Supercritical Fluid Science and Technology
Volume 8

Synthesis of Nanostructured Materials in Near and/or Supercritical Fluids

Supercritical Fluid Science and Technology

Series Editor – Erdogan Kiran

Vol. 1 Supercritical Fluids and Organometallic Compounds: From Recovery of Trace Metals to Synthesis of Nanostructured Materials.
By Can Erkey

Vol. 2 High-Pressure Fluid Phase Equilibria: Phenomenology and Computation.
By Ulrich K. Deiters and Thomas Kraska

Vol. 3 Phase Equilibrium Engineering.
By Esteban Brignole and Selva Pereda

Vol. 4 Introduction to Supercritical Fluids: A Spreadsheet-based Approach.
By Richard Smith, Hiroshi Inomata, and Cor Peters

Vol. 5 Hydrothermal and Supercritical Water Processes.
By Gerd Brunner

Vol. 6 Particle Formation with Supercritical Fluids: Challenges and limitations.
By Michael Türk

Vol. 7 In situ Spectroscopic Techniques at High Pressure.
By Andreas Braeuer

Vol. 8 Synthesis of Nanostructured Materials in Near and/or Supercritical Fluids.
By Can Erkey and Michael Türk

Supercritical Fluid Science and Technology
Volume 8

Synthesis of Nanostructured Materials in Near and/or Supercritical Fluids

Methods, Fundamentals and Modeling

Can Erkey
Department of Chemical and Biological Engineering, Koc University, Istanbul, Turkey

Michael Türk
Institute for Technical Thermodynamics and Refrigeration, Karlsruhe Institute of Technology, Karlsruhe, Germany

ELSEVIER

Elsevier
Radarweg 29, PO Box 211, 1000 AE Amsterdam, Netherlands
The Boulevard, Langford Lane, Kidlington, Oxford OX5 1GB, United Kingdom
50 Hampshire Street, 5th Floor, Cambridge, MA 02139, United States

Library of Congress Cataloging-in-Publication Data
A catalog record for this book is available from the Library of Congress

British Library Cataloguing-in-Publication Data
A catalogue record for this book is available from the British Library

ISBN: 978-0-444-64089-5

ISSN: 2212-0505

For information on all Elsevier publications
visit our website at https://www.elsevier.com/books-and-journals

Publisher: Susan Dennis
Acquisitions Editor: Anita Koch
Editorial Project Manager: Aleksandra Packowska
Production Project Manager: Bharatwaj Varatharajan
Cover Designer: Alan Studholme

Typeset by STRAIVE, India

Contents

Foreword

It is with great pleasure that I introduce Volume 8 in the *Elsevier Book Series on Supercritical Fluid Science and Technology.*

This volume entitled "*Synthesis of Nanostructured Materials in Near and/or Supercritical Fluids*" is written by Professor Can Erkey of Koç University in Istanbul, Turkey, and Professor Michael Türk of the Karlsruhe Institute of Technology in Karlsruhe, Germany. They are internationally recognized experts in their fields and thus need no further introduction other than pointing out that they each authored previous volumes in this book series. Professor Erkey wrote the first volume "*Supercritical Fluids and Organometallic Compounds. From Recovery of Trace Metals to Synthesis of Nanostructured Materials*" which was published in 2011. Professor Türk was the author of Volume 6 "*Particle Formation with Supercritical Fluids. Challenges and Limitations*," which was published in 2014. They have now combined their experience and expertise in producing this outstanding volume, which is on a technologically important area of rapid developments.

Chapters 1 to 3 provide the authors' motivation for undertaking this new book project along with broad perspectives on coordination compounds, porous support materials ranging from silica to metal organic frameworks, and various applications such as catalysts preparation, extractive metallurgy, purification and processing of nuclear waste, preparation of dyes, coatings, and solar energy storage systems.

Chapter 4 is devoted to fundamental aspects of supercritical fluids. A detailed overview of the volumetric, thermal, and transport properties of pure fluids with a greater focus on carbon dioxide and water is provided. Promising applications and advantages of high-pressure technologies are briefly discussed.

Chapter 5 discusses the high-pressure phase behavior and solid-liquid-gas (SLG) equilibrium in binary mixtures consisting of a supercritical fluid and a substance of low volatility. Solubility and binary diffusion coefficients of metal complexes in supercritical fluids is discussed. Modeling of solubility and transport properties of metal complexes in supercritical fluids and discussions on high pressure phase equilibria, measurement methods of solubility and transport property are also provided.

Chapter 6 covers the thermodynamics and kinetic aspects of adsorption of metal complexes on surfaces, which are important in the formation of metallic films or nanoparticles. Adsorption of solutes from supercritical fluid mixtures is discussed with emphasis on adsorption of metal complexes from supercritical fluid solutions. The nature of phase separation in porous materials and the challenges pertaining to critical behavior of fluids in confined pores are presented.

Chapter 7 is devoted to the synthesis of nanostructured composites by Supercritical Fluid Reactive Deposition (SFRD). Examples include supported bimetallic and multimetallic nanoparticles, and metallic films.

Chapter 8 is devoted to the hydrothermal synthesis of metal-oxide nanoparticles. The basic principles of Continuous Hydrothermal Synthesis (CHTS) using sub and supercritical water is described. The phase behavior and solubility of inorganic salts in sub and supercritical water are discussed in some detail. Kinetics of particle formation and offline particle characterization techniques are also covered. The concepts are demonstrated with selected experimental results and modeling studies.

Chapter 9 describes the current state of knowledge regarding the theoretical description of the particle formation and characteristics obtained from the SFRD and the CHTS processes. The challenges and further needs concerning the modeling of the SFRD, and CHTS processes are then discussed in detail.

The last chapter provides insights on future direction in the areas of synthesis of mono and bi-metallic nanoparticles, and synthesis of metal oxide nanoparticles.

I trust that this volume will be a valuable resource to researchers both in academia and in industry and will serve as a key reference to those who employ the concepts discussed in the book in their practice at a production setting.

Erdogan Kiran
Series Editor
Blacksburg, VA, USA
May 2021

Preface

The book presents and discusses the scientific and engineering fundamentals for the synthesis of either supported metallic nanoparticles, films or metal-oxide nanoparticles that can be prepared by innovative supercritical fluid (SCF)-based processes. The unique properties of SCFs enable the tailoring of the nanostructure that is not possible using other technologies and results in the creation of novel nano-structured materials with superior properties. The book covers fundamental aspects such as high-pressure phase behavior of complex mixtures, thermodynamics and kinetics of adsorption/desorption from supercritical solutions, phase separation kinetics, mechanism of particle formation and growth phenomena in either supercritical carbon dioxide or water particle synthesis processes. Based on this, the underlying physical and chemical phenomena needed to understand the different process strategies and the relationship between process conditions and product properties, are presented and discussed.

Furthermore, this book bridges the gap between theory and application and imparts the scientific and engineering fundamentals of innovative particle synthesis processes. This interdisciplinary "modus operandi" makes it a valuable tool for chemical engineers, materials scientists, chemists, and physicists from both academia and industry and encourages an intensive cooperation between scientists from different but complementary disciplines.

We would like to thank the editor of the series, Erdogan Kiran from Virginia Tech, Blacksburg, Virginia, who encouraged us to write this book. Since more than 20 years, we have had a large number of stimulating discussions about new and promising research activities and novel applications of SCF-based techniques in the field of material processing. Without doubt, it is his enthusiasm, and tireless commitment for promoting the exchange of knowledge and experience in the field of supercritical science and technology that has been an inspiration to a countless number of colleagues in their research.

M. Türk would especially like to thank the large number of bachelor, master and PhD students who spent endless hours in the laboratories, working in the broad field of the determination of high-pressure phase equilibria data and/or the synthesis of nanoparticles with outstanding properties.

M. Türk would also like to express his thanks to his current collaborators Marlene Crone, Manuel Hohm, and Christian Schüßler and the large number of colleagues with whom he has cooperated successfully in a large number of joint projects.

C. Erkey would especially like to thank Carl Saquing, Ying Zhang, Betül Cangül, Selmi Bozbağ, Nazire Seda Yaşar, Ramazan Oğuz Canıaz, Şansım Bengisu Barım and Hande Güneş who all had their thesis work in some aspects of supercritical deposition to prepare supported nanoparticles.

C. Erkey would also like to thank Prof. Mark Aindow at the University of Connecticut for imaging the supported nanoparticles produced in my laboratory using TEM for the past 20 years. Prof. Aindow has enabled us to see the world of nanoparticles and has been a great collaborator.

Finally, we are grateful for the support of the Elsevier team, in particular Grace Lander, Kostas Marinakis and Anita Koch for all their efforts in ensuring the publication of this book.

Michael Türk
Karlsruhe Institute of Technology, Karlsruhe, Germany

Can Erkey
Koç University, Istanbul, Turkey

Chapter 1

Introduction

Below, some useful conventions and a short overview of the motivation and the basics of the different processes for the production of metallic and metal oxide nanoparticles and their promising fields of application are given.

1.1 Some conventions[a]

At the beginning, before entering into the discussion of the various particle formation processes and their underlying fundamentals, it is useful to introduce and clarify some conventions and definitions. It is well known that there are three common states of matter, namely, solid, liquid, and gas. Note that no difference is made in the notation of "gas" or "vapor" phase. Thus, the following abbreviations are used if a phase or an aggregation state has to be indicated in phase diagrams or equations:

- S (s): solid phase, thereby no distinction is made between "solid substance" and "low volatile substance."
- L (l): liquid phase
- G (g): gas (gaseous) = V (v) vapor phase.

The phase equilibria are indicated by the combination of the different phases, e.g.:

- S = L: solid-liquid equilibrium, i.e., a solid and a liquid phase coexist.
- S = G: solid-gas equilibrium, i.e., a solid and a gas phase coexist.
- L = G: liquid-gas equilibrium, i.e., a liquid and a gas phase coexist.
- S = L = G: solid-liquid-gas equilibrium, e.g., in case of a pure substance, the three phases coexist at the triple *point*, and in case of binary systems, the three phases coexist along the S-L-G, three-phase *line*.

In case of mixtures consisting of "*n*" components, component "1" is the substance with the highest sublimation (in case of solids) or vapor (in case of liquids) pressure at a given temperature. From this follows that component "*n*" is

a. Revised version of Chapter 1.1 in Ref. [1]; Copyright © (2014) Elsevier.

https://doi.org/10.1016/B978-0-444-64089-5.00025-1

the substance with the lowest (sublimation or vapor) pressure or, at constant pressure, the highest boiling temperature.

The composition, unless mentioned otherwise the mole fraction, of the liquid phase is characterized by "x" and of the gas/vapor phase by "y," e.g.:

x_1 is the mole fraction of component "1" in the liquid phase with $\Sigma x_i = 1$.
y_1 is the mole fraction of component "1" in the gas/vapor phase with $\Sigma y_i = 1$.

1.2 Motivation

1.2.1 Metallic materials[b]

A large number of technologies, including coating, electronic, and material and separation science will be enhanced by the ability to prepare nanoparticles (NPs) with controlled particle size (PS), size distribution (PSD), and particle shape. In NPs, i.e., particles which have at least one dimension in the nanometer scale (1–100 nm), the number of surface atoms compared to the overall number of atoms is much higher than in macroscopic materials. This results in an enhanced energetic state and thus a higher reactivity of NPs. Due to the dramatic surface enlargement, NPs deposited on different substrates are, among other applications, currently used extensively as catalysts for chemical transformations. As in case of such applications, the improved catalytic properties are directly related to the core-shell atom ratio. It is worthy to consider, that the decrease of the PS from 10 to 1 nm results in an increase of the amount of the shell atoms from 20% to 99%. Therefore, it is essential to control the PS, PSD, and particle shape to achieve high dispersions of the metal on the various substrates. If the NPs are to be incorporated into a matrix, their concentration and dispersion should also be controlled.

Noble metal catalysts are often prepared by conventional and established methods such as incipient wetness impregnation (IWI), co-precipitation, chemical vapor deposition (CVD), flame spray pyrolysis (FSP), and pulsed laser ablation in liquid (PLAL). Catalysts prepared by these conventional techniques have some disadvantages and limitations such as metal particles with broad size distributions, large volumes of waste water, and, in case of CVD and FSP high process temperatures. To overcome this, innovative PS reduction processes based on supercritical fluids (SCFs) are gaining an importance in material science. The application of the supercritical fluid reaction deposition (SFRD) method is a promising technique to design inorganic nano-structured materials. Note, that the SFRD process is identical with the so-called chemical fluid deposition (CFD), supercritical fluid deposition (SCFD), supercritical deposition (SCD) and supercritical fluid chemical deposition (SFCD) method. Nevertheless, the basic principle of the deposition and particle formation process remains the same.

b. Revised and extended version from Chapter 1.4.2 in Ref. [1]; Copyright © (2014) Elsevier.

The SFRD process is performed usually in the batch mode and this technique involves the dissolution of a metallic precursor in CO_2. After adsorption of the precursor on the substrate, the precursor is transformed to its metal form through either chemical or thermal reduction.

The following section gives a survey (that never can pretend to be comprehensive!) of published knowledge about particle synthesis with focus on the SFRD technique.

From a general point of view and based on numerous experimental results published in literature one can summarize that the PS of monometallic NPs synthesized by SFRD are influenced by pressure, temperature, and precursor concentration in $scCO_2$ in the following way:

- A higher precursor concentration in $scCO_2$ and hence higher adsorbed amount of the precursor (i.e., loading of the substrate) leads to larger particles.
- Increasing the reduction pressure results in smaller PS and narrower PSD, while increasing the reduction temperature leads to an increase in PS and a broader PSD.
- Higher surface area of the substrate results in highly dispersed and therefore smaller NP.
- The reduction method greatly influences the size and distribution of the NP formed.

Results obtained within a framework of a joint project showed for the first time the great potential of the SFRD process for the preparation of highly active catalysts. Experiments performed by Incera Garrido et al. demonstrate that SFRD leads to highly dispersed Pt NP with an average diameter of 3.1 nm and a Pt content of 0.71 wt% [2]. These catalysts showed superior activity toward preferential CO oxidation compared with a catalyst prepared by conventional aqueous impregnation of Pt. Furthermore, the results show that a reduction temperature higher than the melting temperature at the given pressure leads to relatively large particles associated with the lower catalytic activity.

Based on this result, PtCuO NP were deposited on α-Al_2O_3 sponges previously coated with CeO_2 by simultaneous deposition of $Cu(thd)_2$ and Pt(cod) me_2 from $scCO_2$ [3]. Again, these catalysts were tested for preferential oxidation of CO and showed high activity and selectivity. The activity of PtCuO NP was found to be twice as high as the activity of the reference catalyst prepared by conventional wet impregnation.

In an extensive research study, the influence of different synthesis methods, i.e., SFRD, FSP, and IWI on PS and the hysteresis behavior during CO oxidation over Pt/γ-Al_2O_3 catalysts were systematically investigated, and the most important results of this study are depicted in Fig. 1.1 [4]. Obviously, the lowest temperature for 50% CO conversion was observed for the catalyst prepared by SFRD with an average particle size of 2.2 nm and a Pt content of 1.6 wt%. Both, smaller and larger particles prepared by alternative synthesis routes were less

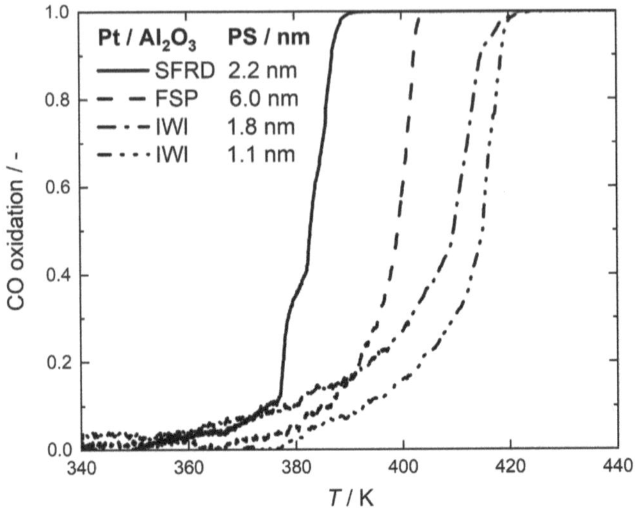

FIG. 1.1 CO oxidation light-off over differently prepared Pt/Al_2O_3 catalysts; reaction conditions: 1000 ppm CO/10% O_2/N_2 at 500 mL/min and heating rate 5 K/min between 323 and 500 K. *(Modified from M. Casapu, A. Fischer, A. Gänzler, R. Popescu, M. Crone, D. Gerthsen, M. Türk, J.-D. Grunwaldt, Origin of the normal and inverse hysteresis behavior during CO oxidation over Pt/Al2O3, ACS Catal. 7 (2017) 343–355. Copyright © 2017, American Chemical Society.)*

active under these reaction conditions. Note that the formation of CO_2 starts at significantly higher temperatures for the two catalysts prepared by IWI using different precursors. The temperature for 50% CO conversion varies between 386 K synthesized by SFRD and 414 K for the catalysts prepared by IWI. The results obtained for the catalysts prepared by different preparation methods clarify the influence of the PS on the light-off temperature and particularly on the hysteresis profile of the CO oxidation. As discussed in detail in Ref. [4], an inverse hysteresis behavior is observed for Pt NPs prepared by IWI which are smaller than 2 nm. The lowest light-off temperature (386 K) was observed for a PS of 2.2 nm, while larger particles (6.0 nm) lead to a light-off temperature of ca. 400 K.

Qiao et al. demonstrated that the catalytic properties of SBA-15 supported Pt/Pd bimetallic catalysts are strongly influenced by the choice of the solvent used for the impregnation of the substrate [5]. The comparison between the catalysts synthesized by SFRD and prepared with the common liquid phase impregnation method showed that SFRD leads to a superior activity for the hydrogenation of levulinic acid to γ-valerolactone with almost perfect selectivity.

Recently, Ogel et al. systematically investigated the influence of the preparation method (IWI, FSP, SFRD, PLAL) and the resulting PSD on catalytic performance during CO/NO oxidation [6]. As one result, three different categories of $Pt/\gamma\text{-}Al_2O_3$ catalysts were obtained:

- Catalysts containing homogeneously distributed Pt NPs with <2 nm diameter, prepared by IWI.

- Catalysts with homogeneously distributed but slightly larger Pt NPs with ~2–4 nm diameter, prepared by SFRD and FSP.
- Catalysts with a heterogeneous PSD containing either a combination of small and large Pt NPs, or an insufficient distribution of the noble metal component, prepared by PLAL.

Further investigations revealed an optimal PS of about 2–3 nm for the CO oxidation reaction, as obtained by SFRD, whereas slightly larger Pt particles with diameters of 3–4 nm led to the highest NO conversion at lower temperatures.

Thus, in summary one can conclude that catalysts prepared by SFRD exhibit an activity higher than reference samples prepared by different conventional methods [1, 4, 6–8]. Furthermore, SFRD allows avoiding undesired organic solvents during catalyst preparation. However, it is worthy to note that a number of established processes (e.g., CVD, IWI, FSP, PLAL) for the production of monometallic NP exist, while synthesis processes for bimetallic and multimetallic nanoparticles are scarce.

1.2.2 Metal oxide materials[c]

The continuous hydrothermal synthesis (CHTS) of metal oxide NPs without the addition of organic solvents, strong bases, or surfactants is of enormous interest for a large number of applications such as catalysis, ceramics, coating, ferrofluid and semiconductor technology, magnetic data storage, sensors, solar energy devices, but also in medical fields such as hyperthermia, bio-imaging, cell labeling, special drug delivery systems and so on. Among others, more details about catalysis and sensing can be found in a concept paper that has been published recently [9].

Various appearances and usages of ceria (CeO_2) NPs are, among others, related to the high oxygen storage capacity and its role as the catalytic component in the oxidation and volatile organic compounds purification catalysts. In addition, CeO_2 can be used as substrate for Pd and/or Pt catalysts for the CO and C_3H_8 oxidation and automotive emission control.

The influence of three different shaped (rod, cube, and octahedron) CeO_2 substrates on the physico-chemical and catalytic properties of Pd/CeO_2 for the CO and C_3H_8 oxidation were examined by Hu et al. [10].

It is depicted in Fig. 1.2 that in case of the CO oxidation, the Pd/CeO_2-rod catalyst showed the highest catalytic activity among the three catalysts. Furthermore, the specific reaction rate of the CO conversion decreases linearly with increasing Pd particle size. In opposite thereto, in case of the C_3H_8 oxidation, the highest reaction rate was obtained with the Pd/CeO_2-octahedron catalyst. Moreover, the reaction rate increases with increasing Pd particle size. These results demonstrate the inverse sensitivity of different shaped CeO_2 substrates to the CO and C_3H_8 oxidation and emphasize that

c. Revised and extended version from Chapter 1.4.3 in Ref. [1]; Copyright © (2014) Elsevier.

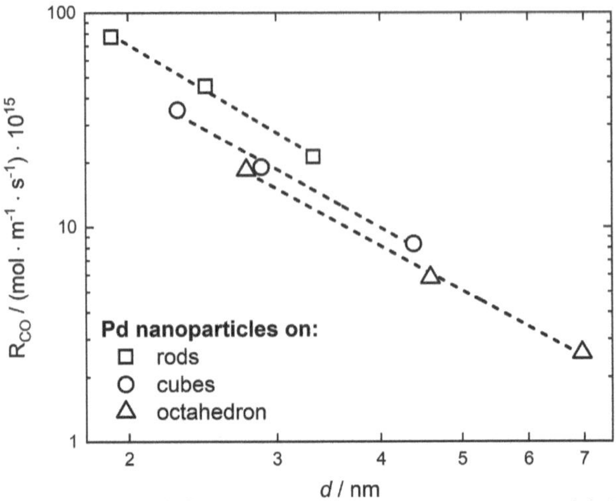

FIG. 1.2 Effect of ceria shape and Pd particle size of different Pd/CeO_2 catalysts on the specific reaction rate of CO conversion at 323 K. The samples were pretreated at 473 K with 10% H_2/Ar (40 mL/min) for 30 min before the test. *(Modified from Z. Hu, X. Liu, D. Meng, Y. Guo, Y. Guo, G. Lu, Effect of ceria crystal plane on the physicochemical and catalytic properties of Pd/ceria for CO and propane oxidation, ACS Catal. 6 (2016) 2265–2279. Copyright © 2016, American Chemical Society.)*

the controlled and reliable synthesis of CeO_2 NPs with uniform and well-defined morphologies is extremely important. This holds especially for the structure-sensitive catalytic reactions which occur on the catalyst surface.

Thus, there is a need for processes which enable the synthesis of different metal oxide NPs with well-defined particle properties such as size and shape and it is of major interest for a large number of applications. Such oxide NPs can be synthesized by conventional solid state reactions at high temperatures (up to 1600 K). However, at such temperatures the particles may sinter and grow together to larger particles. Alternatively, classical wet impregnation methods such as co-precipitation, sol-gel, micro-emulsions, oxidation techniques and hydrothermal synthesis can be used to overcome the drawbacks.

Since the early 1990, hydrothermal synthesis has been used as reaction media for the synthesis of highly crystalline metal oxides. Various experimental investigations from different researches show that the CHTS process enables the synthesis of metal oxide NPs such as Al_2O_3, CeO_2, Fe_xO_y, TiO_2, ZrO_2 [11].

The CHTS process has been pioneered during the 1990s through extensive studies by Adschiri and Arai [12]. CHTS has been recognized as a "green" process and exploits the strong variations in the thermo-physical properties of H_2O around its critical point. Thereby the strong variations in the properties of water, in particular, its dielectric constant around the critical point, are exploited. In this region, sub and supercritical H_2O (scH_2O) has very unusual and tunable properties. Macroscopically, it behaves like a nonpolar solvent but the single molecules are still polar [13].

In CHTS, a pure H_2O stream is heated to the desired supercritical conditions and mixed with a cold aqueous metal salt stream. The reaction mixture then passes through the reactor and the particle-water suspension is cooled down to room temperature either only through an external cooler or also through the introduction of a second cold water stream into the hot, particle-loaded, solution. After passing a filter unit to protect the back-pressure regulator the product dispersion is obtained. It is explained and discussed in detail in Chapter 8 that the CHTS process enables the fast synthesis of highly crystalline particles in an environmentally friendly fluid. Compared to subcritical temperatures, the reaction rate in scH_2O is higher while the solubility of the metal oxides is much lower. This leads to the generation of a higher degree of supersaturation and therewith extremely high nucleation rates, which leads to the formation of nanoparticles. Thereby the PS and particle morphology can be tuned by varying pressure, temperature, and residence time, total mass flow rate (TMFR), and mixing ratio (MR). The influence of TMFR and MR on the PSD of iron oxide NPs is illustrated in Fig. 1.3 [1, 14].

It is obvious that a higher TMFR and MR results in smaller particles and a narrower PSD and, in case of the MR, to a significant decrease of the larger particle fraction.

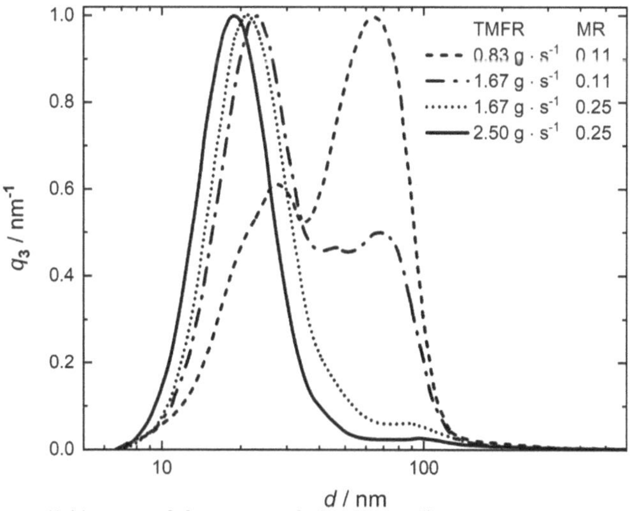

FIG. 1.3 Influence of TMFR and MR on the particle size distribution (q_3) of iron oxide particles. *(Modified from M. Türk, Particle formation with supercritical fluids: challenges and limitations, in: E. Kiran (Ed.), Supercritical Fluid Science and Technology Series, vol. 6, Elsevier, Amsterdam, 2014, Copyright © (2014) Elsevier.)*

The published knowledge about the influence of various process parameters on PS and PSD with a focus on the CHTS technique [1, 11] can be summarized as follows:

- The dielectric constant is a controlling factor of solubility of metal oxides and reaction rate.
- Generally, an increasing reactant concentration leads to larger particles.
- As expected from the (classical) nucleation theory, the size of the synthesized particles decreases with increasing supersaturation.
- The mixer geometry is of extreme importance; an invariant instead of an abrupt changing inner diameter leads to less frequent clogging and ensures a reliable long-term operation and enables the optimization of the process with regard to reproducibility.
- Particles synthesized under subcritical conditions are larger than those produced in scH_2O.
- The Reynolds number increases with increasing flow rate and temperature, resulting in an improvement in turbulence intensity and therewith shorter residence time which results in smaller particles.
- For a given reaction system, the particle size is stable when the Reynolds number is higher than a specific value that depends on the reaction temperature.
- An increasing heating rate results in smaller particles; this holds for different T-shape mixers.

Due to the fact that scH_2O is completely miscible with a large number of organic compounds [15] and that H_2O molecules can act as catalysts for organic reactions, organic surface-modified metal oxide NPs could be synthesized [14]. Such small (<10 nm) organic-inorganic hybrid NPs could be well dispersed in organic solvents and form a lattice structure. Due to their unique properties, these hybrid materials have a huge potential for applications in various technologies related to water and energy supply and efficiency, environmental protection, health, microelectronics, optics, and transportation [1, 16].

In summary one can conclude that scH_2O based particle formations processes such as CHTS enable the formation of well-defined single and bimetallic oxide NPs. However, there is an enormous need for new synthesis techniques which enable the possibility to design advanced functional materials, e.g., core-shell architectures with functionalized surfaces, due to continuous multistep processes as proposed by e.g., Daschner et al. [14, 17, 18] and recently Aymonier et al. [19].

1.3 Scope of the book

This book gives a survey of published knowledge about the synthesis of mono- and bimetallic nano-structured materials with desired product properties. The book is intended for both scientists and engineers who are working in the area of supercritical fluid (SCF) based particle formation processes, as well as for people who are interested in carrying out research in this field.

Thus, one important aspect of this book is to impart the scientific and engineering fundamentals that will allow the reader to understand the relationship between process conditions and the properties of the obtained particles. However, to fulfill this challenge, fundamental knowledge about (complex) high-pressure phase behavior, the transport properties, the thermo and fluid dynamics of the processes, the phase separation kinetics, i.e., the nucleation theory related to these processes and nucleation phenomena, is required.

The interdisciplinary "modus operandi" makes it a valuable tool for chemical engineers, materials scientists, chemists, and physicists and encourages cooperation between scientists and researchers from different but complementary disciplines.

Therefore, the book is organized as follows: A short overview about useful conventions, the motivation, and intention is given in the first chapter. The second and third chapters provide brief introductions to coordination compounds and to porous supports. The fourth chapter gives a detailed account of the thermal and transport properties of SCFs. The fifth chapter is on thermodynamics and transport properties of mixtures of metal complexes and SCFs. The phase behavior of such mixtures is important not only for the design and development of processes for all of the applications covered in this book, but also for understanding the nature of interactions in such systems. The sixth chapter is on thermodynamics and kinetics of adsorption of metal complexes on surfaces in the presence of SCFs, both of which are important in the synthesis of nanostructured materials on surfaces. In this chapter, we tried to also cover some material on adsorption of organics on surfaces in the presence of SCFs due to the applicability of the subject matter to metal complexes. Formation of nanoparticles and films on a wide variety of substrates using supercritical fluid reactive deposition (SFRD) are explained in detail in the seventh chapter. The techniques used to characterize these materials as well as the advantages of this technique are also presented in this chapter. The eighth chapter is on synthesis of metal oxide nanoparticles by continuous hydrothermal synthesis (CHTS) in near/supercritical H_2O. The ninth chapter is on modeling of particle formation in SFRD and CHTS processes while the book ends with a chapter on suggestions for future research needs and development.

References

[1] M. Türk, Particle formation with supercritical fluids: challenges and limitations, first ed, in: E. Kiran (Ed.), Supercritical Fluid Science and Technology Series, 6, Elsevier, Amsterdam, 2014.

[2] G. Incera Garrido, F.C. Patcas, G. Upper, M. Türk, S. Yilmaz, B. Kraushaar-Czarnetzki, Supercritical deposition of Pt on SnO_2-coated Al_2O_3 foams: phase behaviour and catalytic performance, Appl. Catal. A Gen. 338 (2008) 58–65.

[3] S. Lang, M. Türk, B. Kraushaar-Czarnetzki, Novel $PtCuO/CeO_2/\alpha$-Al_2O_3 sponge catalysts for the preferential oxidation of CO (PROX) prepared by means of supercritical fluid reactive deposition (SFRD), J. Catal. 286 (2012) 78–87.

[4] M. Casapu, A. Fischer, A. Gänzler, R. Popescu, M. Crone, D. Gerthsen, M. Türk, J.-D. Grunwaldt, Origin of the normal and inverse hysteresis behavior during CO oxidation over Pt/Al_2O_3, ACS Catal. 7 (2017) 343–355.
[5] Y. Qiao, N. Said, M. Rauser, K. Yan, F. Qin, N. Theyssen, W. Leitner, Preparation of SBA-15 supported Pt/Pd bimetallic catalysts using supercritical fluid reactive deposition: how do solvent effects during material synthesis affect catalytic properties, Green Chem. 19 (2017) 977–986.
[6] E. Ogel, M. Casapu, D.E. Doronkin, R. Popescu, H. Störmer, C. Mechler, G. Marzun, S. Barcikowski, M. Türk, J.-D. Grunwaldt, Impact of preparation method and hydrothermal aging on particle size distribution of Pt/γ-Al_2O_3 and its performance in CO and NO oxidation, J. Phys. Chem. C 123 (2019) 5433–5446.
[7] M. Türk, C. Erkey, Synthesis of supported nanoparticles in supercritical fluids by supercritical fluid reactive deposition: current state, further perspectives and needs, J. Supercrit. Fluids 134 (2018) 176–183.
[8] S. Müller, M. Türk, Production of supported gold and gold-silver nanoparticles by supercritical fluid reactive deposition: effect of substrate properties, J. Supercrit. Fluids 96 (2015) 287–297.
[9] S.A. Müller, D. Degler, C. Feldmann, M. Türk, R. Moos, K. Fink, F. Studt, D. Gerthsen, N. Barsan, J.-D. Grunwaldt, Exploiting synergies in catalysis and gas sensing using noble metal-loaded oxide composites, ChemCatChem (2017), https://doi.org/10.1002/cctc.201701545.
[10] Z. Hu, X. Liu, D. Meng, Y. Guo, Y. Guo, G. Lu, Effect of ceria crystal plane on the physicochemical and catalytic properties of Pd/ceria for CO and propane oxidation, ACS Catal. 6 (2016) 2265–2279.
[11] M.K. Mitchell Lane, J.B. Zimmerman, Controlling metal oxide nanoparticle size and shape with supercritical fluid synthesis, Green Chem. 21 (2019) 3769–3781.
[12] T. Adschiri, K. Kanazawa, K. Arai, Rapid and continuous hydrothermal crystallization of metal-oxide particles in supercritical water, J. Am. Ceram. Soc. 75 (4) (1992) 1019–1022.
[13] A. Kruse, E. Dinjus, Hot compressed water as reaction medium and reactant: properties and synthesis reactions, J. Supercrit. Fluids 39 (2007) 362–380.
[14] M. Daschner de Tercero, Near-Critical and Supercritical Hydrothermal Flow Synthesis of Metal Oxide Nanoparticles and Hybrid Metal Oxide Nanoparticles Presenting Clickable Anchors (Ph.D. thesis), Karlsruhe Institute of Technology (KIT), Germany, 2014, ISBN: 978-3-7315-0262-3.
[15] H. Weingärtner, E.U. Franck, Supercritical water as a solvent, Angew. Chem. Int. Ed. 44 (2005) 2672–2692.
[16] O. Pascu, S. Marre, C. Aymonier, Creation of interfaces in composite/hybrid nano-structured materials using supercritical fluids, Nanotechnol. Rev. 4 (2015) 487–515.
[17] M. Daschner de Tercero, I. Gonzales Martinez, M. Herrmann, M. Bruns, C. Kübel, S. Jennewein, U. Fehrenbacher, L. Barner, M. Türk, Synthesis of in situ functionalized iron oxide nanoparticles presenting alkyne groups via a continuous process using near-critical and supercritical water, J. Supercrit. Fluids 82 (2013) 83–95.
[18] M. Daschner de Tercero, M. Bruns, I. González Martínez, M. Türk, U. Fehrenbacher, S. Jennewein, L. Barner, Continuous hydrothermal synthesis of in situ functionalized iron oxide nanoparticles—a general strategy to produce metal oxide nanoparticles presenting clickable anchors, Part. Part. Syst. Charact. 30 (2013) 229–234.
[19] C. Aymonier, G. Philippot, A. Erriguible, S. Marre, Playing with chemistry in supercritical solvents and the associated technologies for advanced materials by design, J. Supercrit. Fluids 134 (2018) 184–196.

Chapter 2

Compounds of interest

2.1 Introduction

A metal complex, also known as a coordination compound, consists of a central metal atom surrounded by ions and/or molecules. The metal is generally a transition metal, lanthanide or actinide. The ions or molecules that bind to the metal to form coordination compounds are called ligands. The word ligand comes from the Latin word ligare, which means to tie. Coordination compounds are a very important class of compounds. Many reactions in the industry are carried out using highly selective homogeneous catalysts, which are indeed coordination compounds. Such catalysts are also utilized extensively in the synthesis of advanced nanostructured materials. Extractive metallurgy is based on the formation of coordination compounds by the reactions of complexation agents with metal ions with many plants operating on an industrial scale. Complexation reactions are used extensively in the nuclear industry for purification of uranium and also in processing of nuclear waste. Coordination compounds also play an important role in a wide variety of biochemical processes in the body. Hemoglobin, vitamin B_{12} and enzymes that contain a metal atom (metalloenzymes) are examples of such coordination compounds. Chlorophyll is a coordination compound of magnesium. Coordination compounds are used in medical diagnosis and therapy as contrast agents for magnetic resonance imaging and as active compounds in some chemotherapy formulations for the treatment of cancer. Many dyes and pigments such as the blue color of writing ink are metal complexes and they are being used increasingly more in surface finishes and coatings. Metal complex dyes are being developed for use in optical data storage and in electrochromic devices. Coordination compounds are also playing an important role in the development of technologies for conversion and storage of solar energy.

2.2 Types of coordination compounds

The coordination compounds are generally classified according to the nature of the ligands as given below:

Synthesis of Nanostructured Materials in Near and/or Supercritical Fluids
https://doi.org/10.1016/B978-0-444-64089-5.00020-2

1. Classical Complexes: Ligands in classical coordination chemistry bind to metals via their "lone pair" electrons residing on the main group atoms of the ligand. Typical ligands are H_2O, NH_3, Cl^-, CN^- and some examples of complexes are $[Cr(C_2O_4)_3]K_3$ and $[Co(NH_3)_6]Cl_3$.
2. Organometallic Complexes: Ligands are organic such as alkenes, alkynes and alkyls, as well as "organic-like" ligands such as phosphines, hydride, and CO. The difference between an organometallic complex and a classical complex is that an organometallic complex possesses a metal-carbon bond. Among examples of such complexes are $Ni(CO)_4$ and $Fe(C_5H_5)_2$ and $HRhCO(PPh_3)_3$.
3. Bioinorganic Complexes: Ligands are those provided by nature. Examples of such complexes are vitamin B_{12} co-enzyme and hemoglobin.
4. Cluster Complexes: Ligands are all of the above and also include other metals as ligands. Some examples are $Ru_3(CO)_{12}$, $Rh_6(CO)_{16}$ and $Co_4(CO)_{12}$.

2.3 Structures and bonding in coordination compounds

In coordination chemistry, a structure is first described by its coordination number which is the number of ligands attached to the metal. Coordination numbers are normally between two and nine, but large numbers of ligands are not uncommon for the lanthanides and actinides. Typically, the chemistry of complexes is dominated by interactions between s and p molecular orbitals of the ligands and the d orbitals of the metal ions. The maximum coordination number for a certain metal is therefore related to the number of empty orbitals of the metal ion and to the ratio of the size of the ligands and the metal ion. Generally, large metals and small ligands lead to high coordination numbers, e.g. $[Mo(CN)_8]^{4-}$ and small metals with large ligands lead to low coordination numbers, e.g. $Pt[P(CMe_3)]_2$. Due to their large size, lanthanides, actinides, and early transition metals tend to have high coordination numbers. Metal ions may have more than one coordination number. The structural arrangements of the ligands depend on the coordination number. Most structures follow the points-on-a-sphere pattern as if the central atom was in the middle of a polyhedron where the ligands are located at the corners of the polyhedron. There are many complexes that deviate from a regular geometry. The reasons for such deviations may be due to the irregular bond lengths because of the existence of ligands of different types, due to the size of ligands, or due to electronic effects. The most observed geometries and the corresponding coordination numbers are given below:

- Linear for two-coordination.
- Trigonal planar for three-coordination.
- Tetrahedral or square planar for four-coordination.
- Trigonal bipyramidal or square pyramidal for five-coordination.

- Octahedral (orthogonal) or trigonal prismatic for six-coordination.
- Pentagonal bipyramidal for seven-coordination.
- Square antiprismatic for eight-coordination.
- Tri-capped trigonal prismatic (triaugmented triangular prism) for nine coordination.

Some examples of coordination compounds with these geometries are given in Fig. 2.1.

In classical complexes, the ligand is bound to the metal by coordinate covalent bonds in which both of the bonding electrons are supplied by the ligand. In

Coordination Number	Name	Geometry	Polyhedron	Example
2	Linear	—M—		$AuCl(PPh_3)$
3	Trigonal planar	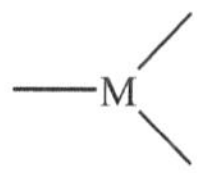		$Pt(PPh_3)_3$
4	Square planar			$RhCl(PPh_3)_3$
4	Tetrahedral	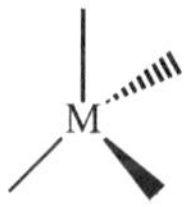		$Ni(CO)_4$
5	Trigonal bipyramidal	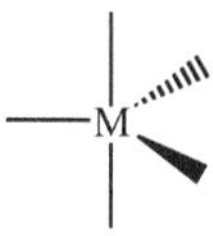	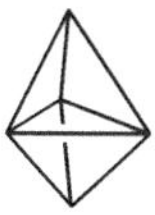	$Fe(CO)_5$
5	Square pyramidal			$[VOCl_4]^{2-}$
6	Octahedral			$Mo(CO)_6$
6	Trigonal prismatic			$W(CH_3)_6$

FIG. 2.1 Examples of coordination compounds with different geometries.

NH_3
H_2O
CO
CO_2
Cl^-
F^-
Br^-
I^-
CN^-
NCS^-
OH^-
H_2NCONH_2
C_5H_5N

FIG. 2.2 Examples of classical ligands and their complexes.

such a complex, the ligand therefore acts as Lewis base and the metal atom or cation acts as a Lewis acid. Some examples of classical ligands and their complexes are given in Fig. 2.2.

A ligand is called a unidentate ligand if it is bound to a metal ion through a single donor atom. Some examples of such ligands are Cl^-, NH_3 and H_2O. If the ligand is bound to the metal ion through two donor atoms, it is called a bidentate ligand. Some examples of bidentate ligands are acetylacetonato, 2,2-bipyridine and 1,10-phenanthroline. If two or more donor atoms are involved in the binding, the ligand is called a polydendate ligand. When a bi- or polydentate ligand uses two or more donor atoms to bind to a single metal ion, it is said to form a chelate complex. The word chelate comes from the Greek word for chelos which means claw. Such complexes tend to be more stable than similar complexes containing unidentate ligands. The structures of some common polydentate ligands are given in Fig. 2.3.

In organometallic complexes, metal and ligand orbitals can interact by sigma and pi interactions as shown in Fig. 2.4.

Most of the ligands can be classified as sigma donor, pi donor and pi acceptor and examples of these ligands are given in Table 2.1. Sigma donor ligands have an electron pair, which can be donated to the empty orbital of a metal atom. Pi donor ligands may donate electrons in a pi fashion using a filled p orbital on a ligand. In some cases, the ligands can also accept electron density from a metal using suitable acceptor orbitals. A good example is the carbonyl ligand which

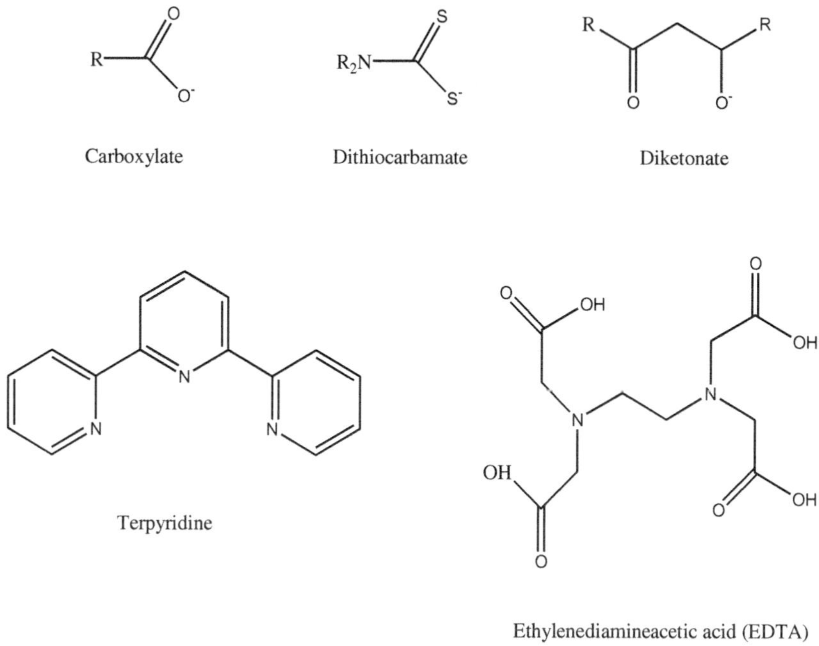

FIG. 2.3 Chemical structures of some polydentate ligands.

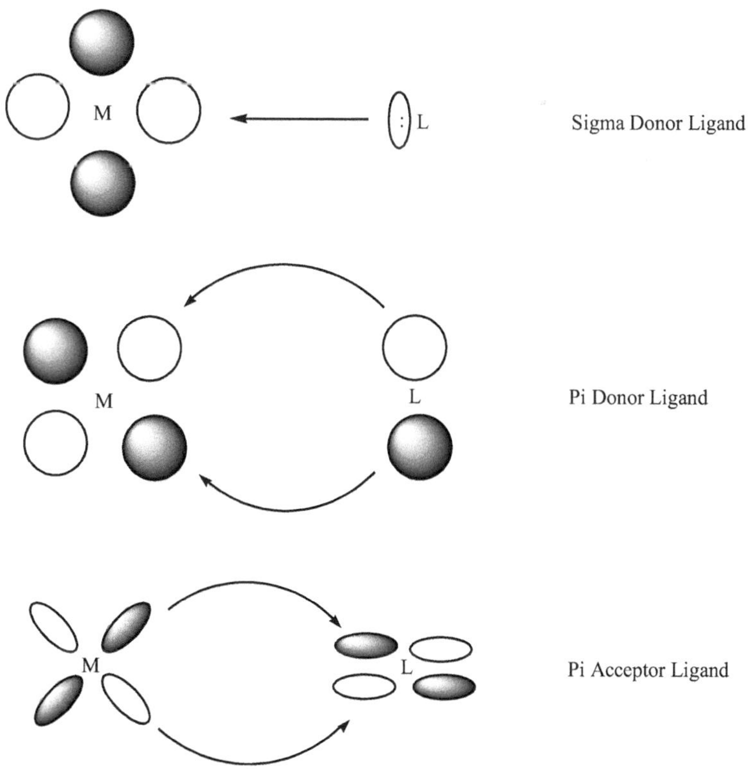

FIG. 2.4 Types of metal-ligand interactions.

TABLE 2.1 Common donor and acceptor ligands.

Sigma donor	Pi donor	Pi acceptor
NH_3	OH^-	CO
H_2O	Cl^-	CN^-
H^-	RCO_2^-	PR_3

has empty pi* orbitals which can accept electron density from the metal. The overall effect of this donation and acceptance is rather strong bonding.

The 18-electron rule is a rule of thumb used primarily for predicting formulas for stable metal complexes. The rule is similar to the octet rule for organic compounds and is based on the fact that valence shells of a transition metal consists of nine valence orbitals, which collectively can accommodate 18 electrons either as nonbinding electron pairs or as bonding electron pairs. The combination of these nine atomic orbitals with ligand orbitals gives rise to nine molecular orbitals which are either metal-ligand bonding or nonbonding. When a metal complex has 18 valence electrons, it is said to have achieved the same electron configuration as the noble gas at the end of the period.

The two methods that can be used to count electrons are the Donor Pair Method and the Neutral Ligand Method. In the first method, ligands donate electron pairs to the metals. To determine the total electron count, one must take into account the charge on each ligand and determine the formal oxidation state of the metal. For example, for $ClMn(CO)_5$, the manganese atom with a charge of +1 has six electrons outside its noble gas core. Each CO is considered to act as a donor of two electrons and each Cl^- would contribute 2 electrons. Thus, the total electron count is 18. The second method uses the number of electrons that would be donated by ligands if they were neutral. For simple inorganic ligands, this usually means that they donate the number of electrons equal to their negative charge as free ions. For example, Cl is one electron donor if it is singly bonded to a metal. To determine the total electron count by this method, one does not need to determine the oxidation state of the metal. For example, for $ClMn(CO)_5$, Mn would contribute 7 electrons, five CO ligands would contribute 10 electrons and Cl^- would contribute 1 electron bringing the total number of electrons to 18.

2.4 Reactions of organometallic compounds on surfaces

The reactions organometallic complexes can go through is given in Table 2.2. Such reactions govern the cycles which organometallic complexes go through as catalysts.

TABLE 2.2 Reactions of organometallic compounds.

Type of reaction		Change in oxidation state of metal	Change in coordination number of metal
Ligand substitution	$L_2 + L_nML_1 \longrightarrow L_nML_2 + L_1$	0	0
Oxidative addition	$L_nM + X\text{-}Y \longrightarrow L_nM(X)(Y)$	+2	+2
Reductive elimination	$L_nM(X)(Y) \longrightarrow L_nM + X\text{-}Y$	−2	−2
1,1-Insertion	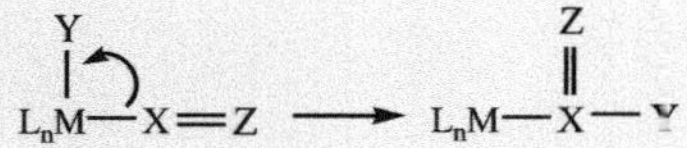	0	−1
1,2-Insertion	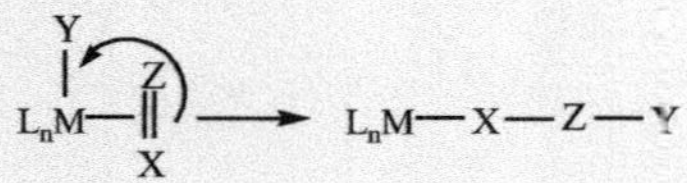	0	−1
Nucleophilic addition	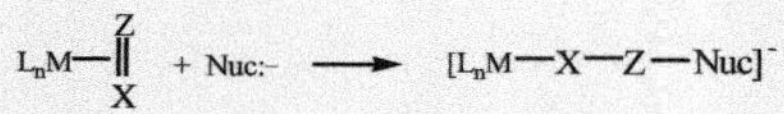	0	0
Nucleophilic abstraction		0	0
Electrophilic addition	$L_nM{-}X{=}Z + E^+ \longrightarrow [L_nM{=}X{-}Z{-}E]^+$	0	0
Electrophilic abstraction	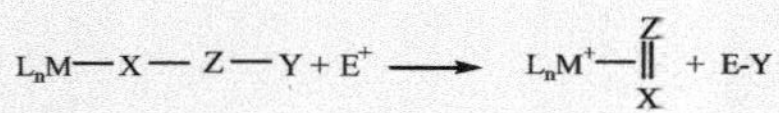	0	0

Chapter 3

Support materials

Supported nanoparticles are extensively used as catalysts in many diverse applications for chemical transformations. Catalytic sites (i.e., active site) are generally composed of metal and/or metal oxide particles with sizes in the nanometer range. The surface atoms are more active compared to bulk atoms due to the presence of fewer neighboring atoms and unsaturated sites or more dangling bonds. Moreover, the surface imperfections of the particle change the electronic structure of the metal atoms that can act as electron or hole trap centers. Hence, metal species are generally dispersed on high surface area materials, termed as supports, to utilize metal species effectively in the form of nanoparticles. Since metal atoms tend to aggregate to minimize their surface energy, support material is used to stabilize the nanoparticles preventing them from agglomerating through metal-support interactions. Pore structure is also important for support materials when we consider heterogeneous catalysis since pore structure is important in terms of the mass transport of reactants to active sites and products from the active site. The support material itself should be stable under reaction conditions and also should meet the requirements for a specific reaction. For example, for electrochemical applications support material should possess adequate electrical conductivity. Support material can also participate in catalysis in some cases.

Many different organic and inorganic materials have been employed as supports in combination with different metal/metal oxides. The supercritical deposition method is suitable for almost any porous support to deposit metal/metal oxide nanoparticles with exceptionally fine dispersion even at a very high metal content. The zero surface tension of supercritical fluid enables the preservation of pore structure during deposition. Most commonly utilized supports are silicas, mesoporous silicas (SBA-15, MCM-41), alumina, silica-aluminas, zeolites, metal organic framework (MOFs), aerogels (inorganic and polymeric), carbonaceous materials (carbon black, graphene/graphene oxide, carbon nanotubes (CNTs), carbon aerogel and graphene aerogel).

Synthesis of Nanostructured Materials in Near and/or Supercritical Fluids
https://doi.org/10.1016/B978-0-444-64089-5.00037-8

3.1 Different support materials

3.1.1 Silicas

Silica is commonly used as a catalyst support due to its good thermal stability as well as tunable porosity and surface area. The physical structures and chemical properties of silicas vary widely in terms of surface area, pore size distribution, particle size, the type and concentration of hydroxyl groups and the acid-base properties according to their preparation method. Siloxane bridges (Si—O—Si) links the corners of the tetrahedral SiO_4 therefore each oxygen atom is shared by two tetrahedra. When this tetrahedra is organized in a definite long-range order to form a crystalline structure in which the silicon and oxygen atoms have defined positions. On the other hand, amorphous forms have only short-range order between neighbors. For industrial amorphous silicas, the size of the primary particles is in the range of 5–30 nm. These primary particles can aggregate to flakes that range in size from microscopic to macroscopic which determines the specific surface area. Amorphous silica is hydrophilic due to the superficial silanol groups (Si—OH) on the silica surface which are also the reactive centers forming hydrogen bonds with each other and with other substances. Hydrogen bonds formed by silanol groups between neighboring aggregates in a liquid form a three-dimensional network of primary particles. Hence the mobility of the liquid molecules is restricted causing an increase in the viscosity of the system (thickening effect). Hydrophobicity can be induced by reacting free silanol groups with dimethyl dichlorosilane. Microporous silica gel can be formed through the reaction of sodium silicate solution and acid through digestion, which is suitable as a desiccant, adsorbent, and catalyst support. Fumed silica has shown some advantages in preparing highly active supported catalysts, due to its unique characteristics such as small particle size in the nanometer range and high surface area.

3.1.2 Mesoporous silicas

Mesoporous silicas are ordered materials with pore sizes between 2 and 50 nm. Mesoporous silicas first emerged in the last decades of the twentieth century with the synthesis of MCM (Mobil Composition of Matter)-type materials [1]. The discovery of the M41S family of ordered mesoporous materials became one the most important milestones in the history of porous nanomaterials creating extensive possibilities for their application as heterogeneous catalyst supports. M41S family of ordered mesoporous silicas has pore dimensions between 2 and 10 nm, very high surface areas ($\sim$1000 $m^2\ g^{-1}$) and large pore volumes ($\sim$1.0 $cm^3\ g^{-1}$). Among the ordered mesoporous silicas MCM-41, MCM-48, MCM-50, FSM-16, FSM-21 and SBA-15 are common and utilized widely in different applications as catalyst support. Synthesis of these materials involves the supramolecular self-assembly of different surfactants (template-directing agents) though there are some differences in conditions. These template directing agents are long chain cationic surfactants such as CTAB (cetyltrimethylammonium bromide) and CPC (cetylpyridinium chloride) or anionic surfactants such as SDS (sodium dodecyl sulfate). At concentrations higher than the critical micelle concentration, co-operative self-assembly of surfactant molecules takes place in solution [1]. A spherical or rod shaped geometry is obtained around which hydrolyzed inorganic silica precursors (e.g., tetraethyl orthosilicate (TEOS) or tetramethyl orthosilicate (TMOS)) arrange themselves to form organic-inorganic metal-template composites in a highly basic medium. During aging at specific conditions, further condensation reactions occur and a highly ordered arrangement such as cubic or hexagonal is obtained. The removal of the surfactants by calcination or solvent-extraction results in the ordered mesoporous silica structure. Depending upon the type of surfactant used, the surfactant-silica ratio and the pH of the solution, the structure of the silica can be varied from hexagonal (e.g., MCM-41), to cubic (e.g., MCM-48) and to lamellar (e.g., MCM-50) [2]. SBA-15 is synthesized using nonionic block co-polymer surfactants under highly acidic conditions. SBA-15 has large mesopores which are arranged as ordered 1D channels and shapes of the primary particles can be varied from rod-like to doughnut-like by changing the copolymer surfactant, silica precursor or aging conditions [3]. Besides the co-operative synthesis pathways, nano-casting is another method developed to synthesize ordered mesoporous structures using already formed ordered mesoporous materials as hard templates. Ordered mesoporous silicas have high surface areas (800–1400 $m^2\ g^{-1}$) and tunable pore diameters (2–50 nm) and therefore attractive as a support material. Introduction of heteroatom dopants or metals into the otherwise inert silica framework while retaining the pore structure and surface area is advantageous. Although ordered mesoporous silicas have various benefits, they suffer from instability issues at elevated temperatures which results in loss of porosity and hence the ordered nanostructure. This drawback limits their application in heterogeneous catalysis.

3.1.3 Aluminas

Alumina is widely used as a basic catalytic support because of its high chemical inertness, strength, and hardness. Mesoporous alumina has excellent properties such as highly uniform channels, large surface area and narrow pore size distribution. Commercial aluminas exhibit specific surface areas between 0.01 and 400 $m^2\ g^{-1}$, pore volumes between 0.1 and 1.4 $cm^3\ g^{-1}$ and average pore size between 2 and 177 nm. Aluminas are also widely used as adsorbents and for other ceramic applications. Alumina can be prepared from bauxite or kaolin in several different phases. There are three different phases of alumina which are α-alumina, β-alumina, and γ-alumina. The α-alumina is also known as nano alumina and it is an inert material with low surface area and high thermal stability generally employed as a ceramic material. β-Alumina is hexagonal, with lamellar structure and the unit cell contains two alumina spinel-based block. γ-Alumina is the most employed type of alumina as support material, because of its reasonably high surface area (up to 400 $m^2 g^{-1}$) owing to the small particle size and good pore parameters. γ-Alumina is generally obtained by the thermal dehydration (i.e., calcination) of aluminum hydroxide and oxyhydroxide precursors. The transformation sequence during this process has been studied for many years, and it also leads to the formation of metastable phases of α-alumina depending on the calcination temperature. γ-Alumina phase transformation occurs at temperatures between 350°C and 1000°C when crystalline or amorphous precursors are used. γ-Alumina is stable at temperatures as high as 1200°C when amorphous precursors are used as the starting material. The structure of γ-alumina is traditionally considered as a cubic defect spinel type. The defective nature derives from the presence of only trivalent Al cations in the spinel-like structure, i.e., the magnesium atoms in the ideal spinel $MgAl_2O_4$ are replaced by aluminum atoms. The oxygen lattice is built up by a cubic close-packed stacking of oxygen layers, with Al atoms occupying the octahedral and tetrahedral sites. To satisfy the γ-alumina stoichiometry some of the lattice positions remain empty (vacancies), although their precise location is still controversial. The partly uncoordinated metal cations and oxide anions that lie at the surface of γ-alumina can act as acids and bases, respectively. Hence γ-alumina has catalytic activity itself for some reactions.

3.1.4 Zeolites

Zeolites are microporous and crystalline aluminosilicates with a framework formed by ordered channels with an opening up to approximately 1 nm. Since they are microporous materials, they have extremely large surface areas with ordered pore structures. Zeolites are naturally occurring materials and have been known for almost 250 years as aluminosilicate minerals. Some naturally occurring zeolite examples are faujasite, mordenite, offretite, ferrierite, erionite and chabazite. Today, these and other zeolite structures are of great interest in catalysis, yet their naturally occurring forms are of limited value, because (i) they generally contain undesired impurity phases, (ii) their chemical composition varies from one deposit to another and even from one stratum to another in the same deposit, and (iii) nature did not optimize their properties for catalytic applications [4]. In 1962, synthetic faujasites (zeolites X and Y) have been utilized for the first time for fluid catalytic cracking of heavy petroleum distillates in industrial scale [5]. These zeolites were much more active than the commercial silica-aluminas and also much more selective for the desired product. Since then utilization of zeolites as catalysts and catalyst supports (co-catalysts) have increased tremendously. The elementary building units of zeolites are SiO_4 and AlO_4 tetrahedra. An oxygen atom links the adjacent tetrahedra corners resulting in a three-dimensional inorganic macromolecule with a structurally distinct framework. The net formula of the tetrahedra is SiO_2 and AlO_2^-, in other words, at each tetrahedron one negative charge resides in the framework which has aluminum in its center. Zeolites contain channels, channel intersections and/or cages with dimensions between 0.2 and 1 nm. Water molecules and small cations reside in these cages or voids that compensate the negative charge in the framework. The chemical composition of a zeolite can hence be represented by the following formula:

$$A^{m+}_{y/m}\left[(SiO_2)_x(AlO_2^{\,-})_y\right]\cdot zH_2O$$

where A is a cation with the charge m, $(x + y)$ is the number of tetrahedra per unit cell and x/y is the so-called framework silicon/aluminum ratio (or simply Si/Al). Many aluminosilicate zeolites can be synthesized with different Si/Al ratios. Aluminum content in zeolites affects the density of negative framework charges, the cation exchange capacity, the density of Brønsted acid sites and their strength, the thermal stability, the hydrophilic or hydrophobic surface properties and the unit cell dimensions [6]. Zeolite crystals are often on the order of one to several micrometers in size. Surface acidity of zeolites is the most important property for their use in heterogeneous catalysis. Both Brønsted and Lewis acid sites occur in zeolites. It has been repeatedly found that the catalytic activity is a result of the Brønsted acid sites rather than the Lewis acid sites. Brønsted acid site density in a zeolite is related to the Al content. The dimensions of the pores of zeolites and molecules interacting with the surface have the same order of magnitude that results in the shape-selective catalytic property of zeolites. Shape-selective catalysis encompasses all effects in which the selectivity of the heterogeneously catalyzed reaction depends unambiguously on the pore width or pore architecture of the microporous catalyst [4]. Shape-selective catalysis using zeolites is widely practiced on industrial scales. A wide variety of zeolites with tunable physical and chemical properties are used as catalysts and catalyst supports for many industrial scale applications.

3.1.5 Silica-aluminas

Amorphous silica-alumina (ASA) is widely used as a solid acid catalyst in various chemical reactions including hydrocracking, isomerization and alkylation. The variable acidity of these materials creates significant interest to their use as catalysts and catalyst supports. The chemical composition of ASA can be expressed as a combination of silica (SiO_2), alumina (Al_2O_3), and residual water (H_2O), depending on the synthesis and pretreatment conditions. Their amorphous nature however makes their structural resolution impossible, contrary to their zeolitic crystalline analogs. They generally exhibit milder acidity than zeolites, which enhances their selectivity toward different products. Similar to zeolites, they have both Lewis and Brønsted acid sites, leading to numerous applications as supports. The acidity of these materials strongly depends on the preparation mode and the Si/Al ratio and hence tunable. Thus, they are widely employed in acid catalysis, from fine chemistry to petrochemistry and refining, and biomass conversion. ASAs are considered as an alternative to zeolites with their limited accessibility of acid sites, inherent to zeolite pore sizes, with respect to bulky reactants, products, or transition states. ASAs can be considered as nonleached de-aluminated zeolites. ASAs have proton conductivity and therefore they can be promising materials as electrolyte in fuel cells.

3.1.6 Metal-organic frameworks

Metal-organic frameworks (MOFs) are a relatively newer class of materials with highly crystalline microporous structures that are comprised of periodically arranged transition metal atoms bonded to organic linkers. Although MOFs were discovered back in 1965, they have attracted great interest in the last two decades. MOFs are materials with well-defined pores, obtained by a method called "reticular synthesis," in which predetermined ordered structures are produced by choosing the appropriate precursors. The spatial organization of these precursors results in a 3D framework consisting of ordered nanometer scale channels and cavities, analogous to that of zeolites. By changing the precursors and the way they are connected allows systematic tuning and modification of the pore structure of MOFs. Due to their extremely high surface area and pore volumes, as well as tailorable structures, MOFs gained significant interest over the last decade. Much research has been conducted to prepare new MOFs and study their applications in different fields including gas storage and separation, catalysis, sensing, drug storage and delivery, and energy storage and conversion. The pores of MOFs can also be considered as microreactors for many kinds of catalytic reactions. MOFs can capture gas molecules or embed catalytically active species inside their cavities, and can also provide catalytic sites, such as metal nodes and modified ligand chains by postsynthesis. There are many different methods to prepare MOFs such as hydro/solvothermal method, microwave method, ultrasonic method, and electrochemical method. High control over the preparation conditions (i.e. metal ions, ligands, solvents, surfactants, temperature, and reaction time) allows tuning of the morphologies of MOFs into the desired dimensional structures from 1D to 3D, such as polyhedron and wires, rods, plates and sheets, or hierarchical structures.

3.1.7 Aerogels

Aerogels are solid colloidal or polymeric networks that are generally ultra-light weight with open porosities as high as 99.9% (v/v). Aerogels are prepared by the drying of the wet gels in which the liquid inside the pores of the network of the wet gel is replaced by air. Aerogels can possess pores from macro to micro range with high surface area and low thermal conductivities and are versatile nanomaterials due to their tunable properties. Since the development and commercialization of silica aerogel-based blankets for thermal insulation, the number of studies that are being conducted at many laboratories around the world on using aerogels for applications other than insulation is growing at an accelerating pace. Aerogels can be organic (e.g., starch aerogel, alginate aerogel, resorcinol formaldehyde aerogel), or inorganic (e.g., silica aerogel, alumina aerogel, titania aerogel). Carbon aerogels is another class of aerogels obtained by pyrolysis of organic aerogels. Aerogels are desirable materials for supporting mono or multicomponent nanoparticles due to their tunable properties such as pore size, surface area and density. Their narrow pore size distribution and high surface areas enables excellent nanoparticle dispersion and enables better control over the rates of diffusion of reactants and products to and from catalytic sites consisting of nanoparticles leading to higher catalytic activity. Combining powerful sol-gel chemistry with different preparation techniques, such as supercritical deposition has led to the development of aerogel supported nanoparticles with superior catalytic properties for any targeted reaction. A variety of organic, inorganic and composite aerogels have been investigated as host materials for a wide variety of metallic nanoparticles for different catalytic applications. Inorganic, alumina and silica aerogels having high hydrothermal and mechanical stability have been investigated as catalyst support materials for high temperature reactions such as hydrogenation/dehydrogenation and Fischer-Tropsch synthesis with promising activities and selectivities. Unlike conventional porous materials, aerogels can also be produced in any form, such as beads, cylinders and monoliths which may be advantageous for various applications, such as optofluidic microreactors. Moreover, many different composite aerogel types can be engineered with desired properties as host materials for a targeted catalytic application. Metal-organic framework (MOF) aerogels composites (MOFACs), for example, is an emerging class of materials that have promising possibilities for catalytic applications as well as drug delivery and gas adsorption.

3.1.8 Silica aerogel

Silica aerogels are extremely low-density, highly mesoporous materials with typical bulk densities which are in the range of 0.05 and 0.150 g cm^{-3}. They have high porosities, typically greater than 99% and average pore sizes in the range of 20–100 nm. Specific surface areas of the silica aerogels are typically between 500 and 1000 and can be as high as 1500 $m^2\ g^{-1}$. As a result of their extremely low densities and nanosized pores, silica aerogels have very low thermal conductivities (~0.014 $W\ m^{-1}\ K^{-1}$), longitudinal sound velocities (~100 $m\ s^{-1}$) and acoustic impedances (~10^{-4}–10^{-5} $kg\ m^{-2}\ s^{-1}$). Due to these properties they are excellent thermal and acoustic insulation materials, un-matched by many other solids. Silica aerogels also exhibit very low dielectric constants and refractive indices, making them suitable for specialty electronic and optic applications. Although silica aerogels have many desirable properties, they are very brittle in nature limiting their application. To overcome this drawback, silica aerogels are frequently combined with other materials producing composite materials to improve mechanical strength while retaining aerogels' desired properties such as low thermal conductivity. Due to their high thermal stabilities and desired structural properties, silica aerogels have also been investigated as catalyst support materials in different catalytic applications such as oxidation, Fischer-Tropsch, gas phase catalysis, hydrogenation and isomerization.

3.1.9 Alumina aerogel

Alumina aerogels are nanoporous materials with low density, high porosity and high surface area. Due to their unique nanostructure, alumina aerogels show relatively high strength, enhanced thermal and chemical stability at high temperatures. Monolithic alumina aerogels have better thermal insulation properties at high temperatures than common silica aerogels and would be a promising candidate for high-temperature thermal insulation, heat-storage and catalysis. Alumina aerogels are commonly synthesized by using two kinds of precursors: hydrated aluminum salts (primarily $AlCl_3 \cdot 6H_2O$ and $Al(NO_3)3 \cdot 9H_2O$) and aluminum alkoxides (primarily aluminum tri-sec butoxide and aluminum isopropoxide). Preparation of alumina aerogel monoliths from aluminum alkoxide can be difficult due to the complex chemical pathways leading to gelation, high reactivity of the aluminum alkoxide and the susceptibility to cracking during drying. Complex solvent mixtures or the addition of chelating agents was usually required to control the hydrolysis and condensation rates of aluminum alkoxide.

3.1.10 Carbonaceous supports

Carbons such as charcoal, diamond, and graphite are some of the commonly utilized naturally occurring carbon materials. High surface area activated carbon and carbon black are widely employed in heterogeneous catalysis as support due their low cost and high availability. Carbon black is an amorphous material obtained by the pyrolysis of organic polymers or hydrocarbon precursors at elevated temperatures (~1500°C). The residual elementary carbon atoms rearrange into stacks of flat aromatic sheets, which are cross-linked randomly with free interstices among them [7]. Resulting carbon materials are composed with roughly planar layers of sp2 hybridized carbons. The materials have a crystalline structure, but they are in the short-range and consequently lack stacking direction [8]. Physical or chemical activation of carbon black results in activated carbon materials. In physical activation carbon materials are treated at 600–900°C in the absence of oxygen and then exposed to oxidizing atmospheres (oxygen or steam) at 600–1200°C. In chemical activation, the carbon precursor is treated either in a strong acid, or a strong base, or a salt, and consequently carbonized at 450–900°C. Chemical activation is more preferred due to lower carbonization temperatures and shorter time needed for activation. Activated carbon has a twisted network of defective carbon layer planes, cross-linked by aliphatic bridging groups. It is an amorphous material with extremely high surface area and high pore volume. Carbon nanotubes (CNTs) are carbon materials consisting of 1D cylindrical hollow channels. They have high surface areas, good electrical conductivity, and excellent mechanical strength and thermal conductivity. Multiwalled CNTs (MWCNTs) are electrically conductive, whereas single-walled CNTs (SWCNTs) are semiconductive, depending on their helicity and diameter which can greatly affect charge transfer processes. CNTs can be considered as ideal and unique structures for anchoring or depositing metallic nanoparticles and therefore attractive supports for heterogeneous catalysis. Moreover, well-defined 1D channels of CNTs enhance the mass transfer of reactants and products. Graphene has emerged as a new kind of catalyst support material having unique physicochemical properties and 2D structure. Graphene has a remarkably high surface area of 2630 $m^2\ g^{-1}$ for a single layer and exhibits a strong interaction with mono or bimetallic nanoparticles. Graphene can also be functionalized with polymers, doped with nitrogen, conjugated with different functional groups to impart varying electronic and acid properties, and highly dispersed with metal nanoparticles that results in high activities.

3.1.11 Carbon aerogels

Carbon aerogels (CAs) are composed of small carbon particles covalently bonded together and were developed first by Pekala in 1989. CAs are usually obtained by pyrolysis of polymeric aerogels at elevated temperatures (500–2500°C). They have high porosity with hierarchical pore structure (micro/meso/macroporous), very high surface areas (400–1500 $m^2\ g^{-1}$), low densities and thermal conductivities. Unlike any other aerogel types carbon aerogels have high electrical conductivity. They can be produced as monoliths, powders, thin films or microspheres. Physical properties of CAs are strongly dependent on the parent polymeric aerogel whose properties can be tuned by changing the parameters in the preparation and gelation steps resulting in a wide variety of CA materials. Because of these favorable properties, CAs have been investigated for different applications such as energy storage, thermal insulation and gas adsorption, broadband nonreflective materials and catalyst supports. CA supported metal nanoparticles have been investigated largely as catalysts for electrochemical reactions due to their high electrical conductivity and electronic structure which can be tuned through pyrolysis temperature and medium.

3.1.12 Graphene aerogels

Graphene aerogels (GAs) are a fascinating class of materials combining the favorable properties of graphene with the high porosity and low density of aerogels. In this regard, mesoporous GAs with high conductivity are potential candidates as support materials. Graphene oxide (GO) is commonly employed as precursor to prepare 3D graphene assemblies because of its' excellent dispersion in aqueous media and its functionality. Also, GO is very useful as a starting material in synthesizing graphene based aerogels (GAs), because the oxygen moieties that are present at both basal planes and edges are able to react covalently with different compounds and thus yield new materials with properties that can be tailored for a specific applications [9]. For example, biocompatibility can be enhanced by immobilizing biopolymers covalently onto GO sheets while retaining its unique properties. A 3D structure with pore sizes in the submicron to micron range can be obtained by the reduction of GO to graphene through a hydrothermal process. Hydrophobic interactions and π-π stacking between the reduced sheets are considered to be responsible for the self-assembly. On the other, GAs generally suffer from extreme brittleness due to their extremely low density and significantly high porosities. Owing to their high surface area, high porosity and functionalities that results in fine dispersion of metallic nanoparticles, GAs gained significant attention as catalyst supports. Furthermore, GAs have high electrical conductivities, high specific capacitance and high cyclic stability. Thus they are commonly employed energy conversion and storage applications such as electrocatalyst supports for fuel cells, electrodes for LIBs, as supercapacitors, and gas sensing applications. They can also be employed as photocatalyst supports, and as phase change materials for better heating-cooling responses. Nitrogen doped GA (N-GA) is a very interesting category of doped GAs due to their mechanical properties as well as their electrochemical properties that can be utilized in different electrochemical applications.

3.2 Characterization of porous materials

3.2.1 Nitrogen physisorption

Nitrogen physisorption is commonly used to characterize porous materials. Physisorption measurements are generally conducted at a constant temperature (77 K for nitrogen) and involve the determination of an adsorption isotherm where amount of nitrogen gas adsorbed is expressed as a function of the equilibrium relative pressure. The features of the isotherm shape are linked to the nature of the gas-solid interactions and the porosity of the material. Pore size is commonly used to classify porous materials. Micropores have sizes less than 2 nm, mesopores between 2 and 50 nm and macropores are greater than 50 nm. Brauner and Emmet conducted the first systematic measurements of physical adsorption of nitrogen at 77 K almost 85 years ago. Since then, nitrogen physisorption became the standard characterization technique to determine the pore structure (macro/meso/microporous), pore volume, pore size distribution and surface area of porous materials. Brauner and Emmet discovered that the adsorption of nitrogen and other gases on a material as a function of pressure follows a similar sigmoidal shape. Experiments indicated that the change from monolayer to multilayer adsorption occurred at a specific point in the isotherm, termed as Point B. Brauner, Emmet and Teller later devised the BET theory to provide theoretical evidence for empirical data which showed that the uptake amount at Point B is in good agreement with the BET monolayer capacity. BET theory is used to determine the specific surface area (BET surface area) from monolayer capacity assuming the closed packed state for the completed monolayer. Although BET is widely used to determine the surface area, it cannot provide reliable data in the case narrow micropores (<2 nm). Pore size determination is also done using nitrogen physisorption data based on Kelvin equation with a correction for the thickness of the adsorbed layer on pore walls. Widely employed calculation scheme for determination of pore size and pore size distribution is devised by Barret, Joyner and Halenda (BJH) and is based on the Kelvin Equation. Conventional technique for nitrogen physisorption employs a discontinuous, point by point procedure in which successive amounts of nitrogen are introduced for each step and the system is allowed certain time to reach equilibrium that corresponds to the points on the adsorption isotherm. Calibration of the pressure gauges, confirmation of adsorption equilibrium and degassing conditions are important for reliable measurement. Degassing is carried out before adsorption measurements to eliminate any physisorbed molecules and impurities such as water as well as to avoid any significant changes due to aging or surface modifications. At degassing, the material is heated to a certain temperature for a certain time under vacuum.

3.2.2 X-ray diffraction

X-ray diffraction (XRD) is a nondestructive technique used to characterize crystalline materials. It is used to determine the crystal structure, crystalline phases and thus gives information about the chemical composition. Crystalline materials have atoms arranged in periodically repeating order. When incident X-rays hit the material, they scatter into many different directions. Some of these scattered X-rays cancel each other (destructive interference), some on the other hand add constructively in specific directions as explained by Bragg's Law.

$$n\lambda = 2d\sin\theta$$

where d is the spacing between diffracting planes, θ is the incident angle, λ is the wavelength of the beam and n is an integer. These specific directions that the X-rays are scattered as the samples are slowly rotated at a specific rate appear as peaks at the diffraction pattern called reflections corresponding to the reflection of X-rays from one set of evenly spaced planes within the crystal. XRD can be used to characterize supports such as aluminas, mesoporous silicas, zeolites, MOFs and crystalline carbon materials such as graphene.

References

[1] C.T. Kresge, M.E. Leonowicz, W.J. Roth, J.C. Vartuli, J.S. Beck, Ordered mesoporous molecular sieves synthesized by a liquid-crystal template mechanism, Nature 359 (1992) 710–712, https://doi.org/10.1038/359710a0.

[2] N. Pal, A. Bhaumik, Mesoporous materials: versatile supports in heterogeneous catalysis for liquid phase catalytic transformations, RSC Adv. 5 (2015) 24363–24391, https://doi.org/10.1039/C4RA13077D.

[3] H.I. Lee, J.H. Kim, G.D. Stucky, Y. Shi, C. Pak, J.M. Kim, Morphology-selective synthesis of mesoporous SBA-15 particles over micrometer, submicrometer and nanometer scales, J. Mater. Chem. 20 (2010) 8483–8487, https://doi.org/10.1039/C0JM00820F.

[4] J. Weitkamp, Zeolites and catalysis, Solid State Ionics 131 (2000) 175–188, https://doi.org/10.1016/S0167-2738(00)00632-9.

[5] P.B. Venuto, E.T. Habib Jr., Fluid Catalytic Cracking with Zeolite Catalysts, Marcel Dekker, New York, NY, 1979.

[6] J. Weitkamp, P. Kleinschmit, A. Kiss, C.H. Berke, The hydrophobicity index—A valuable test for probing the surface properties of zeolitic adsorbents or catalysts, in: R. von Ballmoos, J.B. Higgins, M.M.J. Treacy (Eds.), Proceedings From the Ninth International Zeolite Conference, Butterworth-Heinemann, 1993, pp. 79–87.

[7] E. Lam, J.H.T. Luong, Carbon materials as catalyst supports and catalysts in the transformation of biomass to fuels and chemicals, ACS Catal. 4 (2014) 3393–3410, https://doi.org/10.1021/cs5008393.

[8] E.H. Falcao, F. Wudl, Carbon allotropes: beyond graphite and diamond, J. Chem. Technol. Biotechnol. 82 (2007) 524–531, https://doi.org/10.1002/jctb.1693.

[9] G. Gorgolis, C. Galiotis, Graphene aerogels: a review, 2D Mater. 4 (2017) 032001, https://doi.org/10.1088/2053-1583/aa7883.

Chapter 4

Fundamental aspects of pure supercritical fluids

Our human life takes place at a so-called ambient pressure and temperature conditions (around 0.1 MPa and 298 K). In opposite thereto, a large number of technical applications are carried out at elevated pressures and temperatures, e.g., coffee, hops and flavor extraction, or chemical production processes such as the ammonia and polypropylene synthesis. Under such process conditions, it is often not useful to distinguish between a liquid and a gaseous phase. This holds especially for the supercritical region that is characterized by a pressure and temperature above the fluid's critical values. In this region, only a single homogenous phase exists due to the absence of phase boundaries. Therefore, the expression "fluid" is used for all aggregate states that are not solid.

4.1 Introduction

A schematic *p-T* projection of the phase diagram for a pure substance is shown in Fig. 4.1. The three solid lines divide the diagram into the solid, liquid and gaseous state. At the triple point (TP) the three phases coexist while along the lines the two phases: solid-liquid, solid-gaseous, and liquid-gaseous are in equilibrium with each other. The liquid-gaseous equilibrium $p_S = p_S(T)$ line ends at the critical point (CP), and above the critical pressure p_C and the corresponding critical temperature T_C, the pure substance is in its supercritical (sc) state and is described as a supercritical fluid (SCF).

Fig. 4.2 shows the critical pressures and temperatures of selected pure substances while Table 4.1 lists some common fluids, their molar mass, and the corresponding critical data.

Obviously, the supercritical state can be reached for a large number of substances e.g., CHF_3, CO_2 and C_2H_6 in the temperature range of below 310 K, which is of interest for the processing of heat-sensitive organic substances. Usually, pressures <12 MPa are sufficient to reach the supercritical state. After processing and at ambient pressure CO_2 is in the gaseous state that enables the formation of a solvent-free product without any thermal or mechanical

Synthesis of Nanostructured Materials in Near and/or Supercritical Fluids
https://doi.org/10.1016/B978-0-444-64089-5.00016-0

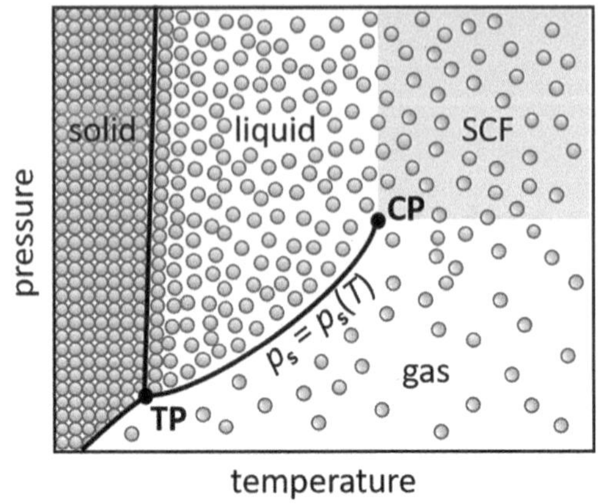

FIG. 4.1 Schematic pressure-temperature diagram for a pure substance (except H_2O).

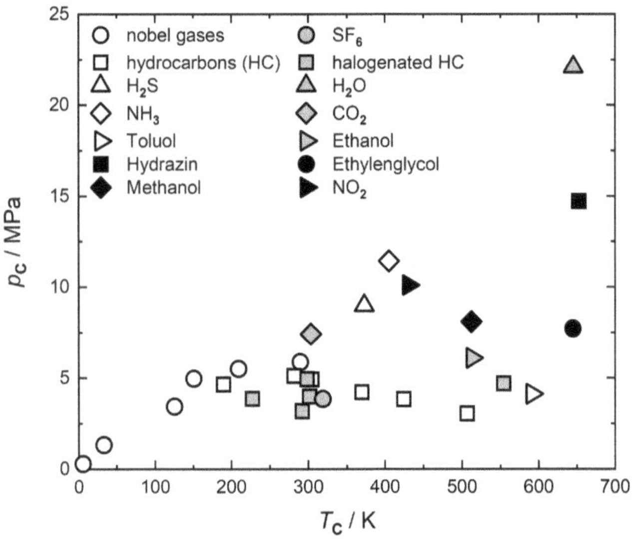

FIG. 4.2 Critical pressure and temperature for selected pure substances.

posttreatment. In summary, it is discussed in various reviews that supercritical fluids exhibit interesting physical and chemical properties.

The specific interest in CO_2 is magnified by its "green" properties e.g., it is nonflammable, relatively nontoxic, and relatively inert. In addition, unlike H_2O, the supercritical region of CO_2 is easily accessible, due to its low critical temperature of 304 K [4, 5]. In addition, $scCO_2$ can act as a solvent and/or separation media and as an in situ protecting group. Due to the low critical temperature $scCO_2$-based particle formation processes can be carried out close to ambient temperatures, which avoids the thermal degradation of the substance which

TABLE 4.1 Molar mass and critical data for some selected substances [1–3].

Substance	M (g mol^{-1})	p_C (MPa)	T_C (K)	ρ_C (kg m^{-3})	v_C (cm^3 mol^{-1})
N_2	28	3.4	126.2	311	90
Xe	131	5.8	289.7	1110	118
CHF_3	70	4.9	299.0	526	133
CO_2	44	7.4	304.2	468	94
C_2H_6	44	4.6	305.4	243	181
N_2O	44	7.3	309.6	454	97
SF_6	146	3.8	318.7	734	199
NH_3	17	11.4	405.6	233	73
H_2O	18	22.1	647.1	322	56

is important for pharmaceuticals, natural substances, and pigments [6, 7]. For example, it is documented in the extensive review written by Danan and Esposito that $scCO_2$-based particle formation processes have been successfully used for inhalable, injectable and oral applications of drugs [8]. The numerous options for the synthesis of nanostructured materials using $scCO_2$ by either physical or chemical transformation have been summarized by Sanli et al. and Bozbag et al. [9, 10]. Recently, a critical survey of published knowledge about synthesis of nanostructured materials and their deposition on different kind of substrates was given by Türk and Erkey and the underlying process basics are discussed [11]. In addition, examples of successful applications of these supported nanoparticles are highlighted and the challenges that need to be overcome to receive an improved understanding of the relationship between process conditions and obtained product properties are stated.

H_2O is certainly the most important substance for all living organisms on Earth. The key role in biological, chemical and technical processes and its various anomalies are well documented in numerous publications. The unique properties of sub and supercritical H_2O are the basis for new innovative technologies in which H_2O acts as a solvent or reactant or catalyst. The solvent behavior of H_2O is controlled by its dielectric permittivity that decreases with increasing temperature and increases with increasing density. Therefore, H_2O changes its character from a solvent for salt at ambient conditions to a solvent for nonionic organic species at supercritical conditions. Thus H_2O is able to form a homogeneous phase at low temperatures with inorganic and in the high-temperature range with organic substances which makes it to an excellent process medium [1, 12–14]. A very extensive overview of the knowledge of the major properties of supercritical water and its solutions with nonpolar, polar, and ionic compounds, and of the underlying molecular processes has been written by Weingärtner and Franck [12]. An overview of the state of the art with regard to hot compressed water (HCW) as reaction medium and reactant has been given by Kruse and Dinjius [13, 15]. Thereby the focus of the work was to summarize results from published synthesis reactions and to understand the basic principles of reactions in HCW. Furthermore, as shown in various overviews, H_2O plays an important role in a large number of technically interesting processes such as the hydrothermal biomass conversion and the oxidative decomposition of organic substances using sub and supercritical H_2O since it acts as solvent, reactant and/or catalyst [14, 16, 17]. Furthermore, the unusual properties of H_2O enable the controlled in situ production of metal oxide nanoparticles by continuous hydrothermal synthesis [18–20]. However, it should be considered that the critical (absolute) temperature of H_2O is about twice and its critical pressure almost three times higher than of CO_2.

The knowledge of reliable p, T and ρ (incl. p_C, T_C and $\rho_C = v_C^{-1}$) data of pure substances and fluid mixtures under sub and supercritical conditions is essential for the optimization of the current and the development of new SCF-based processes. Typically, a static equilibrium apparatus, equipped with a view-cell, can be used for the determination of the critical state by observation of the critical opalescence (see Fig. 4.3).

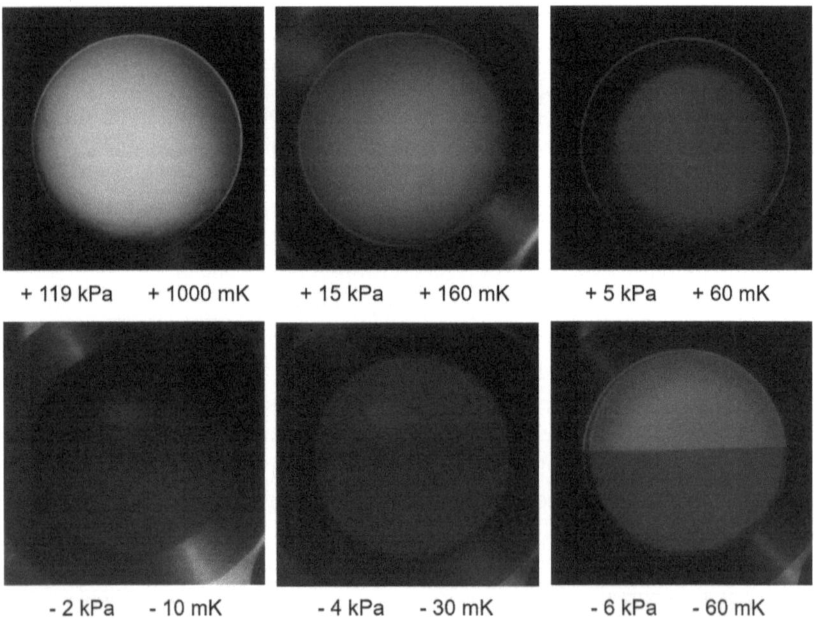

FIG. 4.3 Observation of the critical opalescence of difluoromethane (CH_3F_2). The deviations from the critical data ($p_C = 5.8$ MPa and $T_C = 351.2$ K) are given below the pictures. *(Reprinted with permission from M. Türk, Particle Formation With Supercritical Fluids: Challenges and Limitations, first ed., Elsevier, Amsterdam, 2014, Copyright © (2014) Elsevier.)*

A slow and carefully decrease (0.05 K h^{-1}) of the fluid's temperature from the supercritical to the subcritical state causes that the appearance of the fluid changes from colorless to yellowish and reddish tones. Exactly at the critical point, the fluid is completely opaque (black). Further temperature reduction to the subcritical state causes that the fluid reaches its colorless appearance again. If the density of the fluid is carefully adjusted, the liquid-gaseous interface occurs exactly in the middle of the equilibrium view-cell. In this case, the critical density of the fluid can be determined from the known cell volume and the mass of the used fluid. More details about the apparatus and the experimental procedure, especially the determination of the critical data, can be found elsewhere [21].

In order to estimate the relative amounts of research activities with different types of common solvents a closer look at the papers published in *The Journal of Supercritical Fluids* in the time period 2017–20 will be of interest; the result is shown in Fig. 4.4. Obviously, the researchers in this community prefer to use mainly H_2O, followed by CO_2 and the traditional organic solvents (OS) while ionic liquids (IL) are of minor interest. Among others, the reasons for the preference of H_2O and CO_2 certainly are the low costs and nontoxicity. Furthermore, CO_2 and H_2O can be used as green working media in a wide range of applications, including the synthesis of nanostructured materials in sub and/or supercritical fluids, extraction, dry cleaning, biomass conversion and hazardous waste treatment. According to the Gibbs' phase rule, a supercritical fluid

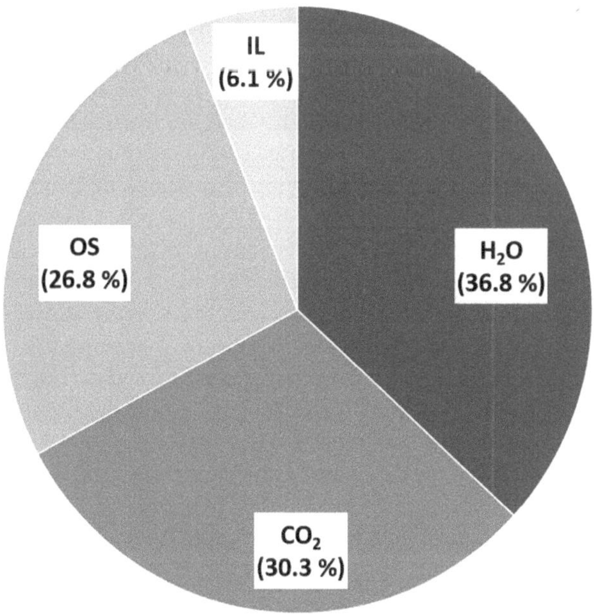

FIG. 4.4 The proportion of papers describing each class of solvent published in *The Journal of Supercritical Fluids* in the time period 2017–20.

has two degrees of freedom and the pressure and temperature can be chosen freely. These two external variables determine the properties of each supercritical fluid, such as its density, heat capacity, viscosity etc. Thus, since CO_2 and H_2O can act as a solvent, reaction and separation medium, the particular properties of both substances will be presented and discussed in detail in the following chapters.

4.2 Thermal and caloric properties

Usually, the thermal (i.e., p, ρ, T) and caloric (i.e., c_p, c_v) properties of fluids can be described by a pressure-explicit thermal equation of state (EoS) $p = f(v,T)$ with $v = 1/\rho$. Nowadays, high-accuracy equations of state such as the fundamental equations explicit in the Helmholtz energy as a function of the dimensionless density ($\delta = \rho/\rho_C$) and temperature ($\tau = T_C/T$) are used for the calculation of the pure fluid properties [22, 23]. The reduced form of such equations is written as:

$$\Phi(\tau, \delta) = \Phi^0(\tau, \delta) + \Phi^R(\tau, \delta) \tag{4.1}$$

where the first expression on the right hand side of Eq. (4.1) is the ideal gas contribution and the second one is the residual (or real) Helmholtz energy that results from intermolecular forces. However, close to the critical point, the isothermal pressure dependence of c_p shows a very pronounced maximum with a c_p value that is by an order of magnitude larger than that extrapolated from data taken far away from the critical point. Even fully empirical equations, such as the IAPWS-95 equation [22, 23] include some nonanalytical terms to obtain accurate descriptions in the near critical region. In applications, the large variations in this region are difficult to control and are usually avoided.

Cubic equations of state are widely used in engineering practice since these equations can describe the phase behavior of mixtures at a reasonable wide range of pressures and temperatures [24]. Common cubic EoS are the Peng-Robinson (PR-EoS), the Redlich-Kwong-Soave (RKS-EoS), and the Trebble-Bishnoi-Salim (TBS-EoS) [25–28]. These EoS are modifications of the van-der-Waals EoS with significant improvements made to the attractive pressure term by the introduction of a temperature-dependent parameter a.

Such an EoS is used due to its simplicity, reliability and the large availability of data bases for various systems. Although highly regarded, a limitation of cubic EoS is that these equations do not represent the behavior of pure fluids near their critical points [24]. Nevertheless, considering their numerous advantages, cubic EoS of the van-der-Waals type can be adjusted in many ways to find acceptable results for most practical applications. These cubic EoS are of the form:

$$p = \frac{R \cdot T}{v - b} - \frac{a}{v^2 + v \cdot (b + c) - (b \cdot c + d^2)} \tag{4.2}$$

The parameters a, b, c and d in Eq. (4.2) of the pure substances can be easily determined from the respective experimental critical data according to the expressions given in the original publications of the respective authors [25–28]. A detailed discussion on the advantages and disadvantages of the numerous cubic EoS can be found in Ref. [24] and in the highly recommended book written by Deiters and Kraska [29].

Note, although these cubic EoS do have some limitations, especially for describing highly polar and/or hydrogen bonding compounds, they are widely used since they give, from an engineering point of view, an adequate thermodynamic model for the representation of complex liquid-gaseous, solid-SCF and solid-liquid-gaseous equilibria. More details about this topic can be found in various papers [30–34].

The density-pressure behavior of pure CO_2 is depicted in Fig. 4.5. In this diagram, $l = g$ denotes the two-phase region in which both, a liquid and gaseous phase exist. The liquid-gaseous saturation curve borders this region and CP stands for the critical point of the fluid at which the distinction between a gaseous and a liquid phase disappears. Above the fluid's critical point, a low-density gas can be compressed into a dense fluid without a phase transition, which is illustrated by the trend of the two supercritical temperatures of 304.2 and 333 K. In opposite thereto at temperatures below the critical one (e.g., 290 K), an increase in pressure leads to dramatically density increase from gaseous to liquid-like as the corresponding saturation pressure $p_S(T)$ is reached.

The comparison between the sub (290 K) and supercritical (304.2 K) isotherm clarifies descriptively that, e.g., at 20 MPa, the density of

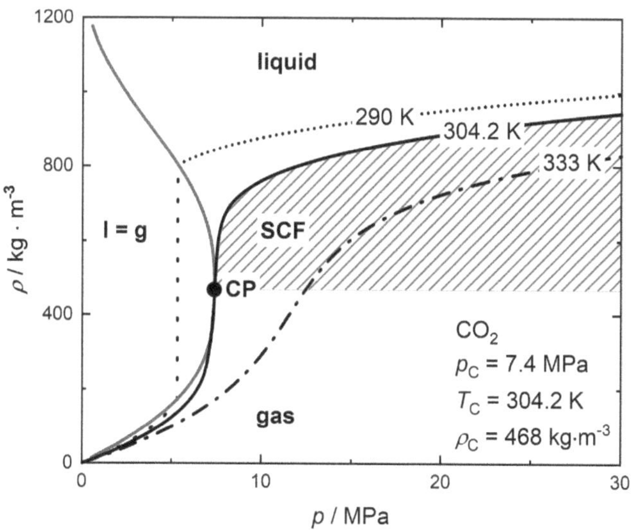

FIG. 4.5 Density-pressure behavior for pure CO_2 [1].

$scCO_2$ (881 kg m^{-3}) is very similar to that of a liquid (950 kg m^{-3}). Note, for all SCFs holds: In the region $1 < T/T_C < 1.1$ the SCF is highly compressible and minor changes in the pressure or temperature result in large changes in the fluid's density and hence, its solvent power. This point is discussed in detail for H_2O below.

From Table 4.2 and Figs. 4.6 and 4.7 follows that close to the critical point, the thermal (e.g., density) and caloric (e.g., specific heat capacity) properties show drastic changes. It is illustrated in Fig. 4.6 that at a subcritical pressure (e.g., 10 MPa) a sudden density jump from the liquid to the gaseous state occurs as the corresponding boiling temperature ($T = 584$ K) is reached.

In opposite thereto, at a constant supercritical pressure the density decreases continuously with increasing temperatures. It is obvious that this behavior is

TABLE 4.2 Selected thermal and caloric properties of H_2O at ambient, sub and supercritical conditions [3, 23, 35].

T	K	298	523	673	673	673
p	MPa	0.1	30	30	50	0.1
ρ	kg m^{-3}	997	826	359	578	0.3
Z	–	8×10^{-4}	0.15	0.27	0.27	0.99
c_p	kJ kg^{-1} K^{-1}	4.2	4.6	25	6.7	2

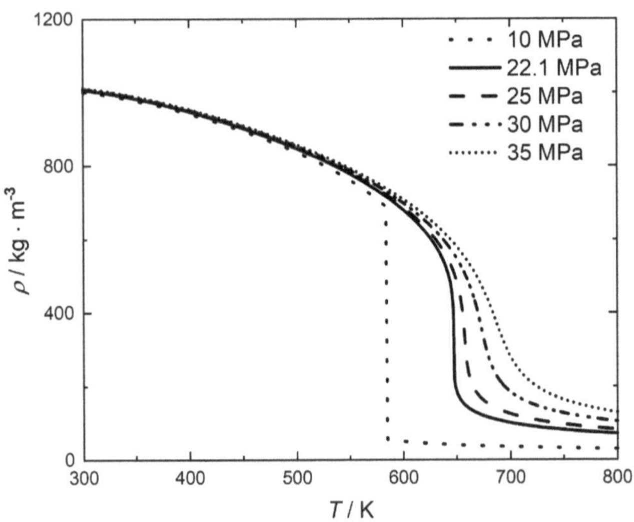

FIG. 4.6 Influence of temperature and pressure on the density of H_2O at sub and supercritical conditions.

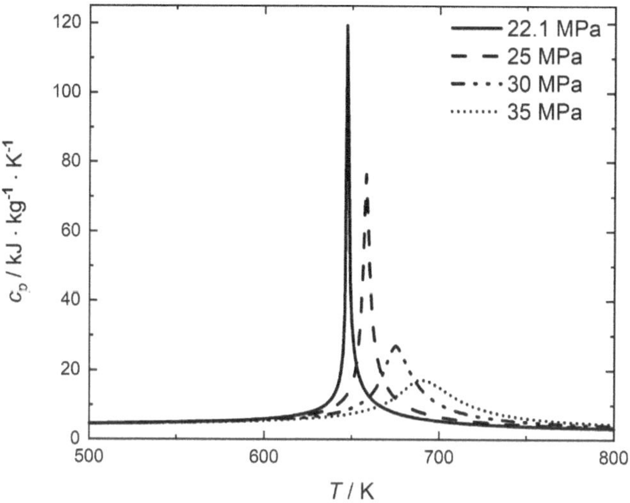

FIG. 4.7 Influence of temperature and pressure on the isobaric specific heat capacity of H_2O at supercritical conditions.

more distinctive at pressures and temperatures close to the critical point. For example, the high density of H_2O at ambient temperature is decreased to a sixth at 30 MPa and 673 K. Thus, the high ability to solve inorganic salts decreases significantly when pressure and temperature are increased.

As depicted in Fig. 4.7 an opposite behavior is observed for the specific heat capacity in the near critical region. At constant pressure, c_p increases with rising temperature up to a maximum that is followed by a decrease with further increasing temperature. In addition, the maximum of c_p decreases and shifts to higher temperatures with increasing pressure. Slightly higher than the critical pressure i.e., 25 MPa, as the critical temperature is approached, the maximum becomes sharper and higher while the drop becomes steeper and larger, resulting in a decrease from 76 kJ kg^{-1} K^{-1} at 658 K to one-tenth at 690 K.

4.3 Transport properties

From the data summarized in Table 4.3 follows that SCFs are characterized by liquid-like densities and transport properties (viscosities, thermal conductivity and diffusivities) lying between those of gases and liquids. Both, the density and viscosity of a gas is typically two orders of magnitude lower than the density and viscosity of liquids. However, in the supercritical region, the density is high and viscosity is low; thus, the ratio of these two quantities, the kinematic viscosity of SCFs, η/ρ, should be extremely low and suggests a high Grashof number.

These specific characteristics are important for various applications in materials processing using a conventional reactor because natural convection effects,

TABLE 4.3 Orders of magnitude for density and transport properties in the gaseous, SCF, and liquid state [1–3].

		Gaseous	SCF	Liquid
ρ	kg m^{-3}	1	$5\times10^{+2}$	10^{+3}
η	μPa s	10	10^{+1}–10^{+2}	10^{+3}
ν	m^2 s^{-1}	10	0.02–0.2	1
D	m^2 s^{-1}	10^{-5}	10^{-7}–10^{-8}	10^{-9}–10^{-10}
λ	mW m^{-1} K^{-1}	3–30	20–80	80–250

i.e., the Grashof number, are inversely proportional to the square of the kinematic viscosity. Furthermore, in case of supercritical particle formation processes such as the continuous hydrothermal synthesis (CHTS, cf. Chapter 8), the Reynolds number, *Re*, should be extremely high for a flow reactor system which causes a strong decrease of the mean residence time and therewith less time available for particle growth that results in small and uniform particles.

Similar to the density, the transport properties such as the viscosity of SCFs vary with changes in pressure and temperature. The accurate knowledge of the viscosity is mandatory because all transport coefficients, e.g., the diffusion coefficient, *D*, strongly depend on the viscosity that decreases significantly with increasing temperature. The low viscosity reflects an improved fluidity and thus high molecular mobility that is attractive for chemical reactions because mass transfer and diffusion-controlled chemical reactions are considerable enhanced. It is depicted in Fig. 4.8 that the dynamic viscosity of CO_2 increases at 313 K by nearly to factor 6 as the pressure increases from 0.1 to 30 MPa. In opposite thereto, the kinematic viscosity decreases by two orders of magnitude as the pressure increases from 0.1 to 7 MPa.

Based on physical and practical purposes, the density and temperature behavior of either the thermal conductivity of a fluid or the dynamic viscosity of a fluid (e.g., CO_2) can be expressed as a sum of three independent contributions [36]:

$$Y(\rho,T)=Y^0(T)+\Delta Y(\rho,T)+\Delta_C Y(\rho,T) \quad \text{with} \quad Y=\lambda \,\text{or}\, \eta \tag{4.3}$$

In Eq. (4.3) the first term, $Y^0(T)$, is the contribution to the transport property in the limit of the diluted gas phase, where only two-body molecular interactions occur. The second term, $\Delta Y(\rho,T)$, the so-called excess property, represents the input of the transport property of the fluid at elevated densities including multiple-body collisions, molecular velocity correlations, and collisional transfer. The final term, $\Delta_C Y(\rho,T)$, is named critical enhancement and describes the contribution which results from the long range fluctuations that arise in a fluid

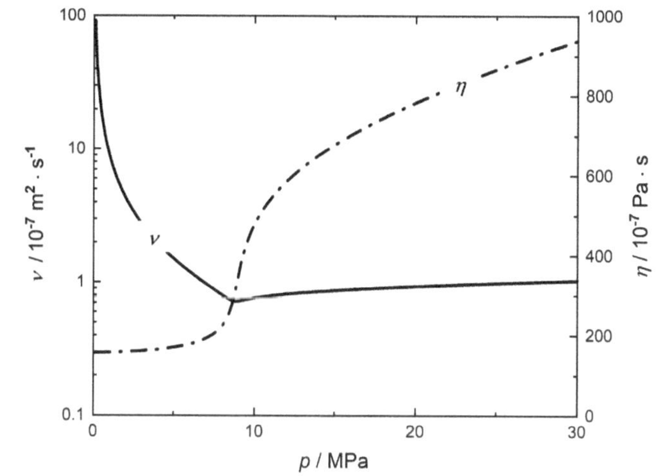

FIG. 4.8 Influence of pressure on the kinematic viscosity (left y-axis) and dynamic viscosity (right y-axis) of CO_2 at 313 K [1, 3]. *(Reprinted with permission from M. Türk, Particle Formation With Supercritical Fluids: Challenges and Limitations, first ed., Elsevier, Amsterdam, 2014, Copyright © (2014) Elsevier.)*

near its critical point. It is worthy to notice that the critical enhancement of the thermal conductivity $\Delta_C \lambda(\rho,T)$ is strong and is observed in a large range of temperatures and densities around the critical point. In opposite thereto, the critical enhancement of the viscosity, $\Delta_C \eta(\rho,T)$, is much weaker and restricted to a small region around the critical point.

From Fig. 4.8 and Table 4.4 follows that the dynamic viscosity of $scCO_2$ (94 μPa s at 30 MPa and 313 K) is about 10 times lower than of classical liquid solvents (e.g., H_2O, 890 μPa s at 0.1 MPa and 298 K). This high fluidity causes lower pressure drops in pipelines and packed bed columns while the enhanced molecular mobility and therefore higher thermal conductivity enables an easier and improved control of exothermal chemical reactions.

TABLE 4.4 Selected transport properties of H_2O at ambient, sub and supercritical conditions [3, 35].

T	K	298	523	673	673	673
p	MPa	0.1	30	30	50	0.1
η	μPa s	890	113	44	68	24
ν	10^{-7} m^2 s^{-1}	8.9	1.4	1.3	1.2	7.6
λ	mW m^{-1} K^{-1}	607	652	328	452	54

In case of H_2O, the temperature and density dependence of the total viscosity can be expressed as a product of three independent contributions [37]:

$$\eta(\rho, T) = \eta^0(T) \cdot \Delta\eta(\rho, T) \cdot \Delta_C\eta(\rho, T) \tag{4.4}$$

while the thermal conductivity of H_2O is represented by the equation [38]

$$\lambda(\rho, T) = \lambda^0(T) \cdot \Delta\lambda(\rho, T) + \Delta_C\lambda(\rho, T) \tag{4.5}$$

Similar to Eq. (4.3), the first factor of Eqs. (4.4), (4.5) represents either the viscosity or the thermal conductivity in the limit of the diluted gas phase and depends only from the temperature. The second expression represents the contribution of the transport property due to increasing density, while the third term reflects the enhancement near the critical point.

From Table 4.4 follows that at ambient pressures the dynamic viscosity of gases and liquids differs by more than one order of magnitude. However, as illustrated in Fig. 4.9, at constant supercritical pressures but subcritical temperatures, the dynamic viscosity decreases continuously by a factor of 10 as the temperature increases from 300 to 600 K. Note, that at 25 MPa a further increase in temperature causes a slight minimum around the critical temperature and a negligible reduction to around 30 μPa s. Thus one can conclude that in opposite to subcritical conditions, the decrease of the dynamic viscosity with increasing temperature is only by the factor 2 at supercritical conditions.

As shown in Fig. 4.10, a different behavior is observed: at supercritical pressures the thermal conductivity increases with increasing temperature, reaches a

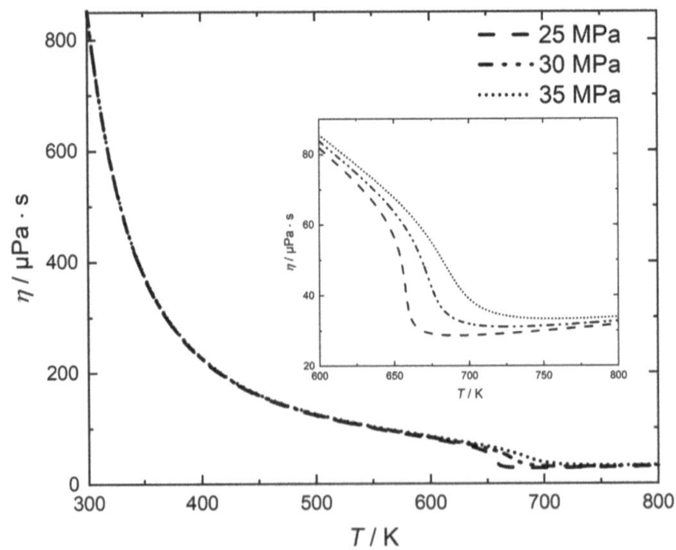

FIG. 4.9 Influence of temperature and pressure on the dynamic viscosity of H_2O at supercritical conditions [35].

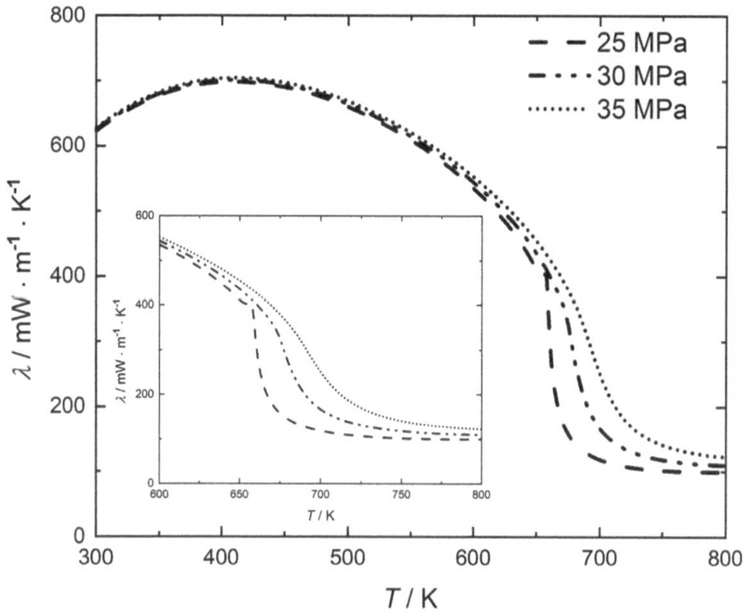

FIG. 4.10 Influence of temperature and pressure on the thermal conductivity of H_2O at supercritical conditions [35].

pronounced maximum at 700 mW m^{-1} K^{-1} and 400 K, and decreases from 400 to 100 mW m^{-1} K^{-1} as the temperature increases from 650 to 750 K.

4.4 Applications of supercritical fluids

This chapter gives a short survey (that never can pretend to be comprehensive!) over some, from the author's point of view, promising applications and advantages of high-pressure technologies. Much more information about the various SCF-based processes can be found in the enormous variety of review articles published in the literature during the last four decades. Furthermore, a very comprehensive overview of research on supercritical fluids and their industrial applications, which has enjoyed a remarkable growth over the past 30 years, is summarized in a number of books which are published during the last decade [39]. A further overview of application of supercritical fluid science and technology is summarized in a collection of current opinion articles written by experts and published in the 30th year special issue of the *Journal of Supercritical Fluids* that is entitled "Perspectives and Opinions on Current State and Future Directions" (details: https://doi.org/10.1016/j.supflu.2018.02.003).

Sub and supercritical fluids offer the possibility to produce new products with special and optimized properties or to create new, environmentally friendly and sustainable processes. Therefore, high pressure technology is a tool which enables the development of distinct processes which result in completely new products. Due to the energetic and environmental advantages of using SCFs in industrial processes, such as low energy consumption during processing in combination with the potential to substitute the damaging conventional organic solvents, SCFs are frequently classified as "green solvents." This applies especially for the use of the two most important SCFs: $scCO_2$ and scH_2O as these solvents are nontoxic, nonflammable and thermodynamically stable. A further advantage is the possibility to adjust the thermal, caloric and transport properties of SCFs by adjusting the process pressure and/or temperature.

A broad overview of the basic principles of processes with SCFs with focus on scH_2O is given by Brunner [40]. Processes which are discussed in this overview are the ROSE process, coal liquefaction and gasification, the production of oil from different sources (e.g., oil shales and oil-sand). All these applications of SCFs benefit from their direct use as a heat transfer medium, their adjustable solvent power, or reactivity and selectivity that lead to improved processes and products. The outcome of the overview emphasizes that simple design, simple operation, higher energy efficiency and understanding the merits of SCFs are the key for continuing and future success.

The high potential of hydrothermal conversion processes which use various product sources from biomass degradation in hot compressed water is described by Kruse and Dahmen [17]. The authors emphasized that biomass has a high reactivity in H_2O and that different products can be synthesized by changing the H_2O properties due to changes of the selectivity. In addition, due to the polar character of the H_2O molecules a new promising reaction pathway is outlined which requires, compared to "dry processes," lower reaction temperatures.

The processes reviewed in the paper written by Cantero et al. are a promising alternative for the efficient and selective use of lignocellulosic biomass as source of bio-fuels, bio-compounds and bio-energy [41]. The authors pointed out that the use of scH_2O as reaction medium has several advantages compared to the traditional methods of enzymatic or acid hydrolysis as well as pyrolysis methods. For example, scH_2O-based hydrolysis and gasification processes avoid the high energy demanding step of H_2O evaporation that is required in the traditional pyrolysis processes.

A very detailed survey of the numerous applications of the SCF technologies used in pharmaceutical, food and textile industries and their energetic and environmental advantages compared to conventional processes are given by Knez et al. [42]. The focus of this overview article is on the extraction of solids and liquids, polymer processing, supercritical drying and cleaning, chemical and biochemical reactions and energy production and energy saving.

Recently, a very comprehensive overview of the promising potential of $scCO_2$ technology as a feasible sterilization technique is published by Soares et al. [43]. This review explores the latest developments on $scCO_2$ sterilization technology, with a strong perspective on the poststerilization performance of biomaterial. In order to receive a better understanding of the potential and limitations of the $scCO_2$ sterilization technique, an overview of the different process parameters, investigated materials and characterization techniques is given.

A very comprehensive overview of knowledge about SCFs in separation and purification processes is given by Manjare and Dhingra [44]. This paper provides a concise review of the applications of SCFs from the perspective of feed materials, type of SCF, co- or antisolvents used, nature of work, operating conditions, findings and limitations of the different processes. In addition, the paper shows clearly that the unique properties of SCFs are well suited for the extraction of solutes, from various plants, fruits, flowers, seeds, leaves, which cannot be easily extracted using conventional extraction processes. Due to the well-known advantages, supercritical fluid extraction (SFE) is becoming an increasing interest in many industrial sectors such as chemistry, energy, food, pharmaceuticals, and petrochemistry. In addition, SCFs are also used for the extraction of metal ions from aqueous solutions and solid and liquid matrices.

Matos et al. pointed out that sub/supercritical CO_2 technology is a promising route for the processing of pharmaceutical compounds and development of different drug formulations by different micronization, coprecipitation, encapsulation and coating techniques [45]. Naturally the selection of the most promising processing strategy requires a deeper understanding of the physico-chemical properties of the active compounds and will be beneficial for the selection of the best possible processing method. The existing techniques as well as the numerous modifications of these techniques for SCF-based processing of thermo-sensitive compounds at low temperatures and using no or low quantities of organic solvents are presented and discussed in detail. In addition, insights into the most suitable formulation and processing strategies for improvement of the dissolution behavior of the active compounds are given too.

In recent years, there has been a growing interest in the processing of biomass-related compounds with ionic liquid mixtures and supercritical CO_2. Among others, one of the key data that are necessary for the development of such processes are reliable pressure-volume-temperature data and gas solubility data. In this context, Hiraga et al. presented and discussed the challenges and knowledge gaps with regard to solvent polarity, mixed ionic liquid acidity, basicity and bipolarity/polarizability [46]. The obtained results and findings of these investigations show that a mixture of two ionic liquids with different functionalities combined with $scCO_2$ can provide a flexible biphasic solvent system for processing biomass and related compounds.

A very fundamental and therewith valuable overview of the advances for the theoretical description of the behavior of SCFs since the late 1980s is given by Yoon and Lee [47]. Thereby the current state of the art theories (such as molecular simulation, fluctuation theory, integral equation theory and others) about the supercritical gas-liquid boundary are investigated and classified. Based on these findings the direction of future research and new useful approaches are given. The obtained results should lead to an improved knowledge of supercritical fluids and an extended utilizing of numerous supercritical fluid technologies.

The impact of accuracy of fluid properties on the efficiency of chemical processes at high pressures and elevated temperatures is highlighted by Mallepally et al. [48]. Thereby the experimental and modeling challenges which are related to the serious determination of the single fluid properties under these conditions are named and discussed. Thereby, the authors demonstrate exemplary for energy related processes that the accurate knowledge of thermodynamic and transport properties minimizes operating costs and enables reliable and robust process development in combination with an environmentally friendly product design.

Thus, as described in brief above, SCF-based particle formation processes enable the sustainable design and production of organic and inorganic materials without contamination or degradation. When SCFs such as $scCO_2$ or scH_2O are involved, particle formation techniques are often classified as physical and chemical transformation processes.

In case of the first process, the driving force for precipitation is a fast depressurization and/or an antisolvent effect which leads to high supersaturations due to the decrease of the fluid's density. Please note that the primary SCF-based techniques for the micronization of organic particles by physical transformation such as RESS (rapid expansion of supercritical solutions), PGSS (particle generation from gas saturated solution), GAS (gas antisolvent) including its numerous modifications (SAS, SEDS, ...) and the large number of supercritical impregnation processes are not considered. The presentation of the essential basics and the discussion of the advantages and disadvantages of these processes would go beyond the scope of this book.

In the chemical transformation technique, the precipitation is caused by a chemical reaction (i.e., hydrolysis, reduction) which induces the nucleation and growth process [1, 10, 11]. The underlying physical and chemical phenomena need to understand the different chemical transformation processes; the relationship between process conditions and obtained product properties are presented and discussed in detail in the subsequent chapters of this book.

Interim summary: As depicted in Fig. 4.11, the thermal (i.e., density), caloric (i.e., heat capacity), as well as transport (i.e., viscosity) properties of a fluid at sub and supercritical temperatures ($0.9 \leq T/T_C \leq 1.1$) are sensitive to changes in pressure. When the pressure is increased across its critical pressure, the density becomes liquid-like while the viscosity remains intermediate between that of a gas and a liquid. The unique ability to adjust tailor-made process conditions enables resource-efficient processing, process intensification and high product selectivity.

In this supercritical state, CO_2 and H_2O can be gradually compressed from gas-like to liquid-like densities. As a solvent, $scCO_2$ can be easily separated from the product just by releasing the pressure without additional time and energy consuming separation and/or drying steps and is of special interest for particle formation processes with regard to pharmaceutical and medical

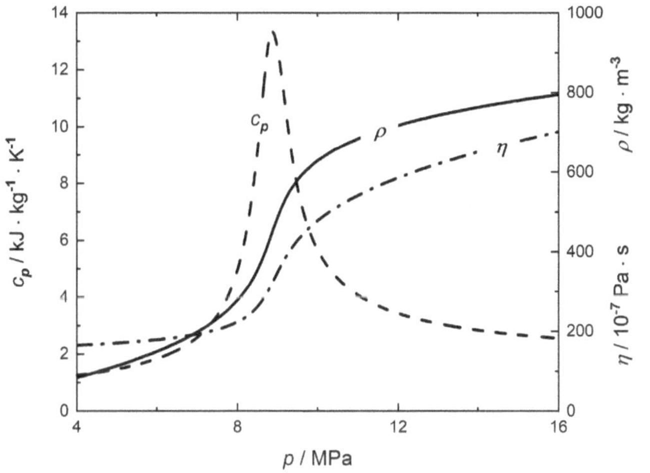

FIG. 4.11 Influence of pressure on selected thermal, caloric, and transport properties for CO_2 at $T/T_C \leq 1.03$.

applications. H_2O changes its character from a solvent for salt at ambient conditions to a solvent for nonionic organic species at supercritical conditions, which makes it to an excellent process and reaction medium. This opens the possibility for controlling the particle formation processes and to control and modify the surface chemistry.

Therefore, SCF-based particle synthesis using "green solvents" can replace undesirable organic solvents.

References

[1] M. Türk, Particle Formation With Supercritical Fluids: Challenges and Limitations, first ed., Elsevier, Amsterdam, 2014.

[2] G.M. Schneider, Physicochemical aspects of fluid extraction, Fluid Phase Equilib. 10 (1983) 141–157.

[3] VDI Heat Atlas, eighth ed., Springer-Verlag, Berlin, Heidelberg, New York, 2008.

[4] E.J. Beckman, Supercritical and near-critical CO_2 in green chemical synthesis and processing, J. Supercrit. Fluids 28 (2004) 121–191.

[5] P.G. Jessop, Homogeneous catalysis using supercritical fluids: recent trends and systems studied, J. Supercrit. Fluids 38 (2006) 211–231.

[6] A. Martín, M.J. Cocero, Micronization processes with supercritical fluids: fundamentals and mechanisms, Adv. Drug Deliv. Rev. 60 (2008) 339–350.

[7] K. Mishima, Biodegradable particle formation for drug and gene delivery using supercritical fluid and dense gas, Adv. Drug Deliv. Rev. 60 (2008) 411–432.

[8] H. Danan, P. Esposito, Critical gases for critical issues: CO_2 technologies for oral drug delivery, Ther. Deliv. 5 (2) (2014) 205–232.

[9] D. Sanli, S.E. Bozbag, C. Erkey, Synthesis of nanostructured materials using supercritical CO_2: part I. physical transformations, J. Mater. Sci. 47 (2012) 2995–3025.

[10] S.E. Bozbag, D. Sanli, C. Erkey, Synthesis of nanostructured materials using supercritical CO_2: part II. Chemical transformations, J. Mater. Sci. 47 (2012) 3469–3492.

[11] M. Türk, C. Erkey, Synthesis of supported nanoparticles in supercritical fluids by supercritical fluid reactive deposition: current state, further perspectives and needs, J. Supercrit. Fluids 134 (2018) 176–183.

[12] H. Weingärtner, E.U. Franck, Supercritical water as a solvent, Angew. Chem. Int. Ed. 44 (2005) 2672–2692.

[13] A. Kruse, E. Dinjus, Hot compressed water as reaction medium and reactant: properties and synthesis reactions, J. Supercrit. Fluids 39 (3) (2007) 362–380.

[14] G. Brunner, Near critical and supercritical water. Part I. hydrolytic and hydrothermal processes, J. Supercrit. Fluids 47 (3) (2009) 373–381.

[15] A. Kruse, E. Dinjus, Hot compressed water as reaction medium and reactant: 2. Degradation reactions, J. Supercrit. Fluids 41 (3) (2007) 361–379.

[16] G. Brunner, Near and supercritical water. Part II: oxidative processes, J. Supercrit. Fluids 47 (3) (2009) 382–390.

[17] A. Kruse, N. Dahmen, A magic solvent for biomass conversion, J. Supercrit. Fluids 96 (1) (2015) 36–45.

[18] A. Loppinet-Serani, C. Aymonier, F. Cansell, Supercritical water for environmental technologies, J. Chem. Technol. Biotechnol. 85 (2010) 583–589.

[19] C. Sanchez, C. Boissiere, S. Cassaignon, C. Chaneac, O. Durupthy, M. Faustini, D. Grosso, C. Laberty-Robert, L. Nicole, D. Portehault, F. Ribot, L. Rozes, C. Sassoye, Molecular engineering of functional inorganic and hybrid materials, Chem. Mater. 26 (2014) 221–238.

[20] J.A. Darr, J.Y. Zhang, N.M. Makwana, X.L. Weng, Continuous hydrothermal synthesis of inorganic nanoparticles: applications and future directions, Chem. Rev. 117 (2017) 11125–11238.

[21] A. Diefenbacher, M. Türk, Critical properties (p_C, ρ_C, T_C) and phase equilibria of binary mixtures of CO_2, CHF_3, CH_2F_2, and SF_6, Fluid Phase Equilib. 182 (2001) 121–131.

[22] E.W. Lemmon, R. Span, Short fundamental equations of state for 20 industrial fluids, J. Chem. Eng. Data 51 (2006) 785–850.

[23] W. Wagner, A. Pruss, The IAPWS formulation 1995 for the thermodynamic properties of ordinary water substance for general and scientific use, J. Phys. Chem. Ref. Data 31 (2002) 387–535.

[24] P.T. Eubank, K.R. Hall, New identities for critical-point constraints with application to cubic equations of state, Ind. Eng. Chem. Res. 43 (2004) 4446–4451.

[25] D.-Y. Peng, D.B. Robinson, A new two-constant equation of state, Ind. Eng. Chem. Fundam. 15 (1976) 59–63.

[26] G. Soave, Equilibrium constants from a modified Redlich-Kwong equation of state, Chem. Eng. Sci. 27 (1972) 1197–1203.

[27] M.A. Trebble, P.R. Bishnoi, Development of a new four-parameter cubic equation of state, Fluid Phase Equilib. 35 (1987) 1–18.

[28] P.H. Salim, M.A. Trebble, A modified Trebble-Bishnoi equation of state: thermodynamic consistency revisited, Fluid Phase Equilib. 65 (1991) 59–71.

[29] U.K. Deiters, T. Kraska, High-pressure fluid phase equilibria, phenomenology and computation, E. Kiran (Series Editor), in: Supercritical Fluid Science and Technology, Elsevier, Amsterdam, 2012.

[30] A. Diefenbacher, M. Türk, Critical (p, ρ, T) properties of CH_2F_2, $\{xCO_2+(1-x)SF_6\}$, $\{xSF_6+(1-x)CH_2F_2\}$, and $\{xCHF_3+(1-x)CH_2F_2\}$, J. Chem. Thermodyn. 31 (1999) 905–919.

[31] A. Diefenbacher, M. Türk, Phase equilibria of organic solid solutes and supercritical fluids with respect to the RESS process, J. Supercrit. Fluids 22 (2002) 175–184.

[32] A. Diefenbacher, M. Türk, (Vapour+liquid) equilibria of binary mixtures of CO_2, CH_2F_2, CHF_3, and SF_6, J. Chem. Thermodyn. 34 (2002) 1361–1375.

[33] M. Türk, T. Kraska, Experimental and theoretical investigation of the phase behavior of naproxen in supercritical CO_2, J. Chem. Eng. Data 54 (5) (2009) 1592–1597.

[34] M. Türk, M. Crone, T. Kraska, A comparison between models based on equations of state and density-based models for describing the solubility of solutes in CO_2, J. Supercrit. Fluids 55 (2010) 462–471.

[35] NIST Chemistry WebBook, NIST Standard Reference Database Number 69, 2018 https://webbook.nist.gov/chemistry. Doi:10.18434/T4D303.

[36] V. Vesovic, W.A. Wakeham, G.A. Olchowy, J.V. Sengers, J.T.R. Watson, J. Millat, The transport properties of carbon dioxide, J. Phys. Chem. Ref. Data 19 (1990) 765–808.

[37] M.L. Huber, R.A. Perkins, A. Laesecke, D.G. Friend, New international formulation for the viscosity of H_2O, J. Phys. Chem. Ref. Data 38 (2009) 101–125.

[38] M.L. Huber, R.A. Perkins, D.G. Friend, J.V. Sengers, M.J. Assael, I.N. Metaxa, K. Miyagawa, R. Hellmann, E. Vogel, New international formulation for the thermal conductivity of H_2O, J. Phys. Chem. Ref. Data 41 (2012) 1–23.

[39] E. Kiran (Series Ed.). Book Series: Supercritical Fluid Science and Technology, vol. 1–7. https://www.elsevier.com/books/book-series/supercritical-fluid-science-and-technology.

[40] G. Brunner, Supercritical process technology related to energy and future directions—an introduction, J. Supercrit. Fluids 96 (2015) 11–20.

[41] D.A. Cantero, M. Dolores Bermejo, M. José Cocero, Reaction engineering for process intensification of supercritical water biomass refining, J. Supercrit. Fluids 96 (2015) 21–35.

[42] Z. Knez, E. Markocic, M. Leitgeb, M. Primozic, M. Knez Hrncic, M. Skerget, Industrial applications of supercritical fluids: a review, Energy 77 (2014) 235–243.

[43] G.C. Soares, D.A. Learmonth, M.C. Vallejo, S. Perez Davila, P. Gonzalez, R.A. Sousa, A.L. Oliveira, Supercritical CO_2 technology: the next standard sterilization technique? Mater. Sci. Eng. C 99 (2019) 520–540.

[44] S.D. Manjare, K. Dhingra, Supercritical fluids in separation and purification: a review, Mater. Sci. Energy Technol. 2 (2019) 463–484.

[45] R.L. Matos, S.A.B. Vieira de Melo, E.C.M. Cabral Albuquerque, N.R. Foster, Dense CO_2 technology: overview of recent applications for drug processing/formulation/delivery, Chem. Eng. Process. Process Intensif. 140 (2019) 64–77.

[46] Y. Hiraga, A. Duereh, R.L. Smith, Aspects of solvent polarity and solvent properties in developing efficient systems for processing biomass with ionic liquid mixtures and supercritical CO_2, J. Supercrit. Fluids 134 (2018) 12–20.

[47] T.J. Yoon, Y.-W. Lee, Current theoretical opinions and perspectives on the fundamental description of supercritical fluids, J. Supercrit. Fluids 134 (2018) 21–27.

[48] R.R. Mallepally, B.A. Bamgbade, A.J. Rowane, H.B. Roknib, M.S. Newkirk, M.A. McHugh, Fluid properties at high pressures and temperatures: experimental and modelling challenges, J. Supercrit. Fluids 134 (2018) 33–40.

Chapter 5

Thermodynamics and transport properties of mixtures composed of metal complexes and supercritical fluids

This chapter gives a detailed overview over the phase behavior of the binary mixtures involved in the different particle formation processes. Furthermore, experimental methods for measuring high pressure phase equilibria and binary diffusion coefficient data are presented. In addition, different approaches for the experimental determination and for modeling the solubility in supercritical media are discussed.

5.1 Introduction

High pressure technologies using supercritical fluids (SCFs) as process media have been considered as an attractive "green" alternative to the conventional chemicals which are utilized in established and new industrial processes. For the design and optimization of SCF-based processes such as extraction, impregnation and dyeing, cleaning, particle formation, reaction, and separation operations, information about high pressure phase behavior is essential. However, since deviations from ideal behavior become larger with increasing pressure, the high-pressure phase behavior becomes more complicated and difficult to predict.

Among others, one of the key properties of interest in e.g., particle formation processes such as SFRD (supercritical fluid reactive deposition) or RESS (rapid expansion of supercritical solutions) is the solubility of the solid in the SCF, since in both processes an insufficient solubility limits the practical applicability [1]. Furthermore, various investigations show, that the properties of the produced particles such as particle size and morphology [2], dissolution behavior [3] or catalytic activity [4], are often strongly influenced by the underlying phase behavior. Therefore, in case of the SFRD process, it would be extremely

Synthesis of Nanostructured Materials in Near and/or Supercritical Fluids
https://doi.org/10.1016/B978-0-444-64089-5.00034-2

helpful to be able to answer questions like "Is the solid compound sufficiently soluble in the supercritical solvent?"

Until today, several reviews of high-pressure phase behavior data are published in the literature [5–11] and a compilation of experimental solubility data in $scCO_2$ covering about 800 different substances was published by Gupta and Shim [12].

5.2 Phase behavior[a]

To estimate the feasibility of the named processes above, and to optimize the choice of operative variables, it is important to understand the phase behavior of the systems involved. Usually the binary mixture consists of a supercritical solvent (1) and a low volatile substance, i.e., the solute (2). The two components of these mixtures differ significantly in mass, size, shape, interaction strength, and polarity. Therefore, it is necessary to take a closer look to the special phase behavior of these highly asymmetric systems.

It is shown in Fig. 5.1 that the main characteristics of such systems are that the triple point temperature of the low volatile substance (TP_2) is much higher than the critical temperature ($T_{C,1}$) of the solvent, and that the solubility of the low volatile substance in the liquid phase of the solvent is quite limited and the vapor pressure curve of the solvent ($l_1 = g_1$) decreases slightly due to the dissolved solid and merges into the S_2LG (I) line. Beyond this, the solubility of the supercritical solvent in the liquid phase of the second component is limited too; that causes a

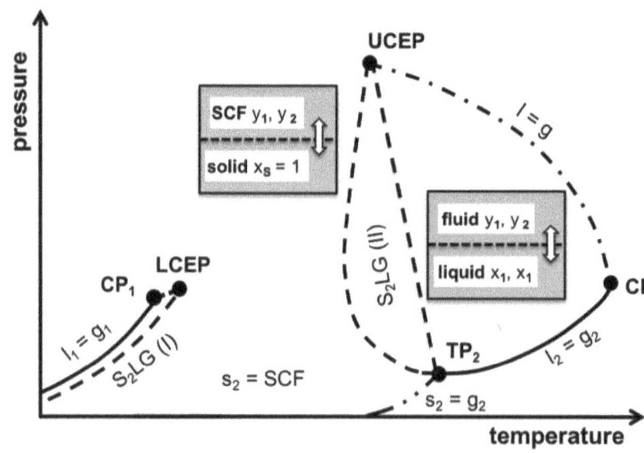

FIG. 5.1 Typical *p-T* projection for an asymmetric mixture consisting of a SCF (1) and a low volatile organic substance (2) [1]. *(Modified from M. Türk, Particle Formation With Supercritical Fluids: Challenges and Limitations, first ed., Elsevier, Amsterdam, 2014, Copyright © (2014) Elsevier.)*

a. Revised and extended version of Chapter 2.2 in Ref. [1]; Copyright © (2014) Elsevier.

lowering of the melting temperature of component 2 (S_2LG line). Depending on the nature and strength of the interaction between the components two different trends of the S_2LG line can be observed. Due to the increasing solubility of the SCF in the liquid phase, the melting temperature decreases either continuously up to the UCEP or to a certain pressure. Above this pressure, the now dominating influence of the hydrostatic pressure induces the increase of the melting temperature up to the UCEP. Thus, the mixture's critical line is interrupted at two distinguished points: the lower critical endpoint (LCEP) and the upper critical endpoint (UCEP). At both, the UCEP and the LCEP; liquid and gas phases merge into a single fluid phase in the presence of excess solid.

At a temperature close to both critical endpoints, small changes in pressure result in a considerable increase of the solubility of the solid in the SCF. Thus, it is desired to take advantage of the increased sensitivity of the solubility with respect to pressure near the UCEP, but to avoid the formation of a liquid phase.

Note that both S_2LG lines (I and II) terminate the region in which a solid-SCF equilibrium ($s_2 = SCF$) exists, i.e., the solid phase consists of only one single component. On the high temperature side of the S_2LG (I) line and on the low temperature side of the S_2LG (II) line a $s_2 = SCF$ equilibrium exists for each pressure. Thus, the S_2LG (I) line shows a lower critical solution temperature (LCST) phase behavior while the S_2LG (II) line shows, at least initially, an upper critical solution temperature (UCST) phase behavior. From the practical point of view, such lines are of great importance as borderline, above or below that line small uniform particles can be produced. In opposite thereto, for a given pressure and a temperature higher than the corresponding melting temperature, a liquid-fluid (i.e., two phase) instead of solid-SCF equilibrium exists and the formation of liquid droplets can already occur. As a consequence, larger particles with a lower dissolution behavior or catalytic activity will be formed which are principally generated from or via the liquid phase [1, 2, 4].

An example which illustrates the importance of a profound knowledge of the phase behavior for the selection of the optimum process parameters with regard to the desired product properties is given below.

Fig. 5.2 illustrates the influence of the mean particle size (d) on the catalytic activity of the preferential CO oxidation. It is obvious that the catalysts synthesized by SFRD (SFRD I and SFRD II) are characterized by a higher catalytic activity compared to the catalyst prepared by the conventional wet impregnation ($d = 27$ nm). Furthermore, the SFRD catalyst synthesized at the same pressure but at a temperature higher than the corresponding melting temperature (SFRD II) shows a mean particle size that is larger than in case of SFRD I ($d = 5.6$ vs 3.2 nm), indicating the presence of a liquid phase after adsorption of the Pt-complex molecules. Such a formation of a liquid phase causes coalescence of the molecules leading to larger Pt particles after reduction and thus affecting the dispersion of the Pt particles and therewith the activity of the catalyst. The observed differences in catalytic activity are caused by the different

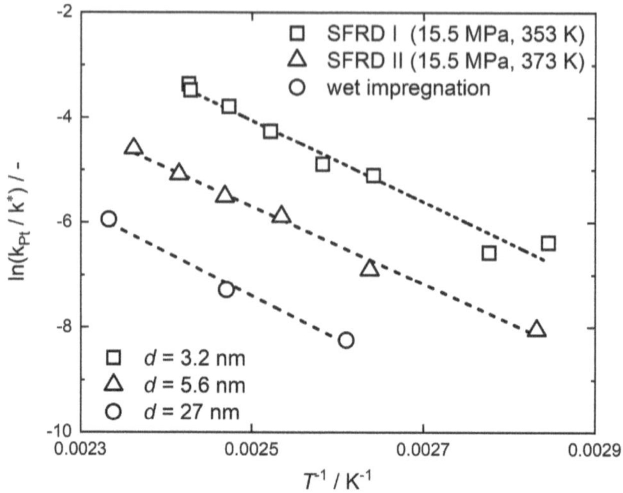

FIG. 5.2 Arrhenius plots of the preferential CO oxidation for catalysts prepared by either SFRD or wet impregnation. *(Modified from G. Incera Garrido, F.C. Patcas, G. Upper, M. Türk, S. Yilmaz, B. Kraushaar-Czarnetzki, Supercritical deposition of Pt on SnO2-coated Al2O3 foams: phase behaviour and catalytic performance, Appl. Catal. A: General 338 (2008) 58–65, Copyright © (2007) Elsevier.)*

degrees of Pt dispersions on the various catalysts, which are the highest in case of SFRD I and the lowest for the catalyst prepared by wet impregnation. Thus, this example demonstrates that the knowledge of the phase behavior of the system involved (e.g., CO_2/Pt(cod)me_2) is a key requirement for the selection of the optimal process conditions.

5.3 Solubility and binary diffusion coefficients of metal complexes in SCF

A typical pressure-solubility diagram corresponding to the temperature range in which a $s_2 = \text{SCF}$ equilibrium exists (cf. Fig. 5.1) is shown in the left part of Fig. 5.3. It is important to notice the significant increase of the solubility with pressure at a temperature close to T_{UCEP} and that no vapor-liquid equilibrium exists in this region. The influence of the pure solvent's density and temperature on the solubility is illustrated in the right part of Fig. 5.3. It is obvious that for a given density, the solubility increases due to rising temperature or, at constant temperature, almost linear with the solvent's density and hence solvent power. Note that such isotherms border the two-phase region of the binary system. On the low-density side a two-phase region exists, while on the right-hand side of these lines a homogeneous phase is formed.

It is discussed in detail in Chapter 4 that the transport properties, e.g., the viscosity, of SCFs vary with pressure and temperature. The viscosity of an

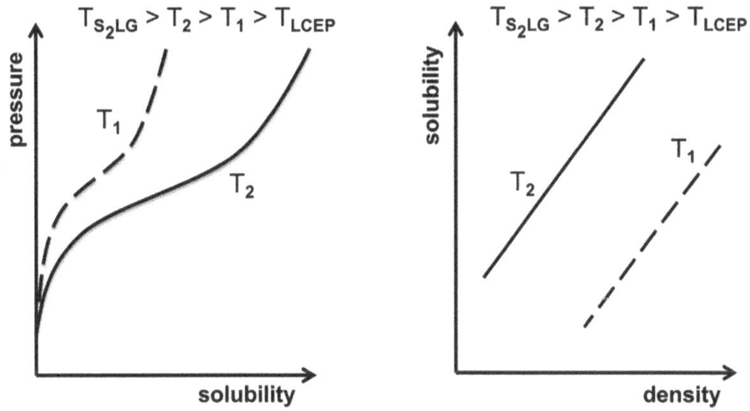

FIG. 5.3 Influence of pressure on the solubility (left diagram) and of density on solubility (right diagram) of an asymmetric mixture, consisting of a supercritical solvent (1) and a low volatile substance (2). *((Right) Modified from M. Türk, Particle Formation With Supercritical Fluids: Challenges and Limitations, first ed., Elsevier, Amsterdam, 2014, Elsevier, Copyright © (2014).)*

SCF is up to two orders of magnitude lower than that of a liquid phase, which leads to lower pressure drops in pipes and packed bed vessels and enables an enhanced mass transfer between different phases. Particularly with regard to the SFRD process another important transport property is the binary diffusion coefficient (D_{12}) of a solute in an SCF. These data are measured and reported at the limit of infinite dilution. A very comprehensive review that includes all the available data of diffusion coefficients in SCFs at infinite dilution has been written by Suarez et al. [13].

Fig. 5.4 shows the influence of pressure and temperature on D_{12} and illustrates that D_{12} of both, Pd(acac)$_2$ and Co(acac)$_3$ in scCO_2, decreases with pressure due to the strong increase of the density in the near critical region. The fact that, at constant temperature and higher pressures, D_{12} decreases slightly with pressure is due to the lower increase of the fluid's density in that region. Furthermore, according to the Stokes-Einstein relation, D_{12} increases with temperature. Obviously, the binary diffusion coefficients of metal precursors in scCO_2 are significantly higher than in (organic) solvents such as Ethanol (EtOH). These higher D_{12} values lead to enhanced mass transfer rates in comparison to organic solvents causing higher fluxes inside the pores of a solid material. Thus, in the near or supercritical region liquid-like density and therewith solubility is combined with gas-like mass transfer rate. Besides these extraordinary properties the quasi nonexistent surface tension of SCFs in combination with the relatively high binary diffusion coefficient of a solute in an SCF enables the deposition and/or the (chemical) modification of porous solids or the deposition of e.g., metallic nanoparticles into porous substrates by SFRD (see Chapter 7).

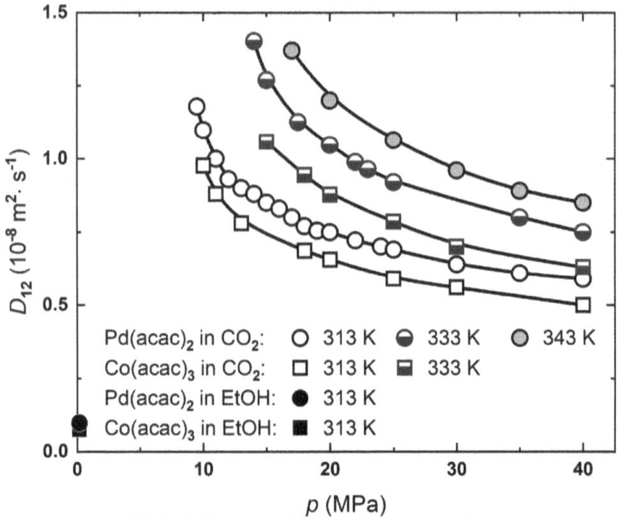

FIG. 5.4 Influence of pressure and temperature on the binary diffusion coefficient, D_{12}, of Pd(acac)$_2$ and Co(acac)$_3$ [14, 15].

An extensive overview of experimental methods for determining the diffusion coefficients in SCFs has been given by Liong et al. [16]. According to this review, the most used techniques for the determination of reliable D_{12} data in SCF systems are: the solid dissolution method, the capillary peak broadening method, the photon correlation spectroscopy method, and the nuclear magnetic resonance spectroscopy method. The essential basic principles and the underlying fundamentals of the different methods are described and their advantages and disadvantages are discussed in detail by the authors. Furthermore, a summary of the available experimental data and some trends in the relationship between the diffusion coefficient and the various solute and solvent properties are given.

A comprehensive review written by Medina summarizes initially the available diffusion coefficients for solutes in pure CO_2 and gives a short overview over the different experimental techniques for determining the diffusion coefficients in SCFs [17]. It is worthy to notice that most of the experimental data reported in literature were obtained by the capillary peak broadening method. Based on the extensive experimental database, the influence of pressure, temperature, density, and viscosity on the binary diffusion coefficients for solutes in SCFs is discussed. Furthermore, correlation methods for modeling the experimental data are presented; thereby equations based on the Stokes-Einstein approach, the Rough-Hard-Sphere theory, and other methods are presented and discussed. Moreover, experimentally determined diffusion coefficients in ternary (i.e., CO_2/co-solvent/solute) systems are presented and the influence of pressure, temperature, the type and amount of a co-solvent, and solute/co-solvent interactions is discussed. Beyond that, several equations which have been proposed for correlating and predicting the diffusion coefficients in ternary systems are presented.

5.4 Determination of high-pressure phase behavior data[b]

Detailed knowledge of binary and/or ternary phase equilibrium data, along with thermodynamic and transport data are needed for a deeper understanding of the technical and natural processes occurring at high pressures. The experimental investigation of phase equilibria remains an important and probably the most reliable source of information about phase equilibria. Especially the measurement of properties of coexisting phases at high pressures is not only of fundamental scientific interest since reliable phase equilibrium data are the key properties of a large number of processes and applications. Some typical examples are the design and optimization of separation techniques, synthesis of nanostructured materials using physical or chemical transformation, as well as biochemical and chemical reactions.

Currently, there are a large number of measurement methods and techniques available for experimental studies of the phase equilibria because usually one certain method is not suitable to determine the different phenomena. For choosing the most suitable method for a given task but also for the evaluation of experimental data, a detailed knowledge and understanding of the different methods is essential. However, it should be considered that it is difficult and expensive to obtain reliable experimental data, especially for multicompound systems at high pressures.

The following section gives a short summary (which does not claim to be comprehensive) of published knowledge about experimental methods for measuring (high pressure) phase equilibrium data. More details about the classification of experimental methods for the investigation of (high pressure) phase equilibria and about systems for which experimental (high pressure) phase-equilibrium data including vapor-liquid equilibria (VLE), liquid-liquid equilibria (LLE), vapor-liquid-liquid equilibria (VLLE), the solubility of high-boiling substances in SCFs, and the solubility of gases in liquids (GLE) are given among others in four up-to-date and very comprehensive review articles [5–7, 11].

Key problems such as sampling steps from the above-mentioned multicomponent equilibrium systems from a (high) pressure cell or technical options for their realization are described in a very extensive review article by Peper and Dohrn [18]. These topics will not be discussed in this work, as they would go beyond the scope of this work.

b. Revised and extended version of Chapter 2.6 in Ref. [1]; Copyright © (2014) Elsevier.

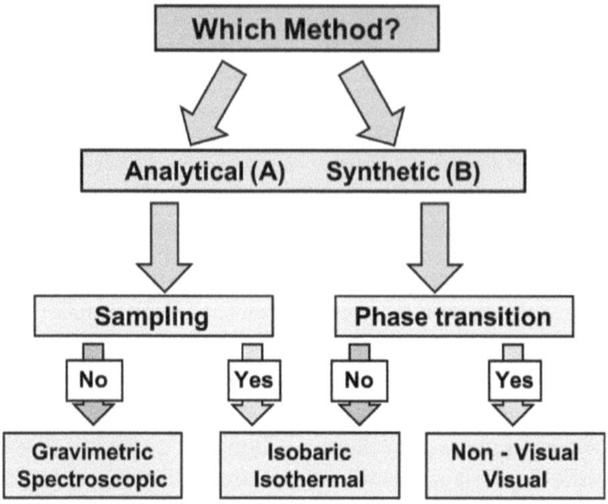

FIG. 5.5 Classification of experimental methods for measuring phase equilibrium data. *(Modified from S. Peper, J.M.S. Fonseca, R. Dohrn, High-pressure fluid-phase equilibria: trends, recent developments, and systems investigated (2009–2012), Fluid Phase Equilib. 484 (2019) 126–224; S. Peper, R. Dohrn, Sampling from fluid mixtures under high pressure: review, case study and evaluation, J. Supercrit. Fluids 66 (2012) 2–15.)*

In general, there are two main classes, depending on how the compositions of the equilibrium phases are determined and whether the mixture to be investigated has been prepared with precisely known composition or not. These experimental methods are known as (A) analytical and (B) synthetic methods (Fig. 5.5). In case of the former one, the phase equilibrium data are determined by analysis of the equilibrium phases while for the latter one the mixture of interest is precisely synthesized and phase equilibrium data are determined without analysis of the different phases.

(A) The determination of phase equilibrium data using analytical methods requires the knowledge of the compositions of the coexisting phases. Usually, at the beginning of such an experiment the equilibrium cell is filled with the mixture of interest and the composition of the mixture is only roughly known. At the desired pressure (p) and temperature (T) the mixture separates into at least two phases, e.g., a liquid and a vapor phase. The composition of these phases can be analyzed either without sampling by using spectroscopic or gravimetric methods inside the equilibrium cell under pressure without taking samples, or by sampling and analysis offline at ambient pressure outside the equilibrium cell (e.g., liquid or gas chromatography, high performance liquid chromatography). Depending on the achievement of the equilibrium condition, the analytical methods with sampling can be divided into isobaric ($p=$ constant, e.g., ebulliometer), isothermal ($T=$ constant, e.g., sampling from each single phase) and isobaric-isothermal (p & $T=$ constant, e.g., continuous-flow methods) methods. These methods are applicable to mixtures consisting of more than two components without significant difficulties, and they can be used where other (e.g., synthetic) methods fail, e.g., when phase boundaries depend strongly on the composition. Thus, the main advantage of analytical methods is the accurate determination of the composition of the different coexisting phases that yield to the complete information about the tie lines. Furthermore, sampling and analysis are separated and analysis can be conducted at ambient pressure.

(B) The determination of phase equilibrium data using a synthetic method requires the preparation of a mixture of precisely known overall composition, to measure pressure and temperature at equilibrium condition and to observe the phase behavior in an equilibrium cell. Since sampling is not necessary, equilibrium cells with a small volume can be used and synthetic methods can be utilized where other (e.g., analytic) methods fail, i.e., similar densities of the coexisting phases close or even at the (mixture's) critical point. However, it must be considered that the difficulty of determining the composition of the mixture is replaced by the problem of accurate synthesis of the mixture. Generally, synthetic methods with phase transition can be divided into visual and nonvisual methods, depending on the method how the phase transition is observed. To cause a phase transition, either pressure or temperature (or composition) inside the equilibrium cell is varied until the new (liquid or gaseous) phase appears or one of the existing phases disappears. Since the overall composition is known, each single experiment yields to one point of the p, T, x phase boundary. In case of synthetic methods with no phase transition, the equilibrium properties such as pressure, temperature, and density are measured. These methods can be divided into isobaric ($p=$ constant, e.g., ebulliometer), isothermal ($T=$ constant, static p, T, x) and others (e.g., measure density and/or volume) methods.

Interim summary: For a given problem, the optimum method depends on twice: the properties of the mixture, respectively, of the single compounds and on the phenomena of interest. For example, when the solubility of a low volatile organic substance (e.g., an organometallic compound) is measured in $scCO_2$, analytical-isothermal methods lead to reliable data sets [18]. A similar hold for the experimental determination of the solid-liquid-gas line of an organometallic compound under CO_2 pressure using a synthetic-visual method [18]. If reliable binary VLE inclusive the critical data are of interest, an analytical-isothermal-visual method is the first choice [11].

However, if multicomponent systems are investigated, synthetic methods yield less information than those with analytical methods because the tie lines cannot be determined without additional experiments. Therefore, synthetic methods are used mainly for binary systems and the main advantages of synthetic methods are that no sampling is required and that they are applicable even when the densities of the coexisting phases are similar.

5.5 Experimental procedure

A schematic diagram of the apparatus used for the determination of phase equilibrium data under high pressure (i.e., melting points of a pure substance, the

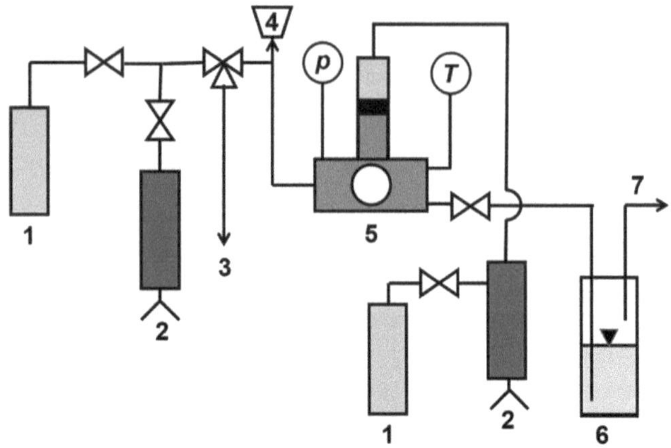

FIG. 5.6 Schematic diagram of the phase behavior set up: 1, gas cylinder; 2, manual piston pump; 3, vacuum connection; 4, pressure-relief valve; 5, high pressure variable volume view cell; 6, collecting vessel; 7, outlet; *p*, pressure indicator; *T*, temperature indicator.

S_2LG line of a binary system, and solubility of a solute in the SCF) is shown in Fig. 5.6. Typically, such studies are determined with a high-pressure variable volume view cell that is discussed in detail elsewhere [19–21]. Therefore, only a brief description of the experimental procedure is given below.

The S_2LG line of asymmetric mixtures (such as CO_2/Pt(cod)me_2 or He/Pt (cod)me_2) is usually measured according to the "first melting point" method. In a typical experiment, an excess of the solid is loaded into the high-pressure equilibrium cell. The cell is equipped with two sapphire windows, which allow the observation of the solid and the fluid under pressure. The melting point of the solid is determined either by finding the melting temperature at a constant pressure (cf. path a, Fig. 5.7; left) or finding the melting pressure at a constant temperature (cf. path b, Fig. 5.7; left).

The solubility of the solid in a SCF (e.g., scCO_2) can be measured using either the cloud-point or the static method. For a typical cloud-point experiment, a certain amount of solid is weighed with a high precise balance and loaded into the high-pressure equilibrium cell. Then the cell is sealed and carefully filled with CO_2. Cloud points (CLPs) are determined visually by noting the pressure when the solid precipitates out of the solute-laden CO_2 solution due to a density decrease and therewith loss of solvent power (cf. path c, Fig. 5.7; right). Furthermore, the solubility of a solid in CO_2 can be determined using a static method coupled with gravimetric analysis described in literature in detail [19–22]. Usually a known amount of solid inside a vial is weighed with a high precise balance and loaded into the high-pressure equilibrium cell and again, the cell is carefully filled with a known amount of CO_2. Once the desired pressure and temperature is reached and after a sufficient time (usually >12 h) to achieve equilibrium, the cell is depressurized, opened, and the vial is

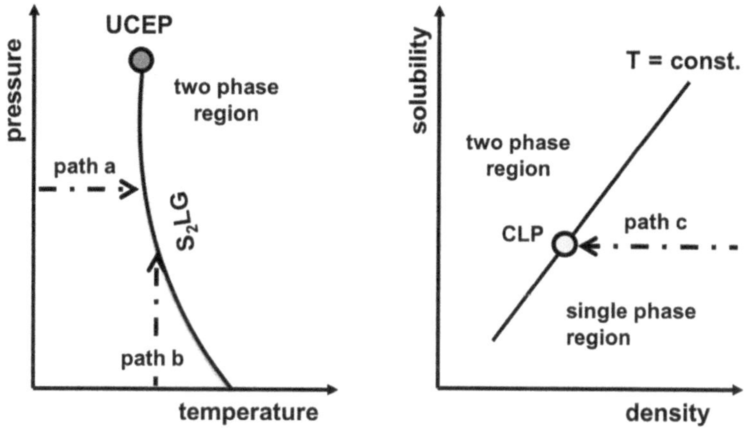

FIG. 5.7 Simplified presentation of the procedure for determining the melting point (left figure) and solubility in a SCF (right figure).

reweighed [21]. In order to minimize the influence of kinetic effects on the onset of melting or precipitation, repeated measurements (e.g., threefold) are performed.

Experimental temperature-pressure data for the S_2LG line of the binary systems CO_2/Cu(thd)$_2$, CO_2/Pt(cod)me$_2$, and He/Pt(cod)me$_2$ are plotted in Fig. 5.8. Due to the relatively high CO_2 solubility in the liquid phase, the S_2LG line

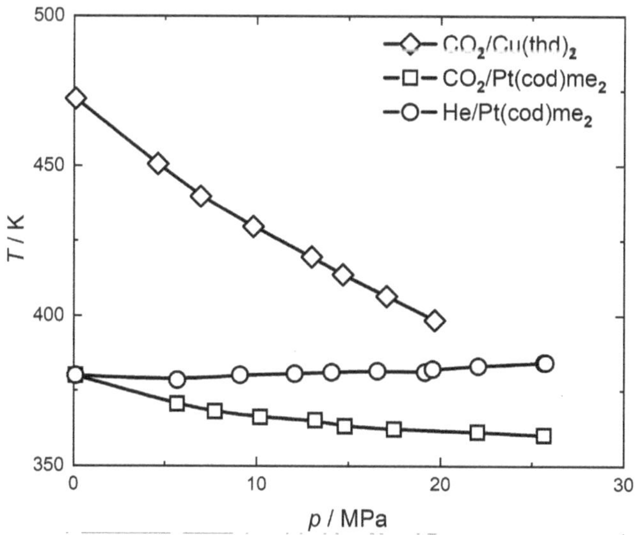

FIG. 5.8 Influence of pressure on the melting temperature of Cu(thd)$_2$ and of Pt(cod)me$_2$. *(Experimental data are taken from M. Türk, M. Crone, G. Upper, Effect of gas pressure on the phase behaviour of organometallic compounds, J. Supercrit. Fluids 58 (2011) 1–6.)*

decreases continuously as the pressure increases in the range investigated. In case of $Cu(thd)_2$ the melting temperature decreases from 472 K at ambient pressure to 399 K at 20 MPa while for $Pt(cod)me_2$ a reduction of the melting temperature from ambient temperature and 380 to 360 K at 26 MPa is observed. Thus, a solid-fluid two-phase equilibrium (s_2 = SCF, cf. Fig. 5.1) exists in the case of the former precursor for pressures below 20 MPa and temperatures from 305 to 399 K and for the latter substance for temperatures between 305 and 360 K at pressures below 26 MPa [21].

An opposite behavior is observed due to the extremely low solubility of He in $Pt(cod)me_2$. In case of this system the solid-liquid phase transition of Pt(cod) me_2 is mainly affected by the increase of volume during melting. Thus, the melting temperature rises slightly from 380 K at ambient pressure with increasing He pressure up to 384 K at 26 MPa due to the dominating influence of the hydrostatic pressure [21].

Note that in case of the mixtures $CO_2/Cu(thd)_2$ and $CO_2/Pt(cod)me_2$ a UCST phase behavior while for the system $He/Pt(cod)me_2$ a LCST phase behavior is observed. Thus, the s_2 = SCF equilibrium exists for the CO_2 systems at temperatures below and for the He system at temperatures above the S_2LG line.

5.6 Modeling solubility of metal complexes in SCFs[c]

Besides reliable experimental measurements of the solubility of a solid in SCFs, the accurate correlation of these data is essential for design of particle formation processes and the determination of their best operating conditions. Until today, approaches based on equations of state (EoS), in combination with the corresponding mixing rules, are most widely used to correlate and predict the solubility in SCFs. However, the importance of the accurate knowledge of the required thermophysical data, such as critical data, acentric factor, molar volume, and sublimation pressure of the solids, has been demonstrated by a large number of investigations, e.g., [19, 20]. The results show that, especially for pharmaceutical compounds, bio-molecules, or organometallic compounds, the common estimation methods which are not based on EoS lead often to inconsistent and unreliable results for p_C, T_C, and sublimation pressure (or acentric factor) of the above-mentioned substances [20]. Therefore, due to the lack of information on these data, empirical density-based models are often used for the correlation of experimental solubility data. A very extensive overview on the various empirical models which are often applied are summarized in Refs. [8–10].

Until today, one of the most common used models has been proposed by Stahl et al. [23] and by Kumar and Johnston [24]. Their approach relates the solubility of a solid compound in a SCF to the temperature and density of the pure SCF:

c. Revised and extended version of Chapter 2.4.3 in Ref. [1]; Copyright © (2014) Elsevier.

$$\ln(y_2) = a + b \cdot (\rho_1) \tag{5.1}$$

where the subscripts 1 and 2 refer to the pure SCF and to the dissolved solid respectively. Thus, y_2 is the solubility of the solid in the SCF, ρ_1 the density of the pure SCF at the equilibrium temperature T and pressure p, and a and b are two empirical constants which depend on the temperature.

A semiempirical density-based model which enables calculating and predicting the solubility of a large number of solids in $scCO_2$ over a wide density and temperature range has been published by Mendez-Santiago and Teja [25]. Thereby it should be considered, that the enhancement factor E:

$$E = \frac{y_2 \cdot p}{p_{2,\mathrm{sub}}(T)} \tag{5.2}$$

Can be regarded as a normalized solubility because it removes the effect of the sublimation pressure and is defined as the ratio of the mole fraction of the solid, y_2, over the solubility in an ideal gas:

$$T \cdot \ln(E) = A + B \cdot \rho_1 \tag{5.3}$$

Since the constants A and B are independent of temperature, the solubility data for binary systems at different temperatures should collapse to a single straight line when plotted in terms of $[T \cdot \ln E]$ vs the CO_2 density. Thus, Eq. (5.3) allows determining the self-consistency of experimental data and allows identifying data sets which are not consistent with other data. Due to the fact that reliable sublimation pressure data for the calculation of E are needed but often not available, Mendez-Santiago and Teja incorporated a Clausius-Clapeyron-type expression for the sublimation pressure:

$$\ln\left(\frac{p_{2,\mathrm{sub}}(T)}{p_0}\right) = a + \frac{b}{T} \tag{5.4}$$

And derived a semiempirical relation for the solid solubility [24]. This equation has three adjustable parameters and is given below:

$$T \cdot \ln(y_2 \cdot p) = A + B \cdot \rho_1 + C \cdot T \tag{5.5}$$

Since the constants A, B, and C are independent of the temperature, again, the solubility data for binary systems at different temperatures should collapse to a single straight line when plotted vs the CO_2 density:

$$T \cdot [\ln(y_2 \cdot p) - C^*] = A^* + B^* \cdot \rho_1 \tag{5.6}$$

Furthermore, the Mendez-Santiago and Teja approach (Eqs. 5.3, 5.5, 5.6) can be used to extrapolate solubility data to other temperatures.

However, it is important to notice that for both approaches (i.e., Kumar and Johnston and Mendez-Santiago and Teja) the lower limit of the linear behavior is about half, while the upper limit is around the twofold of the critical density of the solvent.

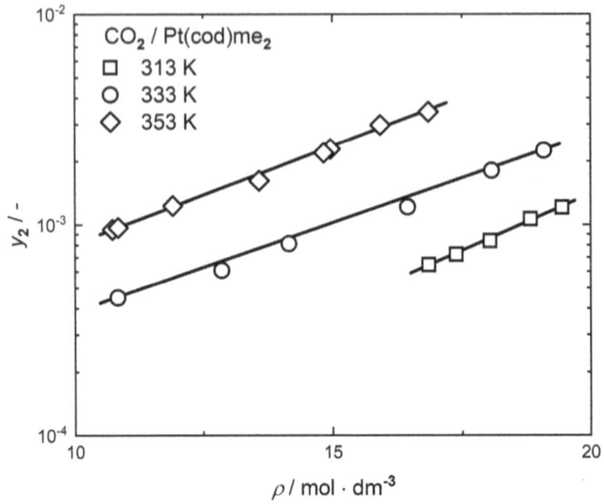

FIG. 5.9 Influence of CO_2 density and temperature on solubility. The *solid lines* represent Eq. (5.1). *(Modified from M. Türk, M. Crone, G. Upper, Effect of gas pressure on the phase behaviour of organometallic compounds, J. Supercrit. Fluids 58 (2011) 1–6, Copyright © (2011) Elsevier.)*

Fig. 5.9 shows the solubility of Pt(cod)me_2 in scCO_2 as a function of the pure CO_2 density for different temperatures. It is obvious that these experimental results show trends which are similar to those observed for other solids in the supercritical region. The continuous lines depicted in Fig. 5.9 are calculated with Eq. (5.1) and confirm the predicted linear relationship between the logarithmic solubility and the density of pure CO_2 for all three isotherms. For a given temperature, the solubility increases almost linear with the solvent density and therewith solvent power. The results shown in Fig. 5.9 illustrate the well-known temperature effect on the solubility for pressures above the retrograde region in which the influence of the temperature on the solid sublimation pressure overlays the effect of the solvent density, resulting in an increase of the solid solubility with increasing temperature.

These experimental results show trends which are similar to those observed for other solids in the supercritical region. Furthermore, the comparison between experimental and calculated solubility data show that the experimental data are satisfactorily correlated with the approach proposed by Kumar and Johnston with an overall ARD[d] of 2.8% [21].

Previous investigations performed by Upper showed, that the sublimation pressure data for Pt(cod)me_2 published in literature lead to inconsistent results [26]. Thus, due to the lack of reliable experimental sublimation pressure data,

---

d. $\mathrm{ARD} = \frac{1}{N}\sum_1^N \frac{|y_{calc} - y_{exp}|}{y_{exp}}$.

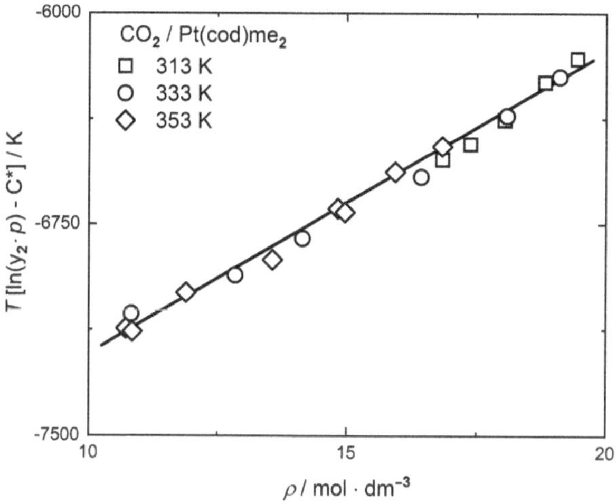

FIG. 5.10 $T\{\ln(y_2\,p)-C^*\}$ vs CO_2 density for the CO_2/Pt(cod)me$_2$ system. The *solid line* represents Eq. (5.6). *(Modified from M. Türk, M. Crone, G. Upper, Effect of gas pressure on the phase behaviour of organometallic compounds, J. Supercrit. Fluids 58 (2011) 1–6. Copyright © (2011) Elsevier.)*

Eq. (5.6) was used to verify the consistency of the experimental solubility data depicted in Fig. 5.9 (more details can be found in literature [21]).

The result is shown in Fig. 5.10; it shows, that all experimental solubility data collapse to a single line and that the straight-line fit is excellent over a temperature range from 313 to 353 K. Furthermore, from the good agreement between the experimental and calculated solubility data one can conclude that the obtained experimental results are independent from the measurement technique (i.e., cloud point or static method). This fact confirms the consistency and reliability of these experimental solubility data.

In addition to the density-based models described above, an approach based on cubic EoS is used for the calculation of the solubility of Pt(cod)me$_2$ in CO_2 as a function of the pressure. Eq. (5.7) is the result of the equifugacity condition between the solid and the fluid phase, under the assumption that:

- the solid phase consists of only one single component
- the volume of the solid phase is pressure and temperature independent.

$$y_2(p,T) = \frac{p_{2,\text{sub}}(T)\cdot\varphi_{2,\text{sub}}(p_{2,\text{sub}},T)}{p\cdot\varphi_2(p,T,y_2)}\exp\left(\frac{v_2(p-p_{2,\text{sub}})}{RT}\right) \tag{5.7}$$

Again, in Eq. (5.7) subscript 2 refers to the pure solid and thus $p_{2,\text{sub}}$ is the sublimation pressure of Pt(cod)me$_2$ at temperature T, $\varphi_{2,\text{sub}}$ is the fugacity coefficient at the sublimation pressure, φ_2 is the fugacity coefficient for Pt(cod)

me_2 in CO_2, and v_2 is the molar volume of the Pt(cod)me_2. Thus, the calculation of the solubility y_2 requires the knowledge of the solid sublimation pressure, $p_{2,sub}$, solid molar volume, v_2, and the fugacity of Pt(cod)me_2 in CO_2, φ_2, which can be calculated with a reliable EoS (e.g., the PR EoS in combination with the van der Waals mixing rules) while $\varphi_{2,sub}$ is equal to 1.

It is described more detailed in the literature [20, 26], that the boiling temperature at ambient conditions, T_B, is the most important parameter for the determination of the above-mentioned parameter of a pure solid, since this parameter allows the estimation of the critical data (i. e. T_C and p_C) and the acentric factor ω [20]. Thus, both the experimental solubility data and the melting temperature of Pt(cod)me_2 were used to fit the binary interaction parameter k_{12} and T_B for the system CO_2/Pt(cod)me_2. The calculated results obtained with Eq. (5.7) are depicted in Fig. 5.11 and lead to ARD values from 6.4% at 313 K up to 21% at 353 K. More details about this approach can be found elsewhere [26].

Interim summary: It can be stated that the cubic EoS in combination with the van der Waals mixing rules, which contain two adjustable parameters, can be successfully applied to binary systems. This approach leads to deviations between experimental and calculated solubility data which are up to one order of magnitude higher than for both empirical density-solubility correlations. However, from an engineer's point of view, it should be considered that the empirical approaches can only correlate the solubility as a function of the

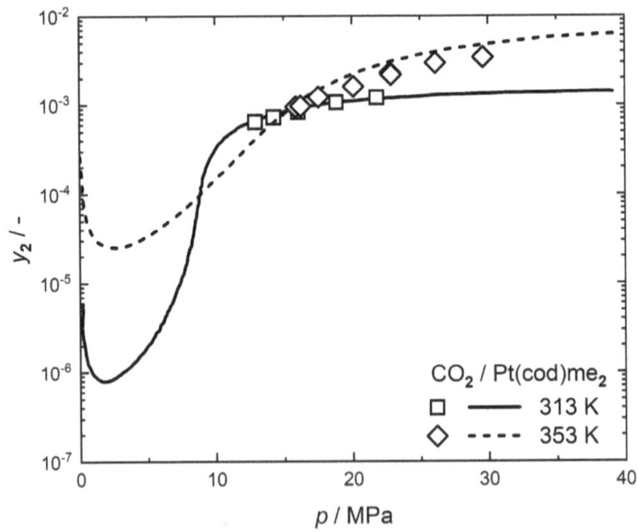

FIG. 5.11 Influence of pressure and temperature on the solubility of Pt(cod)me_2 in CO_2; the *lines* represent the correlation using Eq. (5.7). *(Data from G. Upper, Reaktivabscheidung von Metallen aus überkritischen Fluiden zur Herstellung von funktionsoptimierten Katalysatoren (PhD thesis), University of Karlsruhe (TH), Germany, 2009.)*

density while the EoS approach allow the practically more important correlation between solubility, pressure and temperature. The advantage of cubic EoS is that they provide reliable information on high pressure systems and can be readily used for interpolation and carefully limited extrapolation, which significantly reduces the experimental effort. It is also important to keep in mind that all models which are based on EoS, require substance-specific parameters which can only be calculated if the parameters p_C, T_C, and vapor/sublimation pressure (or acentric factor) of both pure components are known.

5.7 Experimental determination and modeling of the solubility of salts in sub and supercritical H_2O

For the experimental determination of the salt solubility in sub and supercritical H_2O either a batch or continuous flow apparatus can be used [27]. Usually, the measuring and analyzing techniques which are commonly applied include visual technology, neutron ray technology, on- and offline diagnostics, and combination of two or more technologies together.

Typically, in the case of the batch apparatus, a homogeneous solution of a known salt concentration is filled into a cell which is equipped with two sapphire windows and heated up until the salt precipitates. Thereby, the phase transformation can be observed visually or by light extinction.

In the second method, a cold aqueous salt solution is injected into a pure sub or supercritical H_2O stream. Thereby high supersaturations are created which result in a rapid precipitation of small salt particles. Thereby, synchrotron wide angles X-ray scattering (WAXS) in situ analysis can be used to determine the mean size of the particles created [28, 29]. In addition, at the end of the experiment, the cell can be opened and the salt particles are collected for scanning electron microscopy (SEM) and X-ray analysis.

More details about the different measurement methods applied and their further developments can be found in various (review) articles [30–38]. For the current state of the art and common knowledge up to 2018, the paper written by Ding et al. is highly recommended [38]. In this review article, experimental solubility data of KCl, LiCl, NaCl, $CaCl_2$, $MgCl_2$, $Al_2(SO_4)_3$, K_2SO_4, $MgSO_4$, Na_2SO_4, KNO_3, $LiNO_3$, $NaNO_3$, K_2HPO_4, K_3PO_4, NaH_2PO_4, Na_2HPO_4, CuO, PbO, Na_2CO_3 and KOH in sub and supercritical H_2O were correlated by seven empirical and semiempirical approaches (density based-, enthalpy-, c_p-, Flory-Huggins-, ionization-, second and third order polynomial model) published in literature. Depending on the respective salt type (1 or 2), these experimental data cover a temperature range from 330 to 825 K and pressures ranging from 9 to 99 MPa. In addition, the authors determined solubility data of Na_2SO_4 and K_2SO_4 in H_2O at temperatures and pressures ranging from 633 to 723 K and 18 to 26 MPa.

Based on the results of the comparison between the experimental and with the different modeling approaches calculated solubility data, a new four-parameter semiempirical model (cf. Eq. 5.8) was proposed by the authors and used to correlate the above listed solubility data. In summary, the authors conclude that for the description of the solubility behavior of the 20 different salts the new model is more effective than the other approaches named above.

$$\ln(y_2) = \frac{a}{R \cdot T} + \frac{b}{R} + c \cdot (p_1)^2 + d \cdot p_1 \tag{5.8}$$

In Eq. (5.8) the subscripts 1 and 2 refer to the pure H_2O and to the dissolved salt, respectively, while a, b, c and d are adjustable parameters.

The influence of the chemical nature of the salt (Type 1 NaCl, Type 2 Na_2SO_4) on the phase behavior at supercritical temperatures is schematically shown in Fig. 5.12 according to the data published by Armellini and Tester [27]. Note that a Type 1 system exhibits a continuous critical locus while in a Type 2 system the critical locus is interrupted (cf. Fig. 8.5 in Chapter 8). For the H_2O/Na_2SO_4 system the LCEP is close to the critical point of pure H_2O, while the UCEP is at approximately 114 MPa and 713 K [27]. Obviously, in case of the Type 1 system a UCST phase behavior is observed while for the Type 2 system a LCST phase behavior is observed. Thereby is worthy to notice that in case of a Type 1 system and subcritical temperatures an opposite behavior, i.e., the solubility decreases with increasing temperature is observed [39, 40]. Thereby it is worthy to notice that in case of a Type 1 system and subcritical temperatures (< 585 K) and pressures equal or below 25 MPa an opposite behavior, i.e., the solubility of NaCl in H_2O decreases with increasing temperature is observed [39, 40].

However, it must be considered that in case of the Type 1 system the solubility is 5 times smaller at high temperatures, while in case of the H_2O/Na_2SO_4 system the solubility is about three orders of magnitude lower if the temperature is increased from 674 to 773 K [27]. Thus, the H_2O/NaCl system

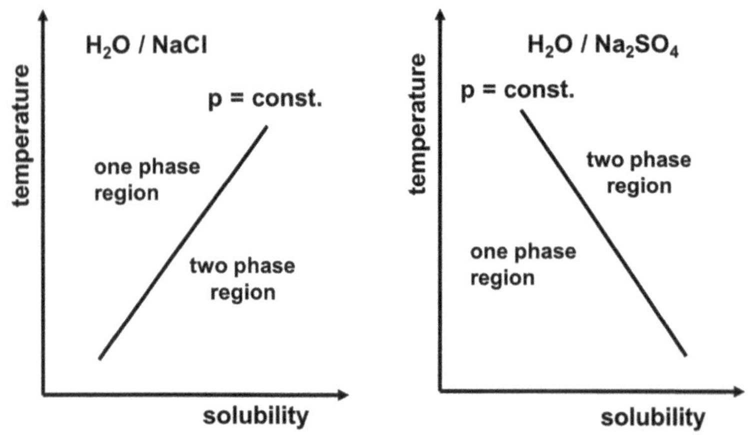

FIG. 5.12 Simplified representation of the influence of temperature on salt solubility for a Type 1 (left) and Type 2 (right) system at supercritical pressures and temperatures.

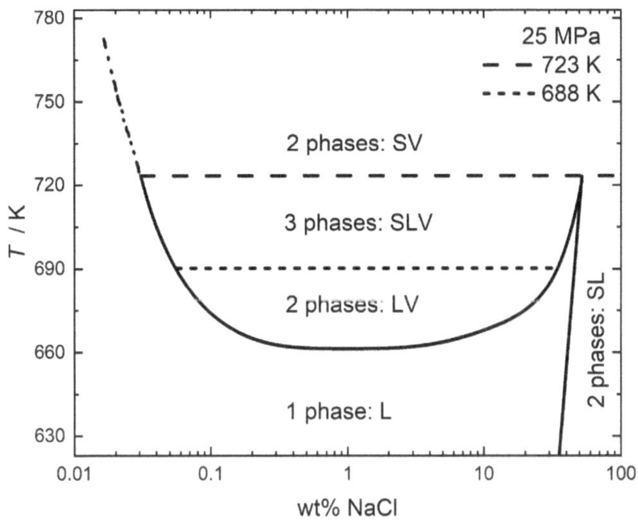

FIG. 5.13 Schematic illustration of the temperature-composition phase diagram of the H_2O/NaCl system at 25 MPa. *(Modified from T. Voisin, A. Erriguible, C. Aymonier, Influence of multiphasic systems on salt(s) solubility in supercritical water: the case of NaCl and NaCl-Na2SO4, J. Supercrit. Fluids 152 (2019) UNSP 104567, Copyright © 2019 Elsevier.)*

exhibits a more complex and much different phase behavior than the H_2O/Na_2SO_4 system at the same pressure.

Fig. 5.13 shows schematically the H_2O/NaCl phase diagram, determined at 25 MPa [29, 37, 40, 41]. Key features of this mixture are the existence of a two-phase liquid-vapor (LV) region at temperatures higher than 663 K, which is above the critical temperature of pure H_2O (647 K) and that for temperatures below 663 K a large one-phase region exists. At a temperature of about 723 K, three phases (solid-liquid-vapor, SLV) are in equilibrium while at temperatures higher than 723 K, a two-phase equilibrium (solid-vapor, SV) is observed. Recently, Voisin et al. have shown that in the temperature range from about 688–723 K a (solid-liquid-vapor, SLV) three-phase region exists, which illustrates the complexity of the binary phase diagram of the H_2O/NaCl system [37]. It needs to be mentioned that such a triphasic region has never been predicted by the earlier studies on the phase behavior of the binary system H_2O/NaCl. Furthermore, between 723 and 973 K, only very low NaCl concentrations (<0.02 wt%) are found in scH_2O.

References

[1] M. Türk, Particle Formation With Supercritical Fluids: Challenges and Limitations, first ed., Elsevier, Amsterdam, 2014.

[2] A. Diefenbacher, M. Türk, Phase equilibria of organic solid solutes and supercritical fluids with respect to the RESS process, J. Supercrit. Fluids 22 (3) (2002) 175–184.

[3] M. Türk, Manufacture of submicron drug particles with enhanced dissolution behaviour by rapid expansion processes, J. Supercrit. Fluids 47 (2009) 537–545.

[4] G. Incera Garrido, F.C. Patcas, G. Upper, M. Türk, S. Yilmaz, B. Kraushaar-Czarnetzki, Supercritical deposition of Pt on SnO_2-coated Al_2O_3 foams: phase behaviour and catalytic performance, Appl. Catal. A: General 338 (2008) 58–65.

[5] R. Dohrn, S. Peper, J.M.S. Fonseca, High-pressure fluid-phase equilibria: experimental methods and systems investigated (2000–2004), Fluid Phase Equilib. 288 (2010) 1–54.

[6] R. Dohrn, S. Peper, J.M.S. Fonseca, Experimental methods for phase equilibria at high pressures, Annu. Rev. Chem. Biomol. Eng. 3 (2012) 343–367.

[7] J.M.S. Fonseca, R. Dohrn, S. Peper, High-pressure fluid-phase equilibria: experimental methods and systems investigated (2005–2008), Fluid Phase Equilib. 300 (2011) 1–69.

[8] M. Skerget, Z. Knez, M. Knez-Hrncic, Solubility of solids in sub- and supercritical fluids: a review, J. Chem. Eng. Data 56 (2011) 694–719.

[9] Z. Knez, D. Cör, M. Knez-Hrncic, Solubility of solids in sub- and supercritical fluids: a review: 2010-2017, J. Chem. Eng. Data 63 (2018) 860–884.

[10] W.H. Teoh, R. Mammucari, N.R. Foster, Solubility of organometallic complexes in supercritical carbon dioxide: a review, J. Organomet. Chem. 724 (2013) 102–116.

[11] S. Peper, J.M.S. Fonseca, R. Dohrn, High-pressure fluid-phase equilibria: trends, recent developments, and systems investigated (2009–2012), Fluid Phase Equilib. 484 (2019) 126–224.

[12] R.B. Gupta, J.J. Shim, Solubility in Supercritical Carbon Dioxide, Taylor & Francis, Boca Raton, FL, 2007.

[13] J.J. Suarez, I. Medina, J.L. Bueno, Diffusion coefficients in supercritical fluids: available data and graphical correlations, Fluid Phase Equilib. 153 (1998) 167–212.

[14] M. Toriumi, R. Katooka, K. Yui, T. Funazukuri, C.Y. Kong, S. Kagei, Measurements of binary diffusion coefficients for metal complexes inorganic solvents by the Taylor dispersion method, Fluid Phase Equilib. 297 (2010) 62–66.

[15] C.Y. Kong, Y.Y. Gu, M. Nakamura, T. Funazukuri, S. Kagei, Diffusion coefficients of metal acetylacetonates in supercritical carbon dioxide, Fluid Phase Equilib. 297 (2010) 162–167.

[16] K.K. Liong, P.A. Wells, N.R. Foster, Diffusion in supercritical fluids, J. Supercrit. Fluids 4 (1991) 91–108.

[17] I. Medina, Determination of diffusion coefficients for supercritical fluids, J. Chromatogr. A 1250 (2012) 124–140.

[18] S. Peper, R. Dohrn, Sampling from fluid mixtures under high pressure: review, case study and evaluation, J. Supercrit. Fluids 66 (2012) 2–15.

[19] M. Türk, T. Kraska, Experimental and theoretical investigation of the phase behavior of naproxen in supercritical CO_2, J. Chem. Eng. Data 54 (5) (2009) 1592–1597.

[20] M. Türk, M. Crone, T. Kraska, A comparison between models based on equations of state and density-based models for describing the solubility of solutes in CO_2, J. Supercrit. Fluids 55 (2010) 462–471.

[21] M. Türk, M. Crone, G. Upper, Effect of gas pressure on the phase behaviour of organometallic compounds, J. Supercrit. Fluids 58 (2011) 1–6.

[22] G. Sherman, S. Shenoy, R.A. Weiss, C. Erkey, A static method coupled with gravimetric analysis for the determination of solubilities of solids in supercritical carbon dioxide, Ind. Eng. Chem. Res. 39 (2000) 846–848.

[23] E. Stahl, W. Schilz, E. Schütz, E.A. Willing, A quick method for the microanalytical evaluation of the dissolving power of supercritical gases, Angew. Chem. Int. Ed. 17 (1978) 731–738.

[24] S.K. Kumar, J.P. Johnston, Modelling the solubility of solids in supercritical fluids with density as the independent variable, J. Supercrit. Fluids 1 (1988) 15–22.

[25] J. Mendez-Santiago, A.S. Teja, The solubility of solids in supercritical fluids, Fluid Phase Equilib. 158–160 (1999) 501–510.

[26] G. Upper, Reaktivabscheidung von Metallen aus überkritischen Fluiden zur Herstellung von funktionsoptimierten Katalysatoren (PhD thesis), University of Karlsruhe (TH), Germany, 2009.

[27] F.J. Armellini, J.W. Tester, Experimental methods for studying salt nucleation and growth from supercritical water, J. Supercrit. Fluids 4 (1991) 254–264.

[28] T. Voisin, A. Erriguible, G. Philippot, D. Ballenghien, D. Mateos, D. Kunegel, F. Cansell, B.B. Iversen, C. Aymonier, Investigation of the precipitation of Na_2SO_4 in supercritical water, Chem. Eng. Sci. 174 (2017) 268–276.

[29] T. Voisin, A. Erriguible, D. Ballenghien, D. Mateos, D. Kunegel, F. Cansell, C. Aymonier, Solubility of inorganic salts in sub- and supercritical hydrothermal environment: application to SCWO processes, J. Supercrit. Fluids 120 (2017) 18–31.

[30] I. Leusbrock, S.J. Metz, G. Rexwinkel, G.F. Versteeg, The solubility of magnesium chloride and calcium chloride in near-critical and supercritical water, J. Supercrit. Fluids 53 (2010) 17–24.

[31] I. Leusbrock, S.J. Metz, G. Rexwinkel, G.F. Versteeg, The solubilities of phosphate and sulfate salts in supercritical water, J. Supercrit. Fluids 54 (2010) 1–8.

[32] M. Schubert, J.W. Regler, F. Vogel, Continuous salt precipitation and separation from supercritical water. Part 1: type 1 salts, J. Supercrit. Fluids 55 (2010) 99–112.

[33] M. Schubert, J.W. Regler, F. Vogel, Continuous salt precipitation and separation from supercritical water. Part 2. Type 2 salts and mixtures of two salts, J. Supercrit. Fluids 52 (2010) 113–124.

[34] M. Schubert, J. Aubert, J.B. Mueller, F. Vogel, Continuous salt precipitation and separation from supercritical water. Part 3: interesting effects in processing type 2 salt mixtures, J. Supercrit. Fluids 61 (2012) 44–54.

[35] G. Lemoine, H.A. Turc, A. Leybros, J.C. Ruiz, Y.S. de Gelicourt, H. Muhr, Na_2CO_3 and K_3PO_4 solubility measurements at 30 MPa in near-critical and supercritical water using conductimetry and high pressure calorimetry, J. Supercrit. Fluids 130 (2017) 91–96.

[36] Q. Liu, X. Ding, B. Du, T. Fang, Multi-phase equilibrium and solubilities of aromatic compounds and inorganic compounds in sub- and supercritical water: a review, Crit. Rev. Anal. Chem. 47 (2017) 513–523.

[37] T. Voisin, A. Erriguible, C. Aymonier, Influence of multiphasic systems on salt(s) solubility in supercritical water: the case of NaCl and NaCl-Na_2SO_4, J. Supercrit. Fluids 152 (2019). UNSP 104567.

[38] X. Ding, T. Zhang, S. Zhang, Z. Shen, T. Fang, Experimental determination and modelling of the solubilities of sodium sulfate and potassium sulfate in sub- and supercritical water, Fluid Phase Equilib. 483 (2019) 31–51.

[39] X. Ding, Y. Lei, Z. Shen, Y. Yu, Q. Zhou, J. Wei, T. Fang, Experimental determination and modeling of the solubility of sodium chloride in subcritical water from (568 to 598) K and (10 to 25) MPa, J. Chem. Eng. Data 63 (2017) 3374–3390.

[40] J.L. Bischoff, K.S. Pitzer, Liquid-vapor relations for the system NaCl-H_2O: summary of the p-T-x surface from 300°C to 500°C, Am. J. Sci. 289 (1989) 217–248.

[41] J.L. Bischoff, R.J. Rosenbauer, K.S. Pitzer, The system NaCl-H_2O: relations of vapor-liquid near the critical temperature of water and of vapor-liquid-halite from 300°C to 500°C, Geochim. Cosmochim. Acta 50 (1986) 1437–1444.

Chapter 6

Thermodynamics and kinetics of adsorption of metal complexes on surfaces from supercritical solutions

A wide variety of applications, which involve metal complexes and SCFs, involve adsorption of metal complexes on surfaces in the presence of SCFs. For example, supercritical deposition for formation of metallic films or nanoparticles on surfaces involves adsorption of metal complexes on such surfaces. The adsorbed metal complexes are then converted to their metallic forms leading to metallic films or nanoparticles. Knowledge of both the thermodynamics and the kinetics of adsorption are necessary for design and development of large-scale processes. This chapter deals with the adsorption phenomena in SCFs starting with a brief introduction to adsorption on surfaces. Subsequently, the thermodynamics and the kinetics of adsorption of metal complexes from SCFs are covered. The high-pressure experimental techniques to measure adsorption isotherms in the presence of SCFs and to study the kinetics of adsorption are described. The models for describing the thermodynamics and the kinetics of adsorption of metal complexes on surfaces in the presence of SCFs are presented.

6.1 Adsorption

Adsorption is the selective transfer of certain components of a fluid phase, called solutes to the surface of an insoluble solid. The adsorbed solute is referred to as an adsorbate, while the solid material is referred to as an adsorbent. When an adsorbent is exposed to a fluid phase, molecules in the fluid phase diffuse to its surface (including its pores if it is a porous adsorbent), where they either chemically bond with the solid surface or are held there physically by weak van der Waals intermolecular forces [1]. When adsorption is caused by van der Waals forces, it is referred to as physical adsorption or physisorption, whereas it is called chemical adsorption or chemisorption if a chemical bond is formed between the adsorbate and the adsorbent.

Synthesis of Nanostructured Materials in Near and/or Supercritical Fluids
https://doi.org/10.1016/B978-0-444-64089-5.00047-0

6.1.1 Physisorption

Physisorption from a pure fluid phase occurs when the intermolecular attractive forces between the adsorbent and pure fluid (solute) molecules are greater than those between the solute molecules themselves. Thus, physisorption is analogous to condensation, an exothermic process and is therefore accompanied by dissipation of heat. The enthalpy change accompanying physical adsorption is given by

$$\Delta H = \Delta G + T\Delta S \tag{6.1}$$

The entropy change of the adsorbate, ΔS, is necessarily negative since the adsorbed state is more ordered than the unadsorbed state due to a loss of at least one degree of freedom. For an adsorption process, which occurs spontaneously, the change in Gibbs free energy, ΔG, is negative, hence the enthalpy change, ΔH, accompanying physical adsorption is always negative (exothermic). An increase in adsorption in temperature then invariably results in a monotonic decrease of uptake. The heat of physical adsorption can be less than or greater than the heat of vaporization and changes with the extent of adsorption. Physical adsorption occurs rapidly because it does not require an activation energy. However, slow uptake may be observed on finely porous adsorbent such as zeolites and some carbons due to diffusion limitations rather than the sorption process itself. For gases at temperatures above their critical temperature, adsorption is confined to a monolayer (unimolecular) since condensation of a second layer on the adsorbed layer is not possible; on the other hand, below the critical temperature, adsorption maybe multilayer (multimolecular) in coverage. Physisorption is reversible and the density of the physisorbed phase is of the order of magnitude of the liquid phase rather than the vapor phase.

6.1.2 Chemisorption

Chemisorption involves the transfer of electrons between the adsorbate and the adsorbent. Because chemisorption occurs through chemical bonding of the adsorbate with the surface of the adsorbent, it often occurs at temperatures above the critical temperature of the adsorbate. As it is the case with most chemical reactions, chemical adsorption requires activation energy. Additionally, the chemisorbed species are more localized on the surface compared to physisorption, and the migration of adsorbate molecules at the surface is more restricted.

6.2 Adsorption isotherms

The amount adsorbed n, on an adsorbent is a function of the absolute temperature, T, pressure, P, and the interaction potential, E, between the adsorbate and the adsorbent. Thus, at equilibrium, the amount of adsorbate per unit adsorbent can be given by

$$n = f(T, P, E) \tag{6.2}$$

Usually, n is measured at constant T so that the equation reduces to

$$n = f(P, E) \tag{6.3}$$

A plot of n as a function of P at constant T provides the adsorption isotherm for a pure gas or vapor. For gas mixtures, P is generally replaced by the partial pressure of the solute in the gas or vapor phase. For liquid or supercritical fluid systems, P is replaced by the concentration of the solute in the liquid or supercritical fluid phase. There exists a vast amount of data in the literature on adsorption isotherms for various adsorbents and adsorbates. Most of these isotherms fit into one of five types of adsorption isotherm as classified first by Brunauer, Deming, Deming and Teller (BDDT) based upon an extensive literature survey [2]. The five isotherms are shown in Fig. 6.1, which also includes the type VI isotherm. The type VI isotherm features a series of steps, thus referred to as the stepped isotherm [3]. This classification of the isotherms is adapted by the International Union of Pure and Applied Chemistry (IUPAC) [4].

The type of the adsorption isotherm of a particular system can provide information on the surface characteristics and pore properties of the adsorbent. The reversible Type I isotherm is obtained when adsorption is limited to at most a few molecular layers. This condition, where the curve is concave to the P/P_o axis and n approaches a limiting value as P/P_o approaches 1, is usually encountered in chemisorption or in physisorption with microporous materials having

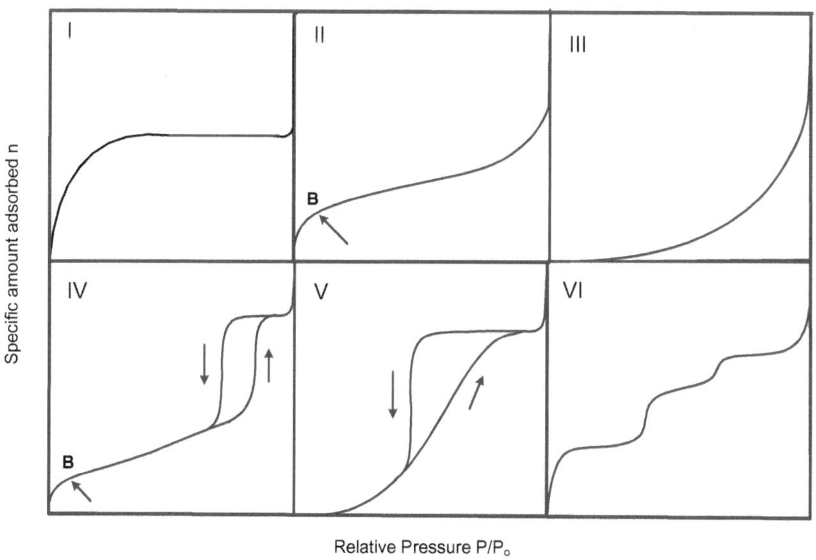

FIG. 6.1 The five types of adsorption isotherms as classified by Brunauer, Deming, Deming and Teller, together with type VI, the stepped isotherm.

small external surfaces such as activated carbons, molecular sieve zeolites and certain porous oxides. The reversible Type II isotherm is typical of nonporous or macroporous adsorbents and represents unrestricted monolayer-multilayer adsorption. The inflection point B at the beginning of the almost linear middle section of the isotherm, is indicative of the completion of the monolayer coverage and the beginning of multilayer adsorption. The reversible Type III is convex to the P/P_o axis over its entire range and consequently, does not display an inflection point B. This isotherm is relatively uncommon, and is characteristic of a system where the adsorbate-adsorbent interaction is weak. It is distinguished principally by heats of adsorption, which are less than the heat of liquefaction of the adsorbate so that additional adsorption is facilitated because the adsorbate-adsorbed layer interaction is greater than the adsorbate-adsorbent interaction. Type IV isotherms exhibit hysteresis loops, which are associated with capillary condensation taking place in the mesopores, and a limiting uptake over a range of high P/P_o. The initial part of the isotherm is attributed to monolayer-multilayer adsorption since it follows the same path as the corresponding part of a Type II isotherm obtained with the given adsorptive on the same surface area of the adsorbent in a nonporous form. Type V isotherm is uncommon and is similar to the Type III isotherm such that the adsorbent-adsorbate interaction is weak. The Type VI isotherm represents the stepwise multilayer adsorption on a uniform nonporous surface. Examples for this type are those obtained with argon or krypton on graphitized carbon blacks at liquid nitrogen temperature.

6.2.1 Adsorption isotherm models

A number of models have been developed to describe mathematically the various adsorption isotherms. These models are helpful in understanding the interactions between the components in the adsorption system such as adsorbate-adsorbent and adsorbate-adsorbate interactions. From an engineering point of view, they are also beneficial in the design and scale up of adsorption processes.

6.2.1.1 The Langmuir model

Using the kinetic theory of gases (KTG), Langmuir [5] was able to describe the Type I isotherm with the assumption that adsorption was limited to a monolayer. The adsorbent surface is conceived as an array of independent and energetically equivalent adsorption sites for localized adsorption (one molecule per site). If θ_o is the fraction of bare adsorption sites, the rate of adsorption is directly proportional to the pressure (or concentration) and the fraction of bare sites according to the KTG:

$$r_{ads} = kP\theta_0 A_1 \tag{6.4}$$

where A_1 is the condensation coefficient and represents the probability of a molecule being adsorbed upon collision with the surface, P is the adsorbate pressure and k is a constant equal to $N_{\mathrm{Avo}}/(2\pi MRT)^{1/2}$ such that N_{Avo} is the Avogadro's number, M is the adsorbate molecular weight, R is the gas constant and T is the absolute temperature. On the other hand, the rate of desorption is proportional to the number of molecules adsorbed and the energy of adsorption, E, given by

$$r_{\mathrm{des}} = n_m \theta_1 v_1 e^{-E/RT} \tag{6.5}$$

where n_m is the number of adsorbate molecules in a completed monolayer, θ_1 is the fraction of occupied sites and ν_1 is the vibrational frequency of the adsorbate normal to the surface when adsorbed. The term $e^{-E/RT}$ represents the probability that an adsorbed molecule possesses adequate energy to overcome the net attractive potential of the surface. At equilibrium, the rates of adsorption and desorption are equal, thus equating Eqs. (6.4), (6.5) and recognizing that $\theta_o = 1 - \theta_1$ yields

$$kPA_1 - \theta_1 kPA_1 = n_m \theta_1 v_1 e^{-E/RT} \tag{6.6}$$

Solving for θ_1 gives

$$\theta_1 = \frac{kPA_1}{n_m v_1 e^{-E/RT} + kPA_1} \tag{6.7}$$

If K is defined as

$$K = \frac{kA_1}{n_m v_1 e^{-E/RT}} \tag{6.8}$$

the Langmuir isotherm given by

$$\theta_1 = \frac{KP}{1 + KP} \tag{6.9}$$

The adsorption energy E is taken constant in invoking Eq. (6.8), which implicitly assumes surface homogeneity (all adsorption sites are energetically identical) and the absence of lateral interactions between adsorbed molecules. A large value of K indicates strong adsorption bonding and the larger the value of K, the greater is the fractional surface coverage at fixed T and P, or the higher is the temperature required for a specified fractional surface coverage at a fixed P. At low values of P, the fraction of the sites covered is directly proportional to P, but at high values of P, the fraction approaches unity asymptotically and becomes essentially independent of P.

6.2.1.2 The Freundlich model

The Freundlich equation is originally an empirical expression that describes an adsorption isotherm where there is a linear correlation between log of adsorption capacity and the log of adsorbate pressure (or concentration). The Freundlich isotherm is represented by

$$n = KP^a \tag{6.10}$$

where n is the amount of solute adsorbed per unit amount of adsorbent used, K and a are empirical constants and P is the equilibrium partial pressure (or concentration) of the adsorbate in the bulk phase. The empirical constants in Eq. (6.10) are determined by fitting the model to adsorption isotherm data. The Freundlich equation is one of the most widely used adsorption isotherms, albeit it lacks a theoretical basis [6]. The value of the Freundlich equation is found in instances where the adsorbate is not actually identified [7]. However, the model is only useful for limited adsorbate concentration (or pressure) and temperature ranges; and it has no predictive capability with regards to adsorption isotherms for similar adsorbates. A new plot must be drawn for each adsorbate-adsorbent system.

6.3 Adsorption of supercritical fluids

The adsorption of supercritical fluids on flat surfaces or on surfaces of porous particles is different in nature from the adsorption of vapors or gases [8]. One of the interesting behaviors of SCF adsorption, which is not observed in gas adsorption, is the occurrence of a maximum or "hump" in the adsorption isotherm when the pressure is sufficiently high and the temperature is close to the critical point. This can be seen in Fig. 6.2 for adsorption of CO_2 on a zeolite. As pressure is further increased beyond the critical, adsorption monotonically

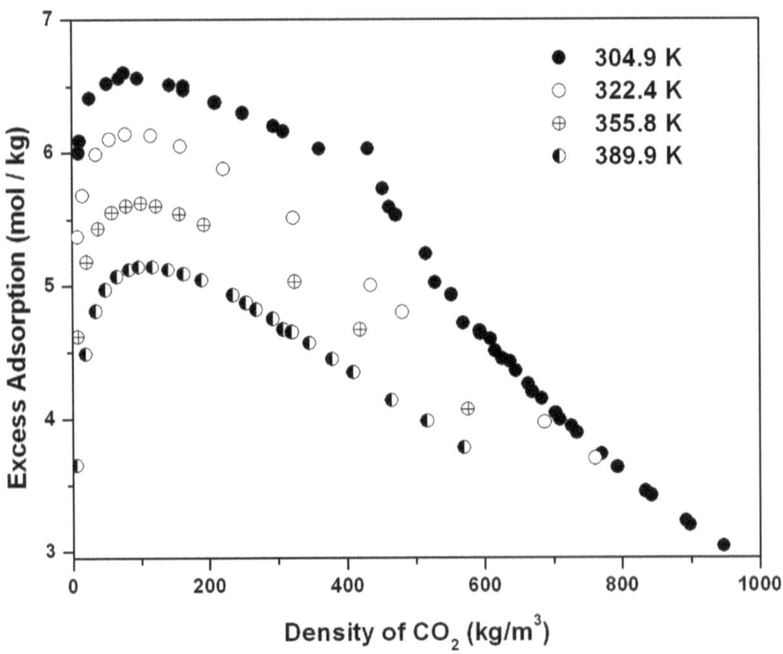

FIG. 6.2 Adsorption isotherm of CO_2 on zeolite.

decreases [9]. This unique behavior is associated to the highly compressible state of the fluid close to the critical point and presently, is difficult to describe quantitatively [10]. Another different feature of SCF adsorption is that the SCF cannot be condensed on flat surfaces and even in the mesopores [11]. It has also been shown that there is enhanced adsorption of supercritical gases (NO, N_2, CH_4) on modified microporous carbons [12–14]. Enhanced adsorption in microporous materials even at very low pressures has also been demonstrated by statistical simulation studies [15, 16]. It is postulated that due to enhanced and overlapping adsorption potential in micropores, both the adsorbate-adsorbent and adsorbate-adsorbate interactions are also enhanced, resulting in increased adsorption. The adsorption of supercritical N_2 and O_2 on pure and Ce/Zr-doped carbon aerogels were recently studied and gas adsorption was correlated with the physico-chemical properties of the selected adsorbents [11]. Although the Ce/Zr-doped carbon aerogels have wider micropores than their pure counterparts, they have higher uptakes, which the authors attribute to specific and nonspecific interactions between the adsorbent and adsorbate molecules. The study demonstrated that the differences in supercritical gas adsorption on various adsorbents can not only be explained on the basis of pore structure.

Understanding the nature of phase transitions of fluids confined in porous media is an important issue. Phase behavior in the critical region in porous systems provides important clues as to the effects of fluid confinement. Wong and Chan [17] measured the coexistence boundary confined in a highly porous aerogel and found a narrow coexistence curve for the confined fluid, enclosed fully within the bulk fluid's coexistence curve. A small shift in the critical temperature was noted, and the value of the critical exponent was calculated [18]. High pressure adsorption studies by Findenegg and coworkers [19–21] on nonporous graphite and controlled pore glass uncovered interesting features of adsorption in the vicinity of the critical region. For adsorption above the critical temperature (T_c) on nonporous homogeneous surfaces the surface excess isotherms generally pass through a sharp maximum, beyond which the surface excess amount decreases. Beneath T_c, excess adsorption increases as the partial pressure approaches the saturation pressure (P_{sat}), beyond which there is a discontinuity in the isotherm as the bulk fluid condenses. At very high temperatures the adsorption resembles gaseous phase adsorption. The phase behavior of confined fluids in a controlled pore size glass with narrow pore size distribution was also studied and it was observed that there was small a shift in the critical temperature in this system as well [22]. The shift was attributed to the weak fluid-pore interactions in the system and the large pore sizes of the material used. Recently, Morishige et al. [23] measured the capillary critical temperature shift for several pure fluids in MCM-41 with much smaller pore sizes, and the authors observed much larger critical temperature shifts. The shifts are much more pronounced at the smaller pore sizes, as theoretically expected. However, the dependence of the confined fluid's critical temperature on the average pore size was not consistent with theoretical predictions, highlighting the current lack of understanding of the critical behavior of fluids in confined pores.

6.4 Adsorption of solutes from supercritical mixtures

Thermodynamics of adsorption of solutes onto surfaces from supercritical mixtures is important for a wide range of applications which involve SCFs. Some of these applications are extraction of solutes from solid matrices, impregnation of inorganic and organic porous matrices, regeneration of catalysts and adsorbents, supercritical fluid chromatography and selective separations by adsorption. Thermodynamics of adsorption from supercritical mixtures are also quantified using adsorption isotherms. Adsorption studies in the literature have been concerned mainly on priority pollutants (such as toluene, benzene [24], nitroaromatic compounds [25], ethyl benzene [26], heavy molecular weight organics [27, 28] and DDT [29]), food such as coffee aroma compounds (ethylacetate and furfural) [30] and terpenes [31], and model pharmaceutical drugs (salycilic acid and benzoic acid) [32, 33]. Activated carbon was the adsorbent commonly used in these studies. The studies in the literature on adsorption isotherms from supercritical solutions are summarized in Table 6.1. By far, most of the existing data have been modeled with the most common adsorption isotherm models—the Langmuir and the Freundlich models.

The enthalpy of adsorption is also a very useful property because it provides directly the strength of the interaction between the adsorbate molecule and the surface of the adsorbent. The common technique is using the van't Hoff plot based on the adsorption isotherms at different temperature [78]. However, the heat of adsorption thus obtained is not accurate. This enthalpy change represents the energy transfer of one mole of solute from the nonideal, supercritical solution onto the adsorbent surface at isobaric and isothermal operations. This quantity, however, is not the heat of adsorption as conventionally defined which represents the energy transfer of one mole of solute from an ideal gas solution onto the adsorbent surface [79]. The energy effects associated with an adsorption process at supercritical conditions should be clearly defined. By equating differentials of the logarithm of the fugacity of the solute in each of the phases [80], the pressure and temperature dependence of K_l is obtained as:

$$\left(\frac{\partial \ln K_2}{\partial T}\right)_P = \frac{(h_2^{IG} - \overline{h}_2^m) + \Delta H_2^{\text{ads}}}{RT^2} + \alpha^m \tag{6.11}$$

$$\left(\frac{\partial \ln K}{\partial P}\right)_T = \frac{\overline{v}_2^m}{RT} - k^m$$

The superscripts, m and IG, represents the mobile phase and ideal gas, respectively. The $\Delta H_2^{\text{ads}} = \overline{h}_2^S - h_2^{\text{IG}}$ is defined as the heat of adsorption of the solute on the solid matrix, $\overline{h}_2^m$ and $\overline{h}_2^S$ are the infinite-dilution partial molar enthalpy of solute in the mobile and stationary phases, respectively, $\overline{h}_2^{\text{IG}}$ is

TABLE 6.1 Adsorption isotherm studies for solute-adsorbent-scCO_2 systems.

Solute	Adsorbent	Temperature (K)	Pressure (MPa)	Correlation	Experimental method	Refs.
α-Tocopherol	Silica gel Silica	313 313	20 16–26	Anti-Langmuir	PC	[34] [35]
δ-Tocopherol	Silica gel Silica	313 313	20 16–26	Anti-Langmuir	PC	[34] [35]
Artemisinin	Silica gel	308–328	14.1–18.1	Freundlich	FAC	[36]
Benzoic acid	Silica gel	318–328	16	Linear	ECP	[37]
Benzene	AC AC	313.2–333.2 308–328	0.99–11.8 6.2–16.3	Dubinin-Astakhov Toth	FAC BA	[38] [39]
Cholesterol	Magnesium silicate	313	24.1	Langmuir		[40]
C.I. Acid Red 52	Silk Wool	313	14	Langmuir	BA	[41]
C.I. Disperse Orange 30	Polyester	368	30	Langmuir	BA	[42]
DDT	AC	313.1–333.1	10.4–16.7	Freundlich	FAC	[29]
DHA-EE	C18-bonded silica	318–338	10–21.9	Linear Langmuir BET	BA	[43, 44]
p-Dichlorobenzene	Silica gel	298–318	7.5–17.8	Linear	ECP	[45]

Continued

TABLE 6.1 Adsorption isotherm studies for solute-adsorbent-$scCO_2$ systems—cont'd

Solute	Adsorbent	Temperature (K)	Pressure (MPa)	Correlation	Experimental method	Refs.
2,6, & 7-Dimethylnaphthalene	NaY-type zeolite	308	9–19.8	Radke-Prausnitz	BA	[46]
Eicosane	AC	324	11.45	Langmuir	FAC	[47]
EPA-EE	C18-bonded silica	318–338	10–21.9	Linear Langmuir BET	BA	[43, 44]
Ethylacetate	AC	304–312	12.5–20	Linear	FAC	[30]
Ethyl benzene	AC	313–338	10–13	Langmuir	FAC	[26]
Furfural	AC	308–312	20–29	Linear	FAC	[30]
Hexachlorobenzene	Soil AC	298–318 308–318	11.3 11.3	Linear Langmuir	FAC FAC	[48] [27]
n-Hexane	ACF	308–328	7.7–14	Langmuir	BA	[49]
1,2-Hexanediol	AC	324	11.45	Langmuir	FAC	[47]
Ketoprofen	Silica aerogels	313	18	Langmuir	BA	[50]
MEK	ACF	308–328	7.7–14	Langmuir	BA	[49]
Miconazole	Silica aerogels	313	18	Langmuir	BA	[50]

Naphthalene	Soil AC Silica aerogels	308–328 308–318 313	10 9.9 18	Linear Langmuir Langmuir	FAC FAC BA	[48] [27] [50]
Pentachlorophenol	Soil AC	298–318 308–318	11.3 11.3	Linear Langmuir	FAC FAC	[48] [27]
Phenanthrene	Soil AC	308–328 308–318	10 10.2	Linear Langmuir	FAC FAC	[48] [27]
Oleic acid	Silica gel Neusilin	313	13–30	N/A	FAC	[51]
Orange oil	Silica gel	313	8.8	Langmuir	FAC	[52]
Sunflower oil	Silica gel Neusilin	313	13–30	N/A	FAC	[51]
Terpene mixtures	Silanized silica	310–320	13–17	Langmuir	FAC	[31]
Toluene	AC ACF Silica gel	308–328 308–328 308–318	7.5–13.8 8–11.4 7.5–17.8	Langmuir Langmuir Linear	FAC BA ECP	[53] [49] [45]
m-Xylene	AC	313–323	10–15	Freundlich	FAC	[54]
Nickelocene	MCM-48	343	11–20	Langmuir Modified Langmuir Freundlich Redlich-Peterson BET	BA	[55]

Continued

TABLE 6.1 Adsorption isotherm studies for solute-adsorbent-$scCO_2$ systems—cont'd

Solute	Adsorbent	Temperature (K)	Pressure (MPa)	Correlation	Experimental method	Refs.
Ruthenocene	MCM-48 AC	333	11–20	Langmuir Modified Langmuir Freundlich Redlich-Peterson BET	BA	[55]
2-Methyl-1, 4-naphthquinone (Vit K_3)	Alginate aerogel	313	15, 20	Langmuir	BA	[56]
Cholecalciferol (Vit D_3)	Alginate aerogel	313	15, 20	Langmuir	BA	[56]
PGMEA	AC	313–353	10–20	Dubinin-Astakhov	BA	[57]
2-Phenylethyl alcohol	SA	323	15	Langmuir	BA	[58]
Toluene	AC	313, 353	4.2–15	Dubinin-Astakhov	FAC	[59]
Acetone	AC	313, 353	4.2–15	Dubinin-Astakhov	FAC	[59]
n-Hexane	AC	313, 353	4.2–15	Dubinin-Astakhov	FAC	[59]
n-Decane	AC	313, 353	4.2–15	Dubinin-Astakhov	FAC	[59]
Methanol	AC	313, 353	4.2–15	Dubinin-Astakhov	FAC	[59]
$Ni(acac)_2$	CA	333	30	Linear	BA	[60]

$Pd(hfac)_2$	SBA-15	313, 353	8.5–14	Langmuir Modified Langmuir Freundlich	BA	[61]
Z-ligustilide	C18-bonded Silica	305–323	12–18	Langmuir Freundlich	ECP	[62]
$Rh(acac)_3$	MCM-41 HMS MSU-H	313–353	15	Langmuir	FAC	[63]
$Pt(cod)me_2$	RFA	308	10.7, 20.7	Modified Langmuir	BA	[64]
Buriti oil	γ-Al_2O_3	333	15–25	Langmuir	FAC	[65]
2-Methyl-1,4-naphthquinone (Vit K_3)	Starch aerogel	313, 333	15	Pseudo-second order model	BA	[66]
α-Tocopherol (Vit E)	Starch aerogel	313, 333	15	Langmuir Freundlich Temkin	BA	[66]
$Rh(acac)_3$	MCM-41 HMS MSU-H	313–353	15	Dubinin-Astakhov	BA	[67]
Toluene	AC Mesoporous carbon Mesoporous silica	313–353	4.2–15	Dubinin-Astakhov	BA	[68]
Acetone	AC Mesoporous carbon Mesoporous silica	313–353	4.2–15	Dubinin-Astakhov	BA	[68]

Continued

TABLE 6.1 Adsorption isotherm studies for solute-adsorbent-scCO_2 systems—cont'd

Solute	Adsorbent	Temperature (K)	Pressure (MPa)	Correlation	Experimental method	Refs.
n-Hexane	AC Mesoporous carbon Mesoporous silica	313–353	4.2–15	Dubinin-Astakhov	BA	[68]
Nimesulide	SA	313–333	21	Freundlich	BA	[69]
Co(acac)$_3$	MCM-41	313–353	12.5–20	Dubinin-Astakhov	BA	[70]
Pt(cod)me$_2$	CA	308	10.6	Langmuir Sips	BA	[71]
CuID6	CA	308	10.6	Langmuir Sips	BA	[71]
Pt(cod)me$_2$ and CuID6	CA	308	10.6	IAST with Langmuir parameters	BA	[71]
Pt(cod)me$_2$	CA4 CA22	353	27.7	Langmuir	BA	[72]
Pt(cod)me$_2$	BP2000	333	20	Langmuir	BA	[73]
Pd(acac)$_2$	BP2000	333	20	Linear	BA	[73]
Ru(cod)(tmhd)$_2$	CA4	353	27.6	Langmuir	BA	[74]
Ru(cod)(tmhd)$_2$	CA22	353 333–353	27.6 19.3	Langmuir Modified Langmuir	BA	[74] [75]

n-Octane	AC	313–353	10–15	Dubinin-Astakhov	BA	[76]
Propylene glycol monomethyl ether	AC	313–353	10–15	Dubinin-Astakhov	BA	[76]
Ethanol	AC	313–353	10–15	Dubinin-Astakhov	BA	[76]
2-Propanol	AC	313–353	10–15	Dubinin-Astakhov	BA	[76]
Benzaldehyde	Polymeric resin	313, 323	10.20–20.68	Langmuir	FAC	[77]
Benzyl alcohol	Polymeric resin	323, 333	20.68	Langmuir	FAC	[77]
Citric acid	Cyano column (ZorBaxSB-CN)	313, 321	13, 15.05	Anti-Langmuir (Hill)	ECP	[77]

AC, activated carbon; *ODS*, octadecyl-bonded silica; *PC*, pulse chromatography; *TIBC*, time integration of the breakthrough curves; *MSG*, molecular sieving carbon; *BA*, batch adsorption; *ACF*, activated carbon fiber; *EPA-EE*, eicosapentaenoic acid ethyl ester; *DHA-EE*, docosahexaenoic acid ethyl ester; *MEK*, methyl ethyl ketone; *PGMEA*, propylene glycol monomethyl ether acetate; *acac*, acetylacetonate; *CA*, carbon aerogel; *ECP*, elution by characteristic points; *cod*, cyclooctadiene; *me*, methyl; *RFA*, resorcinol-formaldehyde aerogel; *IAST*, ideal adsorbed solution theory; *MSA*, maize starch aerogel; *CAA*, calcium alginate aerogel; *N/A*, not available.

the enthalpy of the solute in the ideal gas state, $\overline{v}_2^m$ is the infinite-dilution partial molar volume of the solute in mobile phase, and α^m and κ^m are the volume expansivity and the isothermal compressibility of the mobile or supercritical phase, respectively. The partial molar volume of the solute in the stationary phase has been neglected in Eq. (6.12) because $\overline{v}_2^m$ takes a large negative value near the critical point of the solvent [81, 82]. If the binary interaction parameters between the solute and the solvent are known, $\overline{v}_2^m$, κ^{m}, α^{m} and $h_2^{\mathrm{IG}} - \overline{h}_2^m$ can be calculated using EOS-based methods [83], and therefore $\Delta H_2^{\mathrm{ads}}$ can be calculated from Eq. (6.11).

One significant difficulty encountered when considering supercritical adsorption processes is that, in addition to the fluid phase solute concentration, both the system temperature and the density of the supercritical fluid influence the adsorption equilibrium. This compares with conventional adsorption processes in which temperature is the only significant variable apart from fluid phase concentration. Generally, the parameters in isotherm models vary as a function of both temperature and density that limits the applicability of the equations when they are fitted to experimental data. To date, the density and temperature dependence of the isotherm parameters has not been established due partly to insufficient data.

The supercritical solvent may play an important role in adsorptive processes, and that the interaction between the SCF and the adsorbent surfaces and its effect on the solute adsorption may not be negligible [84, 85]. It is a common practice to treat the supercritical solvent as a continuum, with solutes simply partitioning between the surface and the fluid. In order to have a better understanding of adsorption and other heterogeneous processes involving supercritical fluids, it is important to obtain information about the adsorption-desorption behavior of SCFs with the adsorbent of interest. Unfortunately, when an adsorbate gas mixture is contacted with an adsorbent, the size and structure of the adsorbed phase as well as the actual density and composition profiles of the adsorbed phase cannot be experimentally measured [86]. What are measured are the bulk gas properties such as density, temperature and pressure. The complexity of the real physicochemical structures of commonly used microporous/mesoporous adsorbents also adds to the complexity of the problem The amorphous adsorbents, such as activated carbons, silica and alumina gels, and the polymeric adsorbents consist of an interconnected network of micro- and mesopores of various sizes and shapes. The walls of these pores generally have different surface chemistries from region to region. It is not possible by current techniques to quantitatively characterize the heterogeneous pore structure and surface chemistry of these materials with precision [81].

According to Squires and Paulaitis [87] the adsorption of CO_2 at experimental conditions far removed from the critical point of CO_2 could be neglected. Competition between the SCF solvent and the adsorbate for the adsorbent adsorption sites may also be possible especially for systems involving low solubility solutes or weak adsorbing solute species [32]; however, at conditions away from the critical point such competing effects may be diminished [31].

6.5 Adsorption of metal complexes on surfaces

Supercritical fluid aided incorporation technologies have been extensively exploited to deposit thin metal films onto a wide range of surfaces and incorporate metallic particles into different inorganic and organic substrates for microelectronic, optical and catalytic applications. The adsorption of the metal complex from SCF phase into the substrate phase is a crucial step for the SCF incorporation technology. The equilibrium adsorption isotherm data of the metallic complex (MC)-scCO_2-substrate system at the impregnation conditions of constant temperature and pressure provide an understanding of the interaction between the components involved and how the metal precursor partitions between the SCF and substrate. The isotherm reflects the difference between the affinities or strength of interaction of the MC to the adsorbent and the fluid phases. Moreover, the data also reflect the amount of metal that can be incorporated into the substrate. To design large-scale supercritical adsorption processes, it is necessary to understand in which way process variables, mass transfer effects and equilibrium considerations affect adsorption dynamics. Since the thermodynamic equilibrium between the solid and the fluid phases plays a crucial role in understanding the dynamics of the processes mentioned above, it is imperative to measure and study the adsorption isotherms of the solute of interest in these SCF-adsorbent systems. From an engineering standpoint, they are essential for the accurate design and modeling of supercritical adsorption processes [88, 89].

6.5.1 Experimental techniques for determination of adsorption isotherms

The adsorption isotherms can be measured by using 3 techniques.

1. Batch adsorption
2. Frontal analysis chromatography
3. Pulse chromatography

6.5.1.1 Batch adsorption

In batch adsorption, a certain amount of substrate is contacted with a supercritical fluid solution containing the dissolved metal complex of interest in a closed volume. The metal complex diffuses into the pores of the substrate and adsorb on the surface of the substrate. The system reaches adsorption equilibrium after a certain period of time where the concentration of the metal complex in the fluid phase does not change with time anymore. Measurement of the uptake

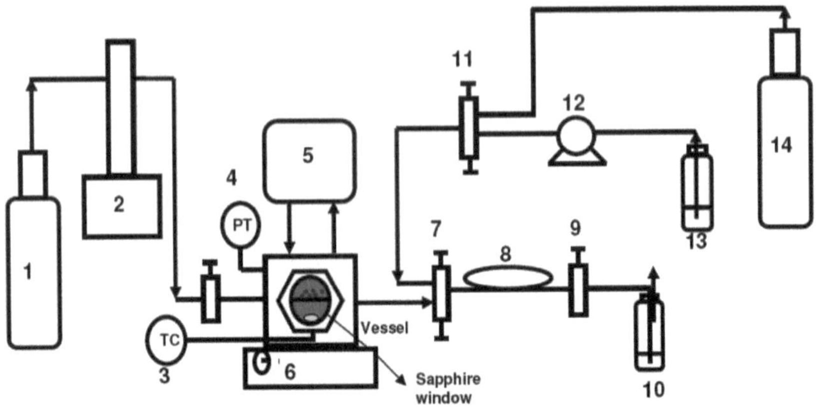

1.CO_2 Cylinder 2. Syringe Pump 3. Thermocouple Assembly 4. Pressure Transducer 5. Circulating Heater 6. Magnetic Stirrer 7. Two-way Valve 8. Sample loop 9. One-way valve 10. Collection Vial 11. Two-way valve 12. Peristaltic pump 13. Ethanol reservoir 14. N_2 Cylinder

FIG. 6.3 Set-up for determination of adsorption isotherms by batch adsorption [71].

of the substrate and the fluid phase concentration at equilibrium gives one a point on the adsorption isotherm. Carrying out experiments with different starting concentrations of the metal complex enables one to construct the whole adsorption isotherm. The experimental set-up located in our laboratory for batch adsorption measurements is given in Fig. 6.3.

The 57 mL stainless steel vessel was custom-made and equipped with two sapphire windows (2.5 cm in diameter (Sapphire Engineering, Inc., Pocasset, Massachusetts), a T-type thermocouple assembly (Omega Engineering), a pressure transducer (Omega Engineering), a vent line, and a rupture disk assembly (Autoclave Engineers). The vessel is sealed with poly(ether ether ketone) O-rings. For each experiment, a certain amount of metal complex is placed in the vessel along with a stirring bar, and a certain amount of substrate. A stainless steel screen is placed in the middle of the vessel in order to separate the substrate and the stirring bar. The system is sealed and heated to the desired temperature by a circulating heater/cooler (Cole Parmer, Model 12108-15). The system is pressurized with CO_2 using a syringe pump (ISCO, 260D) up to the desired pressure and kept at these conditions for the desired exposure time. During this process, the metal complex dissolves in the $scCO_2$ and adsorbs onto the substrate from the fluid phase. After reaching equilibrium, the system is depressurized slowly (0.7 MPa/min) using a needle valve (Autoclave Engineers), and the substrate is removed from the system.

The adsorbed amount of the metal complex is determined by measuring the concentration of the metal complex in the fluid phase. At a particular time, a small volume of the solution in the vessel is transferred and trapped in a sample loop (with a volume of 0.20 mL) by first opening and then closing a two-way valve (7 in Fig. 6.3). Subsequently, the sample loop is depressurized slowly by opening a one-way valve (9 in Fig. 6.3). The CO_2 from the loop is bubbled through an organic solution to trap any metal complex that may be in the gas stream. Then, the sample loop is flushed with a solvent using a peristaltic pump (Gilson, Model Minipuls 3) to collect any complex that precipitated during depressurization. The solutions are collected in vials and analyzed using HPLC (Agilent Technologies, 1200 Series). After each sampling, the vessel is charged with CO_2 to the original pressure to compensate for the pressure drop that occurs during the sampling process. To prepare the sample loop for the next cycle, any solvent left in the sample loop is removed by passing N_2 through the loop. At each particular time, this procedure is carried out twice in order to first remove the amount left in the short tube between the two-way valve (7 in Fig. 6.3) and the vessel during the previous sampling. To determine the fluid phase concentration and uptake of Pt(cod)me$_2$ at a particular sampling time, the following procedure is used.

First, the amount of metal complex in the sample loop for the ith sample at time t, $(m_{SL})_i$, is determined by

$$(m_{SL})_i = (m_{et})_i \times C_{HPLC} \tag{6.12}$$

where $(m_{et})_i$ is the amount of solvent (kg) that is used to dissolve the metal complex which precipitated in the sample loop and C_{HPLC} is the concentration (kg metal complex/kg solvent) of the solvent + metal complex solution which is obtained using HPLC. Then, the amount of CO_2 (kg) in the sample loop, $(m_{CO2})_{SL}$, is calculated by

$$(m_{CO_2})_{SL} = V_{SL} \times \rho_{CO_2} \tag{6.13}$$

where V_{SL} is the volume of the sample loop and ρ_{CO2} is the density of CO_2 in the sample loop.

Using the amount of metal complex and the amount of CO_2 in the sample loop, the mass fraction of the complex in the sample loop for the ith sample, $(w_{SL})_i$, is obtained by the equation:

$$(w_{SL})_i = \frac{(m_{SL})_i}{(m_{CO_2})_{SL}} \tag{6.14}$$

Since $(w_{SL})_i$ is equal to the mass fraction in the vessel at time t, $(w_{vessel})_i$, the fluid phase concentration in the vessel (mol metal complex/kg CO_2) at time t, C_i, is given by

$$C_i = \frac{(w_{vessel})_i}{MW_{metalcomplex}} \tag{6.15}$$

where $MW_{\text{metal complex}}$ is the molecular weight of the metal complex.

To calculate the uptake concentration of the metal complex by the substrate (mol metal complex/kg substrate) at equilibrium, the following equation can be used:

$$q = \frac{m_{\text{initial}} - \sum_{k=0}^{i-1} (m_{\text{SL}})_k - (m_{\text{vessel}})_i}{m_{\text{substrate}} \times MW_{\text{metal complex}}} \tag{6.16}$$

where m_{initial} is the amount of metal complex that is placed in the vessel at the beginning of the experiment, $m_{\text{substrate}}$ is the amount of substrate that is placed in the vessel and $(m_{\text{vessel}})_i$ is the amount of metal complex in the vessel at a particular time when the ith sample was taken (all amounts in kg). The value of $(m_{\text{vessel}})_i$ is calculated using the following equation:

$$(m_{\text{vessel}})_i = (w_{\text{vessel}})_i \times (m_{\text{CO}_2})_{\text{vessel}} \tag{6.17}$$

where $(m_{\text{CO2}})_{\text{vessel}}$ is the amount of CO_2 in the vessel, which was calculated by multiplying the volume of the vessel by the density of CO_2 in the vessel. In Eq. (6.16), the summation term accounts for the cumulative amount of metal complex removed from the vessel with all the samples.

To determine the adsorption isotherms, it is necessary to ensure that the system has reached equilibrium. Therefore, three or more samples can be taken for determination of the concentration in the fluid phase. When the last two fluid phase concentrations are found equal, the sampling can be stopped. In order to minimize the errors associated with replacing the removed sample volume with pure CO_2, the total volume of the samples removed from the vessel should not exceed about 5% of the total volume of the vessel. The adsorbed amount of the precursor can also be determined from the weight change of the substrate using an analytical balance after the depressurization of the vessel.

6.5.1.2 *Frontal analysis chromatography*

In this technique, a step change in the concentration of the solute is imposed at the inlet of the adsorbed bed. The response of this bed to the step change is monitored to obtain a "breakthrough curve." Analysis of these breakthrough curves enables the construction of the adsorption isotherm.

The schematic diagram of such an experimental setup is presented in Fig. 6.4.

Two syringe pumps (ISCO 260D and ISCO LC-2600) are initially filled with liquid carbon dioxide and compressed up to 100 bar using nitrogen. The stream from the first pump, 5, is passed through a preheater in the bath, where it is brought to the bath temperature, and through a column, 8, filled with the metal complex of interest. Glass wool is inserted at the ends of the column to prevent entrainment of the solute. A pure carbon dioxide stream from the second pump, 4, is mixed with the equilibrated carbon dioxide stream.

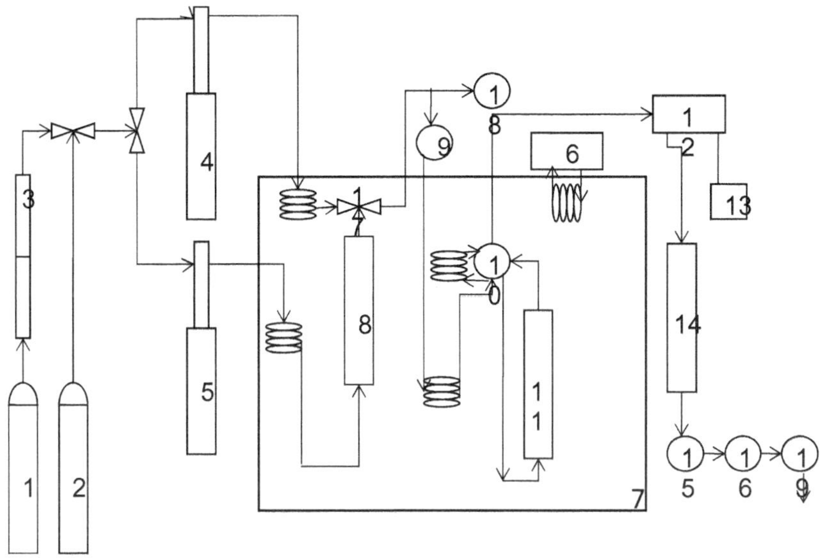

1. Nitrogen cylinder
2. Carbon dioxide cylinder
3. Piston cylinder
4. Syringe pump
5. Syringe pump
6. Immersion circulator
7. Water bath
8. Saturation
9. Sampling valve
10. Six port switch valve
11. Adsorbent bed
12. UV detector
13. Data acquisition
14. Activated carbon trap
15. Back pressure regulator
16. Back pressure regulator
17. Three way valve
18. Pressure transducer
19. Dry gas meter

FIG. 6.4 Set-up for determination of adsorption isotherms by frontal analysis chromatography.

This technique enables one to achieve different fluid concentrations by adjusting the flow rates of the two pumps. This is particularly convenient for conducting experiments at different inlet concentrations and also avoiding condensation in the lines. This metal complex-carbon dioxide mixture is passed through the sample injection valve, 9 (Rheodyne 70101, brought to the conditions of the bath with the aid of a preheater, and passed through the switching valve, 10 (Rheodyne 70001, and the high-pressure UV detector (LDC Analytical, critical extraction monitor). Finally, the metal complex is stripped from carbon dioxide by adsorption onto activated carbon, and the pure carbon dioxide stream is vented through two back pressure regulators placed in series, 15 and 16 (Tescom Inc. 26-1722-24 and Grove SD91-W). The flow rate is verified by the dry gas meter (DTM115, Singer), placed after the back pressure regulators. The temperature of the bed is controlled to 0.02°C using an immersion circular

(Haake Dl), and the pressure is measured accurately to 0.7 atm by a factory-calibrated pressure transducer (Hydronic5 A 006220 TH-1V). The response from the detector is recorded on disk through a MBC488 Metra-byte board which reads the detector voltage from a digital multimeter (Solartron 7150). Once a steady voltage is reached in the detector, the metal complex-carbon dioxide solution is diverted into the bed packed with substrate (11) using a switching valve. The effluent concentration from the bed is monitored online using the UV detector, and the breakthrough profile of the solute is recorded. The column is then isolated and the inlet concentration is increased by adjusting the flow rates of the pumps. After a steady voltage reading is achieved in the detector, the stream is diverted again to the activated carbon bed. The solute eventually breaks through, and the same procedure is repeated until the saturation concentration is reached. Thus, a set of breakthrough curves, each at different initial concentrations, is obtained. The voltage readings at each different inlet concentration are used to construct a calibration curve which is used to convert voltage vs time data to concentration vs time data. The void volume of the carbon bed is determined by a stimulus response technique, using a non-adsorbing tracer, ethylene, introduced to the system through the sample injection valve, 9. The adsorbed mass is given by a mass balance on the system as:

$$m_{\text{ads}}(C_e) = m_{\text{ads}}(C_i) + \int_0^{t^{fa}} [(C_i - C_e(t))\upsilon(t)dt]M - C_iVM\in$$

where C_i is the inlet concentration, C, is the effluent concentration, $u(t)$ is the volumetric flow rate, V is the volume of the bed, M is the molecular weight of the contaminant, c is the combined particle and bed porosity, and t^{fa} is the time when the adsorption breakthrough is reached. Adsorbed values as gram of metal complex per gram of adsorbent can be obtained by the division of the adsorbed mass by the weight of the adsorbent in the column (determined gravimetrically). Analysis of the breakthrough curves at various inlet concentrations by the above procedure enables the construction of the adsorption isotherm. The saturation concentration of the metallic complex in supercritical carbon dioxide at the operating conditions can be determined by depositing the complex on the carbon bed over a period of time. The ratio of the difference in the weight of carbon and the amount of carbon dioxide passed through the column determines the solubility of the organic in carbon dioxide.

6.5.2 Adsorption isotherms for metal complex-adsorbent-$scCO_2$ systems

The adsorption isotherm of an organometallic complex on the surface of a porous substrate was first measured by Saquing et al. [72]. The adsorption isotherms at 80°C and 27.7 MPa for Pt(cod)me_2 in $scCO_2$ were obtained for CA4 and CA22, two carbon aerogels (CAs) with an average pore size of 4 nm and with an average size of 22 nm, respectively. The data are presented in Fig. 6.5.

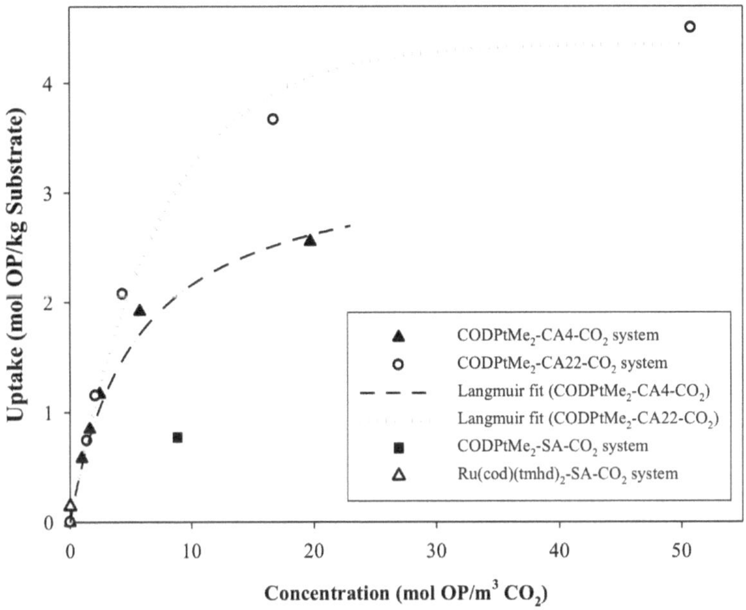

FIG. 6.5 Adsorption isotherms for the Pt(cod)me_2 – carbon aerogel CO_2 system [74].

The slope of the isotherm (i.e., the change in uptake per unit change in Pt(cod)me_2 concentration in scCO_2) at lower concentrations is large. In fact, below the concentration of 5 mol Pt(cod)me^2/m^3 CO_2 the slope is almost infinite and at about 15 mol Pt(cod)me_2/m^3 CO_2, the maximum uptake is almost reached. The amount of Pt(cod)me_2 that can be dissolved in the scCO_2 phase at 80°C and 27.7 MPa is around 50 mol/m^3 of CO_2. The isotherms shown in Fig. 6.5 can be represented accurately using a form of the Langmuir adsorption isotherm which is expressed by the following equation,

$$q = \frac{K_l Q_o C}{1 + K_l C} \tag{6.18}$$

where K_l is the Langmuir adsorption or adsorption equilibrium constant and is correlated to the heat of adsorption, Q_o is the adsorption capacity, and $K_l Q_o$ is a measure of the relative affinity of the solute toward the surface of the adsorbent. The values of $K_l Q_o$ for both CAs are high indicating that Pt(cod)me_2 has a strong interaction with the CA substrate and a higher affinity for CA than scCO_2. CA22 though has a stronger interaction with Pt(cod)me_2 than that of CA4. In the case of silica aerogel as substrate, in comparison to CA at the same Pt(cod)me_2 concentration in the scCO_2 phase, the uptake is about three times lower. This is an indication that the interaction between Pt(cod) me_2 and silica aerogel is weaker than that of the Pt(cod)me_2 interaction and that the nature of the Pt(cod)me_2-substrate interaction is different in the two systems.

As shown in Chapter 5, the solubility of most organometallic solutes in $scCO_2$ is low when compared to other organic solvents. This low solubility reflects the low affinity of $scCO_2$ to the precursor and coupled with strong solute-adsorbent interaction (whether short-range or long-range), explains the solute's biased partitioning to the adsorbent. Along similar lines, Ishii and Ooi [90] compared the intercalation of 4-phenylazoaniline (PAA) into a pillared clay mineral (TMA-Mnt) using $scCO_2$ and three common organic solvents. They found that the uptake of PAA into TMA-Mnt is highest for $scCO_2$, which is the solvent in which PAA hadthe lowest solubility. One of the disadvantages of using $scCO_2$ as a process solvent is the relatively low solubilities of many solutes as compared to those in common organic solvents; however, for the supercritical fluid impregnation of solutes into highly porous materials, this apparently becomes an advantage. Using $scCO_2$, one may possibly control the metal loading over a wide range without using multiple impregnation steps as is required for other impregnation methods.

The surface area of the adsorbent covered by the precursor can be estimated by using the adsorptive capacity determined from the Langmuir model and by assuming monolayer adsorption. The molecular structure of Pt(cod)me_2 can be estimated using the computational chemistry software Gaussian, Inc. which allows for the determination of the center-to-center distances of the atoms in the molecule. From the molecular geometry, there are three major spatial projections in which the Pt(cod)me_2 molecule may adsorb on the CA adsorption sites: (1) the cyclooctadiene projection, (2) the two methyl groups' projection and (3) Pt(cod)me_2 standing projection. Using the measured maximum adsorption capacity of CA22 for Pt(cod)me_2 and the three estimated projected areas (i.e., the surface areas corresponding to each projection) are 984.8, 874.4 and 1463.0 m^2/g CA22, respectively [72]. It is likely that Pt(cod)me_2 is adsorbed either through (1) or (2), or a mixture of both, because of the closeness of the estimated surface area to the BET S_{tot} obtained using N_2 physisorption. Although CA4 has a higher BET S_{tot} and lower mesopore size than CA22, (and with the average pore size of the micropores of both CAs being similar), its adsorption capacity for Pt(cod)me_2 is lower. This is unexpected from the vantage point of enhanced adsorption by pores with smaller diameters. Similar results were obtained in the adsorption of dibenzothiophene from hexadecane on CAs [91]. At low Pt(cod)me_2 concentrations in the $scCO_2$ phase, however, the adsorption is not significantly affected by the difference in the surface chemistry of the substrates because of the similar uptakes in this region. It is believed that adsorption is more influenced by the enhanced adsorption potential in the micropores at low Pt(cod)me_2 concentration in the SCF phase. CA4 and CA22 have similar micropore area and volumes, thus the adsorption of Pt(cod)me_2 in the two CAs at low Pt(cod)me_2 is similar.

The adsorption isotherms of Pt(cod)me_2 in $scCO_2$ for resorcinol-formaldehyde aerogel (RFA) at 10.7 and 20.7 MPa, and at 35°C were investigated by Bozbag et al. [92]. RFA used in this study had an average pore diameter of 8 ± 1.55 nm, a BET surface area of 759 ± 47 m^2/g, a total micropore area of 73.4 ± 5.15 m^2/g and a total pore volume of 2.5×10^{-3} ($\pm 0.09 \times 10^{-3}$) m^3/kg. The adsorption isotherms and model constants found by nonlinear regression are presented in Fig. 6.6 and Table 6.2, respectively.

In both isotherms, the initial slopes of the curves were steep and indicated strong adsorption. The uptake of Pt(cod)me_2 on RFA was not linearly proportional to Pt(cod)me_2 concentration in the $scCO_2$ phase. The uptake of Pt(cod)

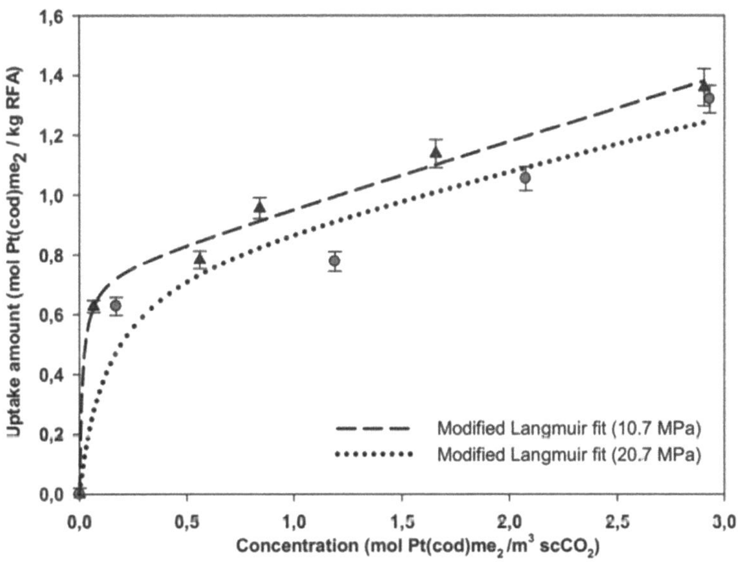

FIG. 6.6 Adsorption isotherms of Pt(cod)me_2/RFA-$scCO_2$ system. (p) 10.7 MPa, 35°C ($\rho = 730$ kg/m^3); (●) 20.7 MPa, 35°C ($\rho = 871$ kg/m^3) [92].

TABLE 6.2 Adsorption isotherm for Pt(cod)me2/RFA-$scCO_2$ at 35°C.

Modified Langmuir isotherm parameters	10.7 MPa	20.7 MPa
K_1 (m^3 $scCO_2$/mol Pt(cod)me_2)	63.84	55.60
K_2 (m^3 $scCO_2$/kg RFA	0.745	0.580
K_3 (m^3 $scCO_2$/mol Pt(cod)me_2)	0.22	0.240

From S.E. Bozbag, N.S. Yasar, L.C. Zhang, M. Aindow, C. Erkey, Adsorption of Pt(cod)me2 onto organic aerogels from supercritical solutions for the synthesis of supported platinum nanoparticles, J. Supercrit. Fluids 56 (2011) 105–113. 10.1016/j.supflu.2010.10.045, Copyright © 2011, Elsevier.

me_2 onto RFA slightly increased with decreasing pressure for a particular Pt(cod)me_2 concentration. The adsorption isotherm was described by the modified Langmuir model as shown below

$$q = \frac{K_1 Q_0 C}{1 + K_1 C} + K_3 C \tag{6.19}$$

where K_1 (m^3 $scCO_2$/mol Pt(cod)me_2) and Q_0 (mol Pt(cod)me_2/kg RFA) are the Langmuir adsorption constants, K_3 (m^3 $scCO_2$/mol Pt(cod)me_2) is the modified Langmuir adsorption constant. In this equation, K_3C accounts for the heterogeneity of the surface and indicates the existence of nonselective sites (the linear term) and some competitive sites (the first term).

Thermodynamics of adsorption of Pt(cod)me_2 and Pd(acac)$_2$ onto carbon black (Black Pearl 2000, BP2000) in $scCO_2$ at 20 MPa and 60°C were investigated by Cangül et al. [93]. BP2000 used had a BET surface area of 1450 m^2/g, an external surface area of 515 m^2/g, a micropore volume of 0.454 cc/g and a total pore volume of 1.75 cc/g. Adsorption isotherms for Pt(cod)me_2 and Pd(acac)$_2$ at these conditions are given in Figs. 6.7 and 6.8, respectively.

It was shown that while adsorption isotherm for Pd(acac)$_2$/BP2000-$scCO_2$ system was linear, Pt(cod)me_2/BP2000-$scCO_2$ system had a nonlinear adsorption isotherm and it could be represented by the Langmuir model given by Eq. (6.18). Using nonlinear regression analysis, K_1 and Q_0 were found to be 0.14 m^3 $scCO_2$/mol Pt(cod)me_2 and 4.5 mol Pt(cod)me_2/kg substrate, respectively.

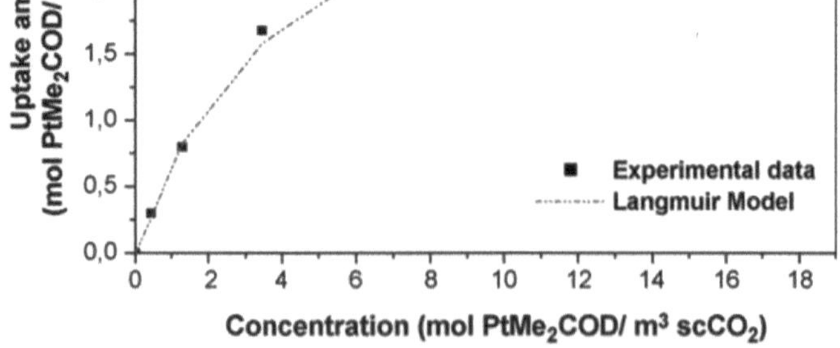

FIG. 6.7 Adsorption isotherm of Pt(cod)me_2/BP2000-$scCO_2$ system at 20 MPa and 60°C [93].

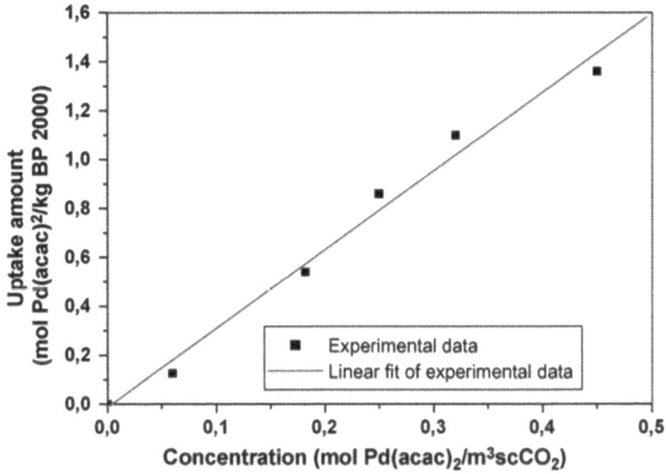

FIG. 6.8 Adsorption isotherm of $Pd(acac)_2$/BP2000-$scCO_2$ system at 20 MPa and 60°C [93].

The difference can be attributed to the higher concentration of $Pt(cod)me_2$ in $scCO_2$ phase. Since the solubility of $Pt(cod)me_2$ (21.7 mol/m^3) is much higher than the solubility of $Pd(acac)_2$ 0.66 mol/m^3) in $scCO_2$, at 20 MPa and 60°C, the range of concentration of $Pt(cod)me_2$ in the fluid phase is much larger. The area covered by the metal precursor molecules on the BP2000 surface can be calculated using:

$$S = Q_o A \pi r^2 \tag{6.20}$$

where S is the covered surface area, Q_o is the capacity, A is Avogadro's number and r is the radius of a precursor molecule. The diameters of the precursor molecules were determined using Marvin Beans software as 0.80 and 0.55 nm for $Pt(cod)me_2$ and $Pd(acac)_2$, respectively. Consecutively, the area covered by $Pt(cod)me_2$ and $Pd(acac)_2$ molecules on BP2000 surface was found as 620 and 640 m^2/g, respectively. This indicates that both metal precursors cover all accessible surface area of BP2000 at maximum uptake. The uptake amount of $Pd(acac)_2$ on BP2000 is higher than that of $Pt(cod)me_2$, which indicates a stronger precursor-support interaction between $Pd(acac)_2$ and BP2000.

The adsorption isotherms of $Ru(cod)(tmhd)_2$ on two carbon aerogels (CA21 and CA4 from $scCO_2$ were measured at 353.2 K and 27.6 MPa [74]. The isotherms are presented in Fig. 6.9. The uptake of $Ru(cod)(tmhd)_2$ by CA-21 was always higher than that by CA4 at the same fluid phase concentration. With the concentration of $Ru(cod)(tmhd)_2$ in $scCO_2$ increasing, the uptake of Ru(cod) $(tmhd)_2$ by both CA21 and CA4 increases. The concentration that can be reached in the fluid phase can be higher than 1.5 wt% in this condition.

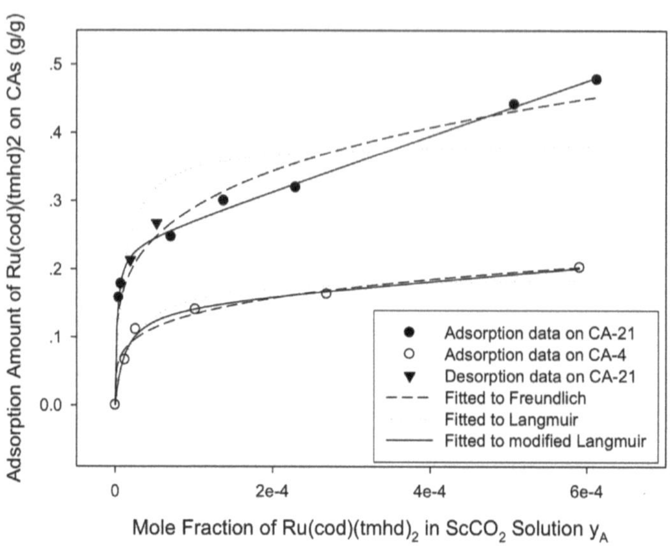

FIG. 6.9 Adsorption isotherms for $Ru(cod)(tmhd)_2$ – carbon aerogel – CO_2 system [75].

The two isotherms were fitted to Langmuir model, modified Langmuir model and Freundlich model and the fitted isotherms are shown in Fig. 6.9 and the isotherm parameters are shown in Table 6.3.

The modified Langmuir model gave the best fit to data for both adsorption isotherms on CA4 and on CA21. Based on the Langmuir isotherm, the maximum uptake of the precursor is given by Q_o, which corresponds to the maximum number of the available sorption sites that the carbon aerogels can supply for the Ru precursor molecules. It is interesting to note that the surface area of CA21 as determined by nitrogen physisorption was smaller than that of CA4, the capacity of $Ru(cod)(tmhd)_2$ on CA-21 was twice of that on CA-4, although the real Q_o for $Ru(cod)(tmhd)_2$ on CA21 was higher than that given by the Langmuir model. This indicates that some of the sorption sites on CA4 are not available, and this may be attributed to both the steric resistance and the accessibility of the sites.

Although the surface area of CA4 is slightly larger than that CA21, the pore volume of CA4 is much smaller than CA-21. Since $Ru(cod)(tmhd)_2$ molecule is very bulky, the small pore volume of CA4 causes more significant steric resistance for adsorption than CA21 does. The accessibility of the sorption sites is determined both by the structure of the pores in the adsorbents and by the molecular size of adsorbate. If one considers that the molecule is a sphere with a diameter of 1.0 nm (estimated approximately from the bond lengths and the geometry of this molecule), the area which is covered by $Ru(cod)(tmhd)_2$ molecules can be calculated using Eq. (6.20). S values were found to be 320 and 158 m^2/g for CA-21 and CA-4, respectively. About 233 m^2/g of the surface area for CA-21 resides in the

TABLE 6.3 Parameters of Ru(cod)(tmhd)2 adsorption isotherms on CA-4 and CA-21 from $scCO_2$ solution at 353.2 K and 27.6 MPa [74].

	Langmuir			Modified Langmuir				Freundlich		
Adsorbent	R^2	K_l	Q_o	R^2	K_l	Q_o	k	R^2	K_f	1/n
CA-4	0.968	4.19×10^4	0.191	0.994	7.61×10^4	0.148	99.5	0.982	1.174	0.236
CA-21	0.828	8.43×10^4	0.387	0.998	3.97×10^5	0.235	403.0	0.979	2.778	0.245

micropores which have an average pore diameter of 0.68 nm. Therefore, it is highly unlikely for Ru(cod)(tmhd)$_2$ molecules to enter the micropores. The low surface area of Ru(cod)(tmhd)$_2$ on CA-4 obtained using Eq. (6.20) indicates that in addition to the micropores, the small mesopores of CA-4, which are comparable in size with the Ru(cod)(tmhd)$_2$ molecules, also result in a substantial fraction of the surface not accessible. The reduction of the cross-sectional area of the pore mouth by adsorption of bulky precursor may also be more pronounced for CA-4, thus limiting access [94].

The desorption of isotherm of Ru(cod)(tmhd)$_2$ for CA21 was also determined at the same condition as the adsorption isotherm [75]. One monolith of CA21 was weighed and impregnated with Ru(cod)(tmhd)$_2$. After the uptake was measured, the impregnated monolith was put into the high-pressure vessel and CO_2 was charged in. The color of scCO_2 changed to light yellow (the color of the precursor), which means some of the adsorbed Ru(cod)(tmhd)$_2$ redissolved in scCO_2 phase. The desorption was continued for 24 h. The solution in the vessel was then drained, the system was allowed to cool down, and the monolith was removed. The equilibrium concentration of the precursor in scCO_2 and uptake was calculated from the measured change in the mass of the material. The result is shown in Fig. 6.9. The desorption isotherm coincides with the adsorption isotherm within experimental error which indicates that the adsorption of Ru(cod)(tmhd)$_2$ onto CAs is reversible.

Additional four adsorption isotherms of Ru(cod)(tmhd)$_2$ on CA21 were obtained under certain temperatures, pressures and densities [61]. The isotherms are presented in Fig. 6.10 together with the one obtained at 353.2 K and 27.6 MPa.

The general shape for each isotherm is similar. As the concentration of Ru(cod)(tmhd)$_2$ in scCO_2 increased, the uptake of Ru(cod)(tmhd)$_2$ by CA21 increased. Each isotherm rises very steeply at low concentrations, which indicates that Ru(cod)(tmhd)$_2$ favored partitioning to the adsorbent. This reflects the low affinity of scCO_2 to the precursor and the strong solute-adsorbent interaction. At both constant density and constant pressure, at 343.5 K the Ru(cod)(tmhd)$_2$ displays the highest adsorption amount on CA21, then followed by 333.5 and 353.2 K. At the same temperature, adsorption amount decreases with the increasing CO_2 density or pressure. This indicates that at higher density (or higher pressure), the interaction between the solute and the solvent increase as compared to that between the solute and the adsorbent. This result is consistent with the results from the literature [31, 88].

Tenorio et al. [64] measured the adsorption isotherms of Pd(hfac)$_2$/SBA-15-scCO_2 system at 40°C, 8.5 MPa; 40°C, 14.0 MPa; 60°C, 10.9 MPa; 60°C, 8.5 MPa; 70°C, 12.0 MPa; 80°C, 8.5 MPa. The adsorption isotherms are given in Fig. 6.11.

At constant temperatures of 40°C and 60°C, the amount of metal complex adsorbed onto support decreased as pressure and density increased whereas at constant pressure of 8.5 MPa, the amount adsorbed increased as the temperature

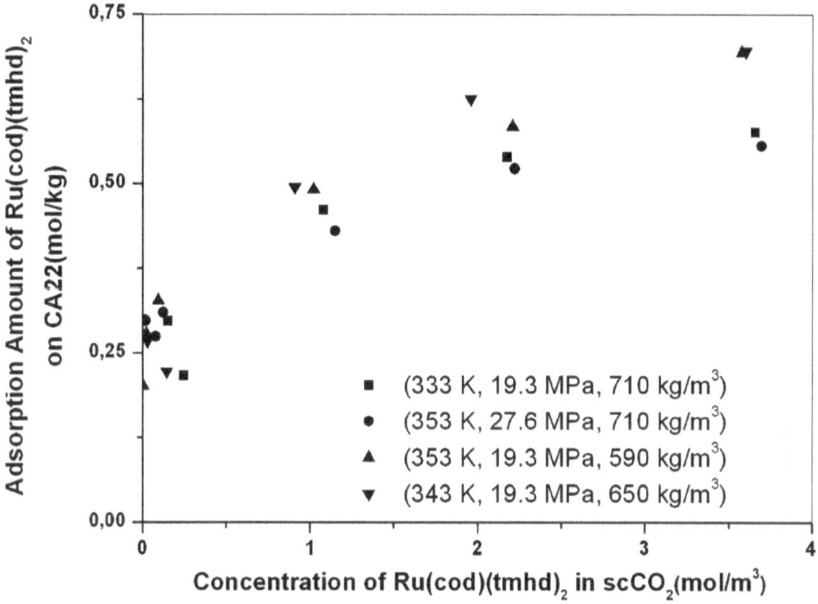

FIG. 6.10 Adsorption isotherms for the $Ru(cod)(tmhd)_2$ – carbon aerogel – CO_2 system at different temperatures and pressures.

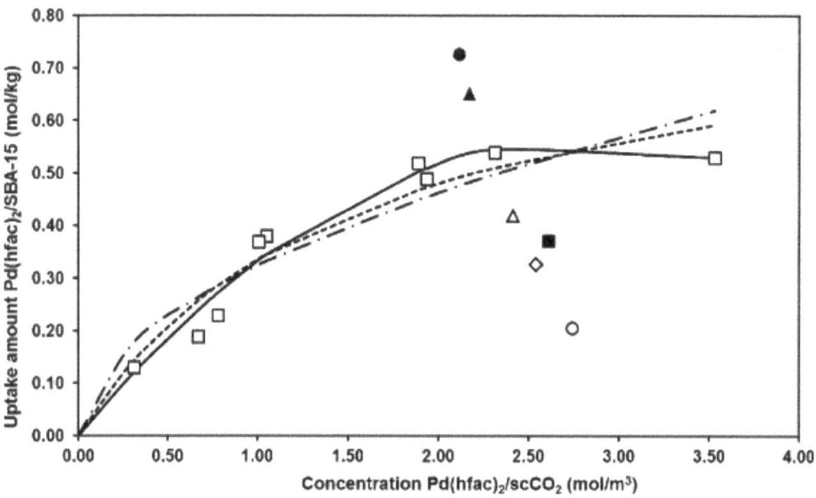

FIG. 6.11 Adsorption isotherms for $Pd(hfac)_2$ on SBA-15 at different temperature, pressure and CO_2 density conditions: (□) 40°C, 8.5 MPa, 350 kg/m^3; (■) 40°C, 14 MPa, 760 kg/m^3; (△) 60°C, 10.9 MPa, 350 kg/m^3; (p) 60°C, 8.5 MPa, 220 kg/m^3; (◊) 70°C, 12.0 MPa, 350 kg/m^3; (○) 80°C, 13.3 MPa, 350 kg/m^3 and (●) 80°C, 8.5 MPa, 170 kg/m^3. Fit of the adsorption isotherm at 40°C and 8.5 MPa to different models: (– –) Langmuir; (—) modified Langmuir; (-··-) Freundlich [64].

increased from 40°C to 80°C. At constant CO_2 density of 350 kg/m^3, adsorption decreased as the temperature increased from 40°C to 80°C. the maximum adsorption was obtained at 80°C and 8.5 MPa. Adsorption isotherm data collected at 40°C and 8.5 MPa was fitted to three different adsorption models: Langmuir (Eq. 6.18), modified Langmuir (Eq. 6.19) and Freundlich (Eq. 6.21).

$$q = K_f c^{1/n} \tag{6.21}$$

where K_f is the Freundlich adsorption constant and $1/n$ is the Freundlich exponent. Parameters regressed from experimental data are given in Table 6.4 along with their standard deviations.

The best fit was obtained when modified Langmuir model was used. For Langmuir and modified Langmuir models, K_1 values were much smaller than K_1 the value of Pt(cod)me$_2$/RFA-scCO_2 system [92] but similar to Ru(cod)(tmhd)$_2$/AC-scCO_2 [75]. K_1Q_0 term gets very small in both Langmuir and modified Langmuir models, indicating weak adsorption of Pd(hfac)$_2$ onto the SiO_2 support. Values of Freundlich model parameters were very similar to those found for Ru(cod)(tmhd)$_2$/AC-scCO_2 system [75]. The authors concluded that adsorption of Pd(hfac)$_2$ onto SBA-15 in scCO_2 was very weak.

Adsorption of Ni(acac)$_2$ precursor onto CA in scCO_2 at 60°C and 30 MPa was investigated by Bozbag et al. [60]. The CA had an average pore radius of 15.1 nm, a BET surface area of 635 m^2/g, a total micropore area of 253 m^2/g and a total pore volume of 4.95 cm^3/g. The adsorption measurements were conducted up to the saturation concentration of Ni(acac)$_2$ in scCO_2 at 60°C and 30 MPa, which corresponds to 6×10^{-5} mol Ni(acac)$_2$/mol CO_2. The adsorption isotherm was linear (Fig. 6.12) and represented by Eq. (6.22).

$$q = KC \tag{6.22}$$

The adsorption equilibrium constant, k, was regressed as 0.466 m^3/kg which indicated a favorable partitioning of the metal complex molecules to the CA.

TABLE 6.4 Fitting parameters of the Pd(hfac)$_2$ adsorption isotherms on SBA-15 at 40°C and 8.5 MPa to the different models [64].

Parameters	Langmuir	Modified Langmuir	Freundlich
K_1 (m3 scCO_2/mol Pd(hfac)$_2$)	0.65	0.15	–
Q_0 (mol Pd(hfac)$_2$/kg SBA-15)	0.84	5.8	–
K_3 (m^3 scCO_2/kg SBA-15)	–	−0.40	–
K_f	–	–	0.32
$1/n$	–	–	0.51
σ (mol/kg)	0.048	0.037	0.063

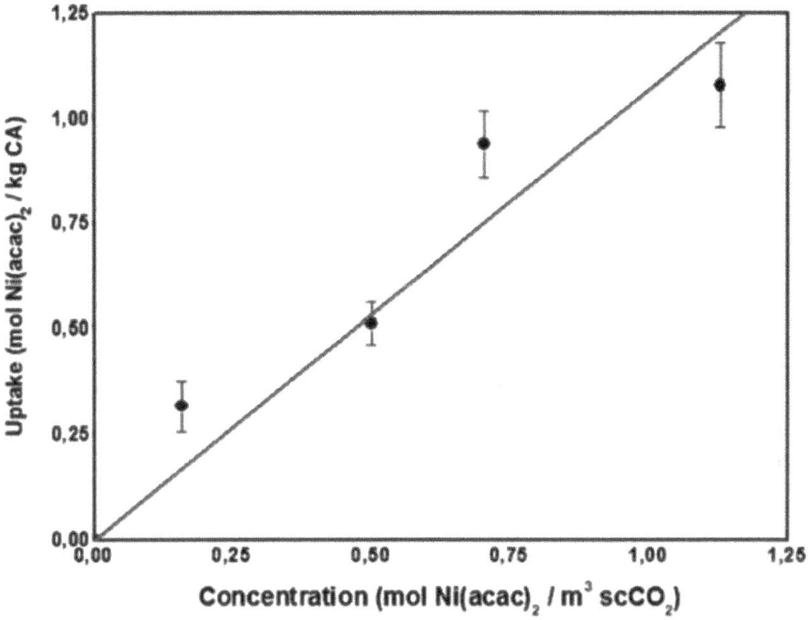

FIG. 6.12 Adsorption isotherm of $Ni(acac)_2$/CA-$scCO_2$ system at 30 MPa and 60°C [60].

Ushiki et al. [67] studied adsorption of $Rh(acac)_3$ onto MCM-41, MSU-H and HMS mesoporous silica substrates in $scCO_2$ with methanol co-solvent at 313, 333 and 353 K and at 15.0 MPa. The adsorption isotherms were represented by Dubinin-Astakhov adsorption isotherm model which are given by Eqs. (6.23), (6.24).

$$W = W_0 \exp\left[-\left(\frac{\varepsilon_{\text{prec}}}{E_{\text{prec}}}\right)^n\right] \tag{6.23}$$

$$q = \frac{\rho_{\text{prec}} W}{MW_{\text{prec}}} \tag{6.24}$$

where W and q are the amount of adsorbed metal complex in volume and molar basis, respectively. W_0 is the saturated adsorption capacity of the adsorbent for metal complex in $scCO_2$, E_{prec} is the adsorption energy between the metal complex and the adsorbent in $scCO_2$, ρ_{prec} is the density of the metal complex, $\varepsilon_{\text{prec}}$ is the adsorption potential of the metal complex, which is given by

$$\varepsilon = RT \ln\left[\frac{p_i^s(T)}{p_i}\right] \tag{6.25}$$

where p_i is the partial pressure of the adsorbate i in the bulk phase and p_i^s is the saturated vapor pressure of adsorbate i at operating temperature T.

MCM-41 had a specific BET surface area of 1011 m^2/g, mean pore diameter of 2.7 nm, pore volume of 1.15 cm^3/g. MSU-H had a specific BET surface area of 585 m^2/g, mean pore diameter of 8.2 nm, pore volume of 0.8 cm^3/g. HMS had a specific BET surface area of 759 m^2/g, mean pore diameter of 3.2 nm, pore volume of 1.6 cm^3/g. Adsorption isotherms along with model predictions are given in Fig. 6.13. The amount of metal complex adsorbed was the highest on MCM-41 due to its high specific BET surface area. The amount of adsorbed metal complex was similar on MSU-H and HMS regardless of differences in their specific surface area. The amount of metal complex adsorbed increased with increasing temperature at 15.0 MPa. The solid lines shown in Figure VI are the Dubinin-Astakhov adsorption isotherm model predictions for the adsorption of the adsorption equilibria of Rh(acac)$_3$ on the three types of silica supports in scCO_2 with methanol co-solvent. Table 6.5 summarizes the fitting parameters of the model with an average relative deviation (ARD). All of the isotherms correlate well to Dubinin-Astakhov adsorption isotherm model within 7% of ARD.

In another study, Ushiki et al. measured and modeled adsorption equilibria of Co(acac)$_3$ on MCM-41 mesoporous silica in scCO_2 and methanol as co-solvent [70]. The MCM-41 had a specific BET surface area of 993 m^2/g, mean pore diameter of 2.7 nm, pore volume of 1.15 cm^3/g. The adsorption isotherms were measured at 313, 333 and 353 K and 12.5, 15.0 and 20.0 MPa and are given in Fig. 6.14 along with Dubinin-Astakhov adsorption isotherm model predictions.

The amount of Co(acac)$_3$ adsorbed on the surface decreased with increasing pressure at 333 K, while the amount of Co(acac)$_3$ adsorbed increased with increasing temperature at 15.0 MPa. Parameters for Dubinin-Astakhov adsorption isotherm model, given in Eq. (6.23) are summarized in Table 6.6 along with

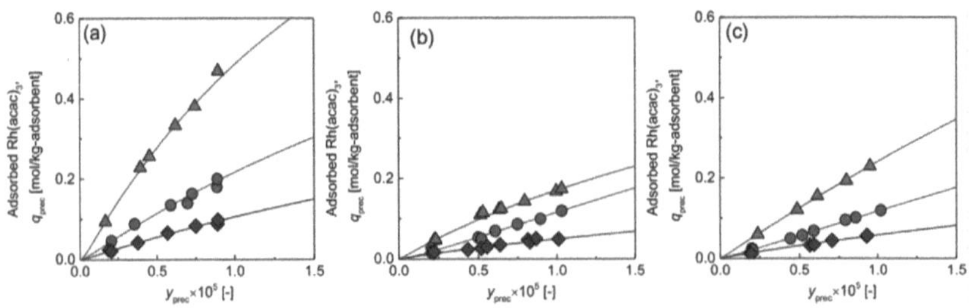

FIG. 6.13 Adsorbent and temperature dependence of adsorption equilibria of Rh(acac)$_3$ on (a) MCM-41, (b) MSU-H and (c) HMS in scCO_2 with methanol co-solvent ($y_{methanol} = 2.60 \times 10^{-2}$) at temperatures of 313 K (◇), 333 K (○) and 353 K (△) and at 15.0 MPa. Solid lines are correlation results of the Dubinin-Astakhov equation [67].

TABLE 6.5 Correlation results of adsorption equilibria of $Rh(acac)_3$ on mesoporous silica in $scCO_2$ with methanol co-solvent at 15.0 MPa by using the Dubinin-Astakhov equation.

SiO_2 type	T (K)	ρ_{CO2} (kg/m^3)	E_{prec} (kJ/mol)	W_0 (cm^3/kg adsorbent)	n	ARD (%)
MCM-41	353	427	7.34	361	1.77	1.7
	333	604	7.77	435	1.77	4.4
	313	780	8.05	415	1.77	5.8
MSU-H	353	427	6.32	164	1.50	3.4
	333	604	5.61	596	1.50	6.8
	313	780	6.91	237	1.50	6.9
HMS	353	427	4.42	432	1.22	0.7
	333	604	4.13	1050	1.22	4.5
	313	780	4.71	694	1.22	6.3

From I. Ushiki, N. Takahashi, T. Shimizu, Y. Sato, M. Ota, R.L. Smith, H. Inomata, Adsorption equilibria of rhodium acetylacetonate with MCM-41, MSU-H, and HMS silica substrates in supercritical carbon dioxide for preparing catalytic mesoporous materials, J. Supercrit. Fluids 120 (2017) 240–248. 10.1016/j.supflu.2016.05.032, Copyright © 2017, Elsevier.

ARD values. The equation correlated to the adsorption equilibria data within 12 % ARD values at all temperatures and pressures.

The first parameter in the Dubinin-Astakhov equation that was regressed was n. It is a constant that corresponds to the ratio between the molecular size of the adsorbate and pore size of the adsorbent. The n value calculated here, 1.74, was very similar to n value calculated for the adsorption isotherm of $Rh(acac)_3$ on MCM-41 in $scCO_2$ and methanol as co-solvent ($n = 1.77$ [67]). The reason for this similarity was attributed to the similarity of the van der Waals diameters of $Co(acac)_3$ ($d_{vdW} = 0.893$ nm) and $Rh(acac)_3$ ($d_{vdW} = 0.897$ nm) molecules.

The dependency of the second parameter E_{prec} to CO_2 density was relatively small, which was also the same for $Rh(acac)_3$ in the previous study [67]. However, E_{prec} values for $Co(acac)_3$ were smaller than those for $Rh(acac)_3$. Since this parameter is related to the interaction between the metal complex and the substrate, it can be deduced that the affinity of $Co(acac)_3$ for MCM-41 was lower than that of $Rh(acac)_3$ in $scCO_2$ and methanol as the co-solvent.

The third parameter, W_0, decreased with increasing density of CO_2. This trend may originate from decreases in the adsorption capacity for $Co(acac)_3$ resulting from an increasing amount of adsorbed CO_2, which competes for absorption sites with the metal precursor.

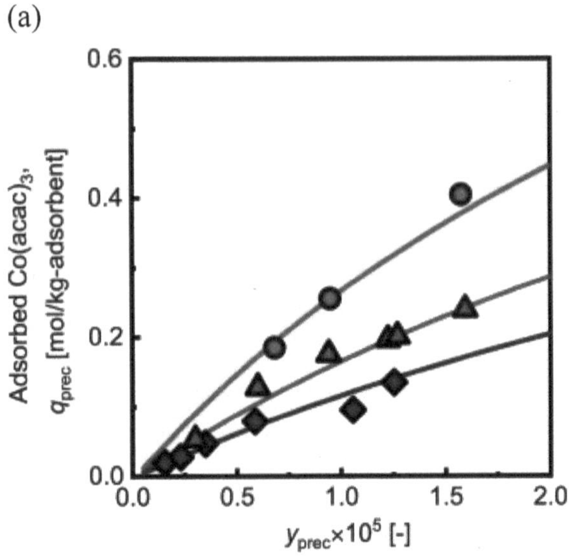

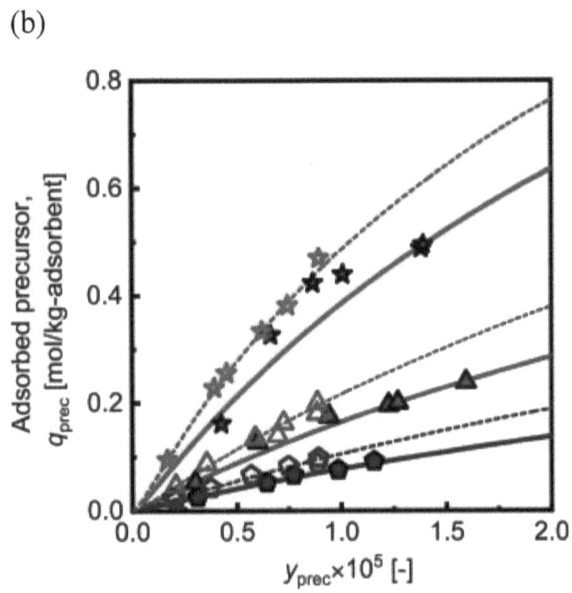

FIG. 6.14 (a) Pressure dependence of the adsorption equilibria of $Co(acac)_3$ on MCM-41 in $scCO_2$ with methanol co-solvent at 333 K (○, 12.5 MPa; △, 15.0 MPa; ◊, 20.0 MPa; solid lines were generated by the Dubinin-Astakhov correlation); (b) Temperature dependence of the adsorption equilibria of $Co(acac)_3$ (closed symbols, solid lines) on MCM-41 at 15.0 MPa in $scCO_2$ with methanol co-solvent in comparison with that of $Rh(acac)_3$ (open symbols, dashed lines) (*, 353 K; △, 333 K; , 313 K; lines, generated by the Dubinin-Astakhov correlation) [70].

TABLE 6.6 Correlation results of the adsorption equilibria of $Co(acac)_3$ on MCM-41 in $scCO_2$ with methanol co-solvent using the Dubinin-Astakhov equation.

	T (K)	P (MPa)	ρ_{CO2} (kg/m^3)	E_{prec} (kJ/mol)	W_0 (cm^3/kg adsorbent)	n	ARD (%)
MCM-41	353	15.0	427.2	7.26	389.4	1.74	10.1
	333	20.0	723.7	7.48	271.8	1.74	11.6
		15.0	604.1	7.54	334.7	1.74	5.8
		12.5	471.5	7.25	359.5	1.74	3.2
	313	15.0	780.2	7.38	242.2	1.74	2.3

From I. Ushiki, M. Koike, T. Shimizu, Y. Sato, S. Takishima, H. Inomata, Measurement and modeling of adsorption equilibria of cobalt (III) acetylacetonate on MCM-41 mesoporous silica in the presence of supercritical carbon dioxide with methanol co-solvent, J. Supercrit. Fluids 140 (2018) 329–335. 10.1016/j.supflu.2018.06.015,

Similar to single metal component SCD systems, adsorption isotherms are important for the control of metal loading and nanoparticle size for the implementation of multimetallic SCD systems. Along this line, the knowledge on the binary adsorption thermodynamics of the metal complexes of the metal complex-substrate systems in $scCO_2$ is crucial for the controlled preparation of supported bimetallic nanoparticles. the binary adsorption isotherm for CuDI6-Pt(cod)me_2-CA systemin $scCO_2$ was determined by Bozbag et al. [71]. To our knowledge, this is the only adsorption isotherm study for a system containing two different metal complexes. In this study, single adsorption isotherms for Pt(cod)me_2/CA-$scCO_2$ and CuDI6/CA-$scCO_2$ were firstly determined at 35°C and 10.6 MPa. Afterwards, binary adsorption isotherm for Pt(cod)me_2-CuDI6/CA-$scCO_2$ system was measured at the same temperature and pressure. Single adsorption isotherms are given in Fig. 6.15 along with model predictions.

At a given concentration, adsorbed amount of Pt(cod)me_2 is higher than that of CuDI6. Single component adsorption data were fitted using Langmuir and Sips adsorption isotherm models. Sips adsorption isotherm is given in Eq. (6.26).

$$q = \frac{Q_s(\alpha C)^{1/n}}{1+(\alpha C)^{1/n}} \tag{6.26}$$

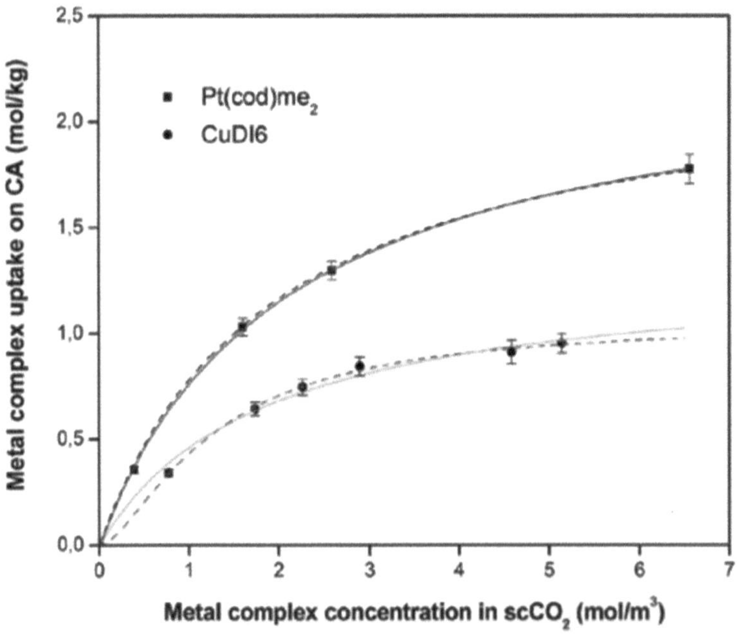

FIG. 6.15 Single component adsorption isotherms of Pt and Cu complexes/CA-$scCO_2$ systems at 10.6 MPa and 35°C. Straight and dashed lines are Langmuir and Sips fits, respectively [71].

where Q_s is associated with the total number of binding sites, α, is the median association constant and $1/n$ is the heterogeneity factor. Values of $1/n$ close to zero indicate heterogeneous adsorbents, while values closer to 1.0 indicate a material with relatively homogeneous binding sites. The parameters for Langmuir and Sips models for single adsorption isotherms calculated from nonlinear regression are given in Table 6.7. In both cases, R^2 values are higher than 0.994, indicating isotherms fit very well to both models. In the Sips model, the n values for Pt(cod)me_2 and CuDI6 are 0.995 and 0.642, respectively. Therefore, on the surface of CA, the adsorption sites available for Pt(cod)me_2 are quite homogeneous but for the adsorption of CuDI6, the sites are rather heterogeneous.

Binary adsorption isotherm is given in Fig. 6.16.

Binary adsorption isotherms were modeled using an extended-Langmuir equation and ideal adsorbed solution theory (IAST) using the respective single-solute isotherm parameters alone. The extended Langmuir adsorption model can be expressed as Eq. (6.27).

$$q_i = \frac{K_{1,i}Q_{0,i}C_i}{1 + \sum_{j=1}^{N} K_{1,i}C_i} \tag{6.27}$$

The IAST adsorption model is usually employed for the prediction of adsorption in multicomponent systems, using the respective single-solute isotherm parameters alone. The model assumes that the solution is ideal which

TABLE 6.7 Single component adsorption isotherm parameters of metal complex/CA-scCO_2 at 10.6 MPa and 35°C.

Model parameters	Pt(cod)me_2	CuDI6
Langmuir		
Q_0	2.343	1.320
K	0.480	0.534
R^2	0.999	0.994
Sips		
Q_s	2.289	1.053
α	0.516	0.789
n	0.995	0.642
R^2	0.999	0.999

From S.E. Bozbag, U. Unal, M.A. Kurykin, C.J. Ayala, M. Aindow, C. Erkey, Thermodynamic control of metal loading and composition of carbon aerogel supported Pt–Cu alloy nanoparticles by supercritical deposition, J. Phys. Chem. C 117 (2013) 6777–6787. 10.1021/jp311641g, Copyright © 2013, American Chemical Society.

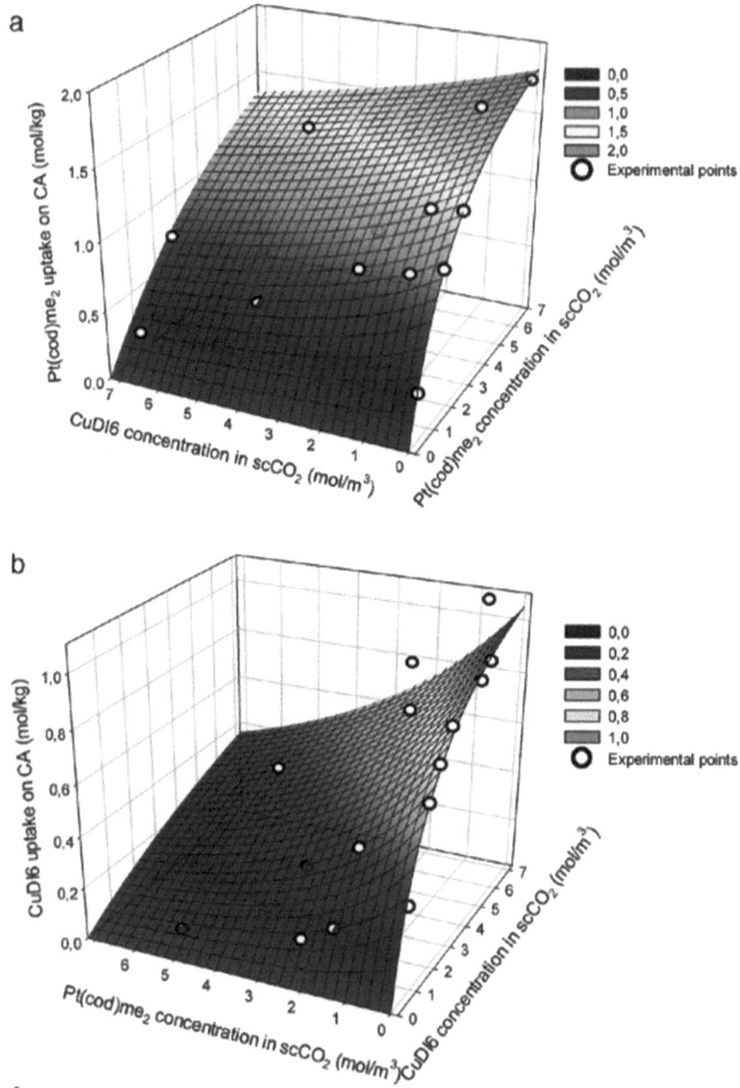

FIG. 6.16 Binary adsorption isotherms of Pt(cod)me_2 − $CuDI_6$ − carbon aerogel system in $scCO_2$ at 35°C and 10.6 MPa. (a) Pt(cod)me_2 uptake, (b) $CuDI_6$ uptake, (c) Total uptake. The 3D surfaces are the IAST model using single component Langmuir parameters [71].

means that no interaction between adsorbent and adsorbate is considered. In IAST, the following equations are used to predict multicomponent behavior:

$$q_t = \sum_{i=1}^{N} qi \tag{6.28}$$

$$z_i = \frac{q_i}{q_t} \tag{6.29}$$

$$q_i^0 = f\left(C_i^0\right) \tag{6.30}$$

$$\frac{1}{q_t} = \sum_{i=1}^{N} \frac{z_i}{q_i^0} \tag{6.31}$$

$$C_i = z_i C_i^0 \tag{6.32}$$

$$\pi_i^* = \int_0^{C_i^0} \frac{q_i^0}{C_i^0} dC_i^0, \pi_1^* = \pi_2^* = \cdots = \pi_N^* \ \text{(at equilibrium)} \tag{6.33}$$

where C_i^0 is the hypothetical single-solute solution concentration in equilibrium with q_i^0, z_i is the mole fraction of the solute in the adsorbed phase, and π_i^* is the modified spreading pressure. This modified spreading pressure is equal to $\pi_i A/RT$, where A is the area of the sites available for adsorption, R is the gas constant and T is the temperature. In IAST, one needs to utilize one of the previously mentioned single isotherm models to end up with binary isotherms. Experimentally obtained adsorption equilibrium data was simulated using IAST, i.e., the nonlinear Eqs. (6.28)–(6.33) were solved in Mathcad using a Levensberg-Marquardt routine and the unknown parameters $(\pi_i, c_i^0, q_i^0, z_i)$ were obtained.

The values of the sum of square errors function between the experimental uptake data (q_{exp}) for a particular metal complex and the model prediction (q_{pred}) are presented in Table 6.8. For the Pt(cod)me_2 uptake, all of the models give reasonable predictions except for the Extended Langmuir model. The best fits were obtained using IAST with Sips single component parameters. For the CuDI6 uptake all of the models fit the experimental data quite well, and the lowest SSE value was obtained using IAST with Langmuir single component parameters. According to these calculations, predicting the binary adsorption behavior of metal complex-substrate systems in scCO2 seems feasible via IAST using either Langmuir or Sips single component parameters.

6.5.3 Kinetics of adsorption of metal complexes on porous substrates from supercritical solutions

Similar to the conventional adsorption process, four steps can be identified for adsorption from the supercritical phase:

1. Mass transfer from the bulk fluid to the adsorbent.
2. Diffusion in the pores of the adsorbent.
3. Adsorption onto the surface.
4. Surface diffusion on the adsorbent surface.

TABLE 6.8 SSE values for the experimental data and the predictions of metal complex uptake on CA.

Model	Pt(cod)me$_2$	CuDI6
Extended Langmuir	0.510	0.063
IAST-Langmuir	0.112	0.061
IAST-Sips	0.091	0.170

From S.E. Bozbag, U. Unal, M.A. Kurykin, C.J. Ayala, M. Aindow, C. Erkey, Thermodynamic control of metal loading and composition of carbon aerogel supported Pt–Cu alloy nanoparticles by supercritical deposition, J. Phys. Chem. C 117 (2013) 6777–6787. 10.1021/jp311641g,

The first step is best described by film diffusion where the flux N_A is proportional to a mass transfer coefficient and the concentration difference between the bulk and surface.

$$N_A = k_m(C_{A,b} - C_{A,s}) \tag{6.34}$$

For perfectly mixed solutions, the mass transfer resistance from the bulk fluid to the exterior surface can be neglected and the concentration at the exterior surface is equal to that measured in the bulk. Solutions that are well stirred can be considered as perfectly mixed and this resistance is assumed to be negligible in the model. Fick's first law best describes the second step, diffusion in the pore volume within the adsorbent in the radial direction:

$$N_A = -D_{e,p}\frac{\partial C_A}{\partial r} \tag{6.35}$$

where $D_{e,p}$ is the effective pore volume diffusivity. The effective diffusivity can be calculated by adjusting the supercritical phase molecular diffusivity, D_{AB}, using the porosity ε_p, tortuosity of the adsorbent τ and a size restriction factor, $F(\lambda)$, where λ is the ratio of the critical solute diameter to the pore size.

$$D_{e,p} = \frac{F(\lambda)\varepsilon_p D_{AB}}{\tau} \tag{6.36}$$

For binary diffusion coefficient estimation, the Schmidt number correlation proposed by Funazukuri et al. [95] can be used since it was found to represent the data in the literature for high molecular weight organic compounds in $scCO_2$ fairly well. The correlation is given by

$$Sc^+ = \frac{Sc}{Sc^*} = 1 + \exp\left[\sum_{i=0}^{5} a_i \left(\frac{v_0}{v}\right)^i\right] \tag{6.37}$$

where

$$Sc^* = \frac{5}{6}\left[\frac{\sigma_1 + \sigma_2}{2\sigma_2}\right]^2 \left[\frac{2M_1}{M_1 + M_2}\right]^{1/2} \tag{6.38}$$

In the equations above Sc is Schmidt number at high pressure, Sc^* is Schmidt number at atmospheric pressure, Sc^+ is the ratio of Schmidt number at high pressure to that at atmospheric pressure, σ is the hard sphere diameter, M is the molecular weight, v is the molar volume of solvent, v_0 is the hard-sphere closest packed volume of solvent molecules. The v_0 values for carbon dioxide can be obtained using the polynomial functions of temperature given by

$$v_0 = \frac{1}{1.384}\left(\sum_{i=0}^{4} c_i T^i\right) \tag{6.39}$$

The constants "a_i" in correlation of Schmidt number with the solvent molar volume (Eq. 6.37) together with the constants "c_i" in correlation of effective hard-sphere packed volume (Eq. 6.41) are given in Table 6.9.

$$Sc = \frac{\nu}{\rho \times D_{AB}} \tag{6.40}$$

The tortuosity factor can be estimated as $1/\varepsilon_p$.

The third step is adsorption onto the surface. This step is generally very fast and one can generally justify using the assumption of local equilibrium within the adsorbent where the adsorbed phase concentration is related to the supercritical phase concentration through the adsorption isotherm. The fourth step is surface diffusion within the adsorbent where the solute is adsorbed onto the surface and

TABLE 6.9 Constants for calculation of effective hard-sphere packed volume.

I	a_i	c_i
0	−4.92519817	4.452×10^{-5}
1	5.45529385×10^{1}	-1.152×10^{-7}
2	$-2.45231443 \times 10^{2}$	2.749×10^{-10}
3	6.07893924×10^{2}	-3.073×10^{-13}
4	$-7.08884016 \times 10^{2}$	1.290×10^{-16}
5	3.29611433×10^{2}	–

From T. Funazukuri, C.Y. Kong, S. Kagei, Impulse response techniques to measure binary diffusion coefficients under supercritical conditions, J. Chromatogr. A 1037 (2004) 411–429. 10.1016/j.chroma.2004.03.043, Copyright © 2004, Elsevier.

then migrates along the surface under the concentration gradient in the adsorbed phase. Surface diffusion is most favorable in cases where the adsorbent has a high adsorption capacity and a weak adsorption bond. When there is a strong adsorption bond between the solute and the substrate, the effects of surface diffusion can be neglected in the model. Subsequently, the unsteady-state adsorption process within the adsorbent particle can be described by the following equation:

$$\varepsilon_P \frac{\partial C_A}{\partial t} + \rho_p \frac{\partial q}{\partial t} = D_{e,p}\left(\frac{\partial^2 C_A}{\partial r^2} + \frac{2}{r}\frac{\partial C_A}{\partial r}\right) \tag{6.41}$$

where $\partial C_A/\partial t$ represents the accumulation term, $\partial q/\partial t$ represents the adsorption term, and $\partial C_A/\partial r$ represents the flux term. The two boundary conditions for Eq. (6.41) are

$$\frac{\partial C_A}{\partial r} = 0 \text{ at } r = 0 \tag{6.42}$$

$$-V\frac{\partial C_A}{\partial t} = mSD_{e,p}\frac{\partial C_A}{\partial r} \text{ at } r = R \tag{6.43}$$

The initial conditions are that at time zero the adsorbent is solute free and the bulk fluid concentration is fixed and are given by the following equations:

$$C_A = 0 \text{ at } t = 0 \text{ for } 0 \le r < R \tag{6.44}$$

$$C_A = C_{A0} \text{ at } t = 0 \text{ for } r = R \tag{6.45}$$

For nonlinear isotherms, Eq. (6.41) must be solved numerically. The accumulation term in Eq. (6.41) can be rewritten in terms of the isotherm relationship and concentration change with time through the use of the chain rule resulting in the following equation:

$$\frac{\partial C_A}{\partial t} = \frac{D_{e,p}}{\varepsilon_p + \rho_p \dfrac{\partial q}{\partial C_A}}\left(\frac{\partial^2 C_A}{\partial r^2} + \frac{2}{r}\frac{\partial C_A}{\partial r}\right) \tag{6.46}$$

Eq. (6.46) can be converted to a set of ordinary differential equations (ODEs) by utilizing the Numerical Method of Lines (NMOL) and solved simultaneously. Eq. (6.46) can be written using finite difference methods as:

$$\frac{dC_{A(n)}}{dt} = \frac{D_{e,p}}{\varepsilon_p + \rho_p \dfrac{dq}{dC_{A(n)}}}\left(\frac{C_{A(n+1)} - 2C_{A(n)} + C_{A(n-1)}}{(\Delta r)^2} + \frac{2}{R-(n-1)\Delta r}\frac{C_{A(n+1)} - C_{A(n-1)}}{2\Delta r}\right) \tag{6.47}$$

with the boundary conditions:

$$C_{A(n)} - C_{A(n-1)} = 0 \text{ at } r = 0 \tag{6.48}$$

$$\frac{dC_{A(1)}}{dt} = \frac{mSD_{e,p}}{-V}\frac{C_{A(1)} - C_{A(2)}}{\Delta r} \quad \text{at } r = R \tag{6.49}$$

Generally, 25 increments provided significant resolution to describe the adsorption dynamics.

The model was tested to predict the batch adsorption of $Ru(cod)(tmhd)_2$ on particles of carbon aerogel with an average size of 21 nm from $scCO_2$ [61]. The experimental data and model fitted by NMOL is shown in Fig. 6.17.

The parameters used in the model and how they were measured and estimated are given in Table 6.10. Considering that there are no adjustable parameters in the model, the good agreement between the model and the data suggest that the model is accurate in describing the kinetics.

The average fitting error of the model to the experimental data is 4.6%. It was observed that the concentration of $Ru(cod)(tmhd)_2$ in supercritical CO_2, decreased from 2.26 to 1.02 mol/m^3 corresponding to the point on the adsorption isotherm. In Fig. 6.18, the concentration values at six different points between the surface and the center of the adsorbent particle carbon aerogel (CA) are presented.

C_1 is the concentration at the outermost point; C_{25} is the concentration at the center of the carbon aerogel particle. The distance between each consecutive points; C_1, C_5, C_{10}, C_{15}, C_{20} and C_{25} is 0.2 mm. It is clearly seen that the

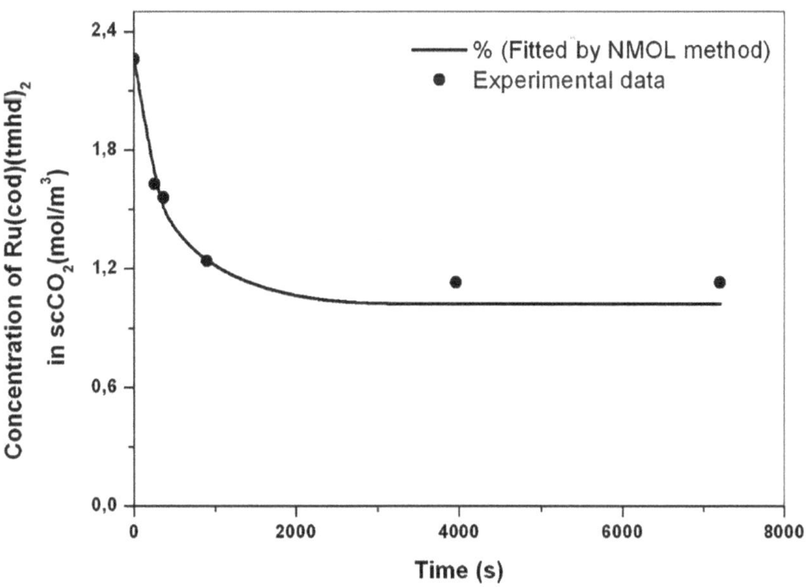

FIG. 6.17 Experimental and predicted adsorption kinetics curves [61].

TABLE 6.10 Parameters for the model for adsorption kinetics.

Parameters	Value	Unit	Sources
Adsorbent properties			
τ	1.1	–	Tortuosity factor
ρ_P	260	kg/m³	Bulk density from measurements
ε_P	0.88	–	Void volume fraction from measurements
R_p	0.001	m	Average radius of the CA particle
S	11.54	m²/ kg	Calculated by $3/(\rho_p R)$
m_{CA}	1.41×10^{-4}	kg	Mass of adsorbent (CA)
Bulk fluid properties			
ρ_{fluid}	650	kg/m³	Fluid density
V_{fluid}	54×10^{-6}	m³	Volume of fluid
C_o	2.26	mol/m³	Initial OM concentration in fluid
Isotherm properties			
Q_o	0.60	Mole OM/kg adsorbent	Modified Langmuir isotherm constants at 19.3 MPa, 343 K and 650 kg/m³ supercritical fluid density
K_1	3.93	m³ $scCO_2$/ mole OM	
k	0.042	m³ $scCO_2$/ kg adsorbent	
Transport properties			
D_{AB}	1.05×10^{-8}	m²/s	Bulk phase liquid diffusivity
d_S	1	nm	Solute molecule diameter
d_P	22	nm	Pore diameter of CA22
λ	0.045	–	d_S/d_P
$F(\lambda)$	0.841	–	$F(\lambda)=1.03e^{-4.5\lambda}$
$D_{e,p}$	7.07×10^{-9}	m²/s	$De,p=\frac{F(\lambda)\varepsilon_p D_{AB}}{\tau}$

From Y. Zhang, B. Cangul, Y. Garrabos, C. Erkey, Thermodynamics and kinetics of adsorption of bis(2,2,6,6-tetramethyl-3,5-heptanedionato)(1,5-cyclooctadiene) ruthenium (II) on carbon aerogel from supercritical CO_2 solution, J. Supercrit. Fluids 44 (2008) 71–77. 10.1016/j.supflu.2007.08.010,

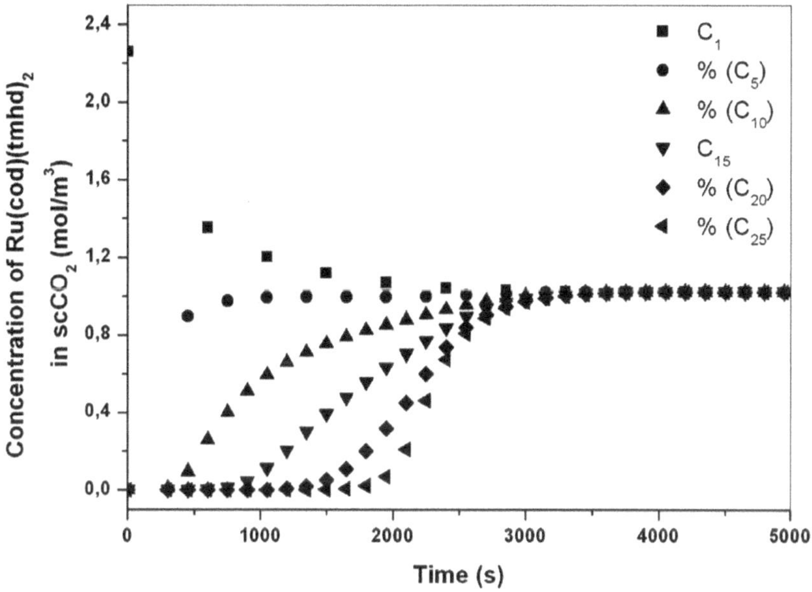

FIG. 6.18 Demonstration of concentration decay curves of Ru(cod)(tmhd)$_2$ on carbon aerogel at 0.2 mm consecutive intervals from the outermost point (C_1) to the center of the adsorbent (C_{25}) [61].

concentration decreases from the surface toward the center of CA at a particular time.

The effects of various parameters on the dynamics of adsorption were also investigated. The constants K_1, Q_0 and k were increased five times to their original values individually while the other two were remained constant. The effect of Modified Langmuir adsorption model parameters can be observed in Fig. 6.19.

As expected, Q_o influences the equilibrium concentration significantly. Increase of Q_o leads to a higher adsorption amount and lower equilibrium concentration. K_l also influences the equilibrium concentration slightly but not as effective as Q_o. However, k does not influence the equilibrium concentration since the second term in the Modified Langmuir isotherm is much smaller than the first term in the concentration range used in the simulation. The particle size of the adsorbent is an important parameter, especially for industrial applications of the technology, since particle size of commercial supported catalysts are generally large. Fig. 6.20 shows the effect of adsorbent particle size on the adsorption dynamics. With increase of the particle size, it takes longer time to reach equilibrium, but the equilibrium concentration remains the same for chosen particle sizes $R = 1$, 3 and 5 mm. In order to prepare particles larger than 5 mm with a radially uniform metallic complex distribution, one has to take into consideration the long diffusion times required.

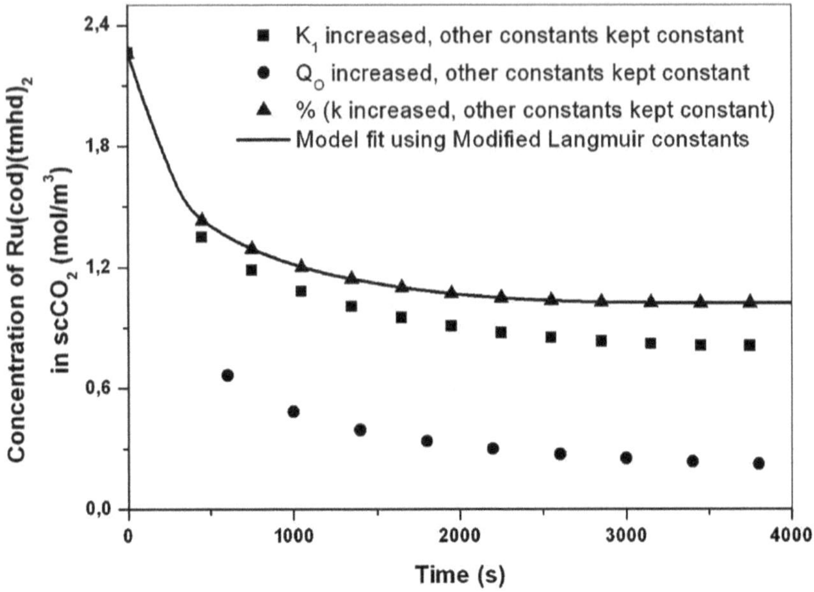

FIG. 6.19 Effect of Modified Langmuir Adsorption isotherm constants on adsorption dynamics [61].

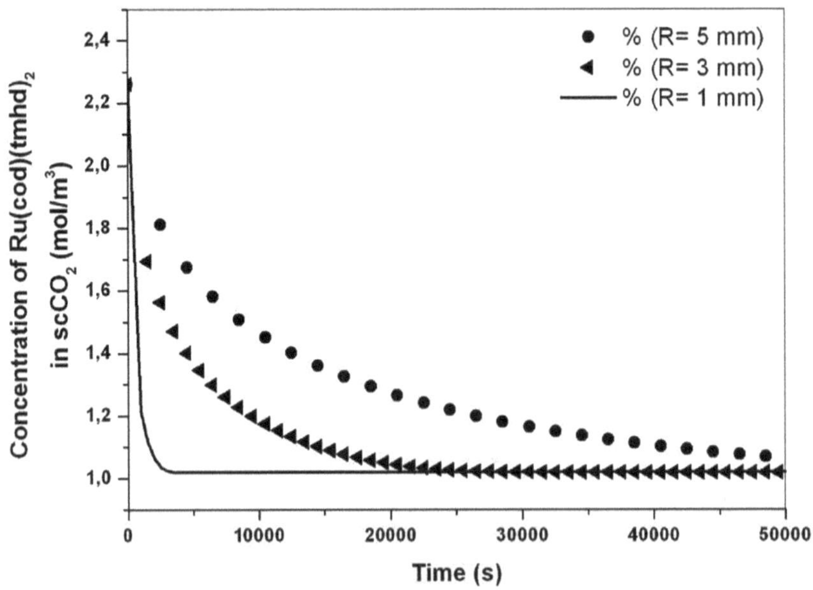

FIG. 6.20 Effect of adsorbent particle size on the adsorption kinetics [61].

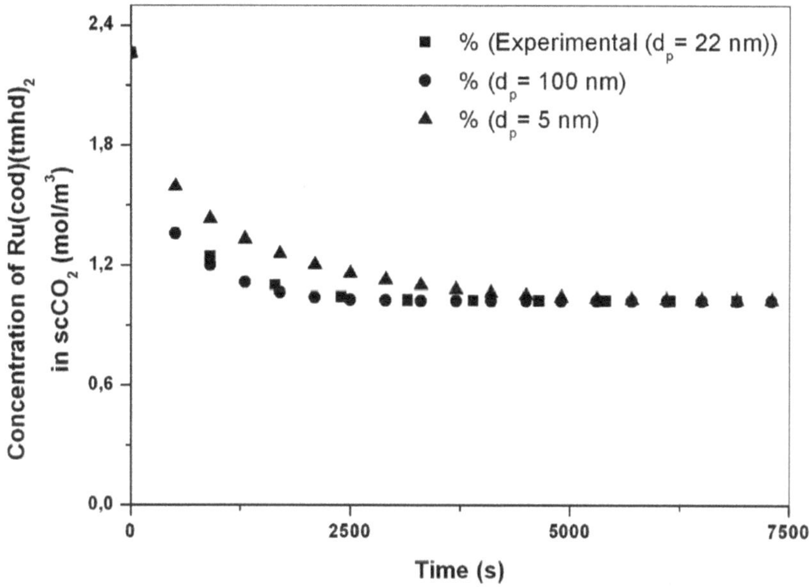

FIG. 6.21 Effect of pore size on the adsorption kinetics [61].

Another important consideration is the size of the pores. The effect of the pore size on the adsorption dynamics can also be simulated. The pore size of the CA used in the experiment was 22 nm. The dynamics of adsorption was simulated for the pore sizes of 5 and 100 nm; the results can be seen in Fig. 6.21. As the pore size starts to approach the solute size, the time to reach equilibrium starts to increase significantly. Therefore, deposition is much faster when λ (solute molecule diameter/pore diameter of CA) is small.

References

[1] J.D. Seader, E.J. Henley, D.K. Roper, Separation Process Principles, third ed., John Wiley Incorporated, USA, 2010. https://books.google.com.tr/books?id=xPfknQEACAAJ.

[2] S. Brunauer, L.S. Deming, W.E. Deming, E. Teller, On a theory of the van der waals adsorption of gases, J. Am. Chem. Soc. 62 (1940) 1723–1732, https://doi.org/10.1021/ja01864a025.

[3] J.R. Anderson, K.C. Pratt, Introduction to Characterization and Testing of Catalysts, Academic Press, Sydney, Australia, 1985.

[4] K.S.W. Sing, D.H. Everett, R.A.W. Haul, L. Moscou, R.A. Pierotti, J. Roquerol, T. Siemienewska, Reporting physisorption data for gas/solid systems with special reference to the determination of surface area and porosity, Pure Appl. Chem. 57 (1985). https://db.degruyter.com/view/IUPAC/iupac.57.0007.

[5] I. Langmuir, The adsorption of gases on plane surfaces of glass, mica and platinum, J. Am. Chem. Soc. 40 (1918) 1361–1403, https://doi.org/10.1021/ja02242a004.

[6] E.L. Cussler, Diffusion: Mass Transfer in Fluid Systems, second ed., Cambridge University Press, Cambridge, 1997.

[7] R.E. Treybal, Mass Transfer Operations, third ed., McGraw-Hill, New York, NY, 1980.
[8] D.D. Do, H.D. Do, Adsorption of supercritical fluids in non-porous and porous carbons: analysis of adsorbed phase volume and density, Carbon 41 (2003) 1777–1791, https://doi.org/10.1016/S0008-6223(03)00152-0.
[9] R. Humayun, D.L. Tomasko, High-resolution adsorption isotherms of supercritical carbon dioxide on activated carbon, AIChE J 46 (2000) 2065–2075, https://doi.org/10.1002/aic.690461017.
[10] D.L. Tomasko, X. Han, D. Liu, W. Gao, Supercritical fluid applications in polymer nanocomposites, Curr. Opin. Solid State Mater. Sci. 7 (2003) 407–412, https://doi.org/10.1016/j.cossms.2003.10.005.
[11] E. Bekyarova, K. Kaneko, Adsorption of supercritical N2 and O2 on pore-controlled carbon aerogels, J. Colloid Interface Sci. 238 (2001) 357–361, https://doi.org/10.1006/jcis.2001.7513.
[12] K. Kaneko, Anomalous micropore filling of nitric oxide on.alpha.-iron hydroxide oxide-dispersed activated carbon fibers, Langmuir 3 (1987) 357–363, https://doi.org/10.1021/la00075a014.
[13] K. Kaneko, K. Shimizu, T. Suzuki, Intrapore field-dependent micropore filling of supercritical N2 in slit-shaped micropores, J. Chem. Phys. 97 (1992) 8705–8711, https://doi.org/10.1063/1.463389.
[14] K. Kaneko, K. Murata, An analytical method of micropore filling of a supercritical gas, Adsorption 3 (1997) 197–208, https://doi.org/10.1007/BF01650131.
[15] A.V. Neimark, P.I. Ravikovitch, Calibration of pore volume in adsorption experiments and theoretical models, Langmuir 13 (1997) 5148–5160, https://doi.org/10.1021/la970266s.
[16] C. Lastoskie, K.E. Gubbins, N. Quirke, Pore size distribution analysis of microporous carbons: a density functional theory approach, J. Phys. Chem. 97 (1993) 4786–4796, https://doi.org/10.1021/j100120a035.
[17] A.P.Y. Wong, M.H.W. Chan, Liquid-vapor critical point of 4He in aerogel, Phys. Rev. Lett. 65 (1990) 2567–2570, https://doi.org/10.1103/PhysRevLett.65.2567.
[18] M.E. Fisher, H. Nakanishi, Scaling theory for the criticality of fluids between plates, J. Chem. Phys. 75 (1981) 5857–5863, https://doi.org/10.1063/1.442035.
[19] G.H. Findenegg, High Pressure Adsorption of Gases on Homogeneous Surfaces, Engineering Foundation, New York, 1983.
[20] S. Blümel, G.H. Findenegg, Critical adsorption of a pure fluid on a graphite substrate, Phys. Rev. Lett. 54 (1985) 447–450, https://doi.org/10.1103/PhysRevLett.54.447.
[21] G.H. Findenegg, J. Specovius, Study of a fluid/solid interface over a wide density range including the critical region.: II. Spreading pressure and enthalpy of adsorption of ethylene/graphite, Ber. Bunsen. Phys. Chem. 84 (1980) 696–701, https://doi.org/10.1002/bbpc.19800840716.
[22] M. Thommes, G.H. Findenegg, Pore condensation and critical-point shift of a fluid in controlled-pore glass, Langmuir 10 (1994) 4270–4277, https://doi.org/10.1021/la00023a058.
[23] K. Morishige, H. Fujii, M. Uga, D. Kinukawa, Capillary critical point of argon, nitrogen, oxygen, ethylene, and carbon dioxide in MCM-41, Langmuir 13 (1997) 3494–3498, https://doi.org/10.1021/la970079u.
[24] B. Turgay, U. Akman, Modeling of equilibrium sorption and fixed-bed desorption for toluene and benzene and supercritical conditions, in: G. Brunner (Ed.), Third Int. Symp. Supercrit. Fluids, Boston, 1994, p. 177.
[25] T. Nakai, Y. Satoh, N. Takahashi, Y. Katoh, Regeneration of activated carbon loaded with nitrobenzene by supercritical carbon dioxide, Nippon Kagaku Kaishi 1992 (1992) 1374–1379, https://doi.org/10.1246/nikkashi.1992.1374.

[26] R. Harikrishnan, M.P. Srinivasan, C.B. Ching, Adsorption of ethyl benzene on activated carbon from supercritical CO2, AIChE J 44 (1998) 2620–2627, https://doi.org/10.1002/aic.690441205.
[27] G. Madras, C. Erkey, A. Akgerman, Supercritical fluid regeneration of activated carbon loaded with heavy molecular weight organics, Ind. Eng. Chem. Res. 32 (1993) 1163–1168, https://doi.org/10.1021/ie00018a022.
[28] G. Madras, C. Thibaud, C. Erkey, A. Akgerman, Modeling of supercritical extraction of organics from solid matrices, AIChE J 40 (1994) 777–785, https://doi.org/10.1002/aic.690400505.
[29] S.J. Macnaughton, N.R. Foster, Supercritical adsorption and desorption behavior of ddt on activated carbon using carbon dioxide, Ind. Eng. Chem. Res. 34 (1995) 275–282, https://doi.org/10.1021/ie00040a029.
[30] S. Lucas, M.J. Cocero, C. Zetzl, G. Brunner, Adsorption isotherms for ethylacetate and furfural on activated carbon from supercritical carbon dioxide, Fluid Phase Equilib. 219 (2004) 171–179, https://doi.org/10.1016/j.fluid.2004.01.034.
[31] P. Subra, A. Vega-Bancel, E. Reverchon, Breakthrough curves and adsorption isotherms of terpene mixtures in supercritical carbon dioxide, J. Supercrit. Fluids 12 (1998) 43–57, https://doi.org/10.1016/S0896-8446(97)00038-7.
[32] I. Kikic, P. Alessi, A. Cortesi, S.J. Macnaughton, N.R. Foster, B. Spicka, An experimental study of supercritical adsorption equilibria of salicylic acid on activated carbon, Fluid Phase Equilib. 117 (1996) 304–311, https://doi.org/10.1016/0378-3812(95)02967-2.
[33] C. Domingo, J. García-Carmona, M.A. Fanovich, J. Llibre, R. Rodríguez-Clemente, Single or two-solute adsorption processes at supercritical conditions: an experimental study, J. Supercrit. Fluids. 21 (2001) 147–157, https://doi.org/10.1016/S0896-8446(01)00088-2.
[34] D. Bolten, M. Johannsen, Influence of 2-propanol on adsorption equilibria of α- and δ-tocopherol from supercritical carbon dioxide on silica gel, J. Chem. Eng. Data 51 (2006) 2132–2137, https://doi.org/10.1021/je060261+.
[35] M. Lübbert, G. Brunner, M. Johannsen, Adsorption equilibria of α- and δ-tocopherol from supercritical mixtures of carbon dioxide and 2-propanol onto silica by means of perturbation chromatography, J. Supercrit. Fluids 42 (2007) 180–188, https://doi.org/10.1016/j.supflu.2007.02.002.
[36] H. Xing, B. Su, Q. Ren, Y. Yang, Adsorption equilibria of artemisinin from supercritical carbon dioxide on silica gel, J. Supercrit. Fluids 49 (2009) 189–195, https://doi.org/10.1016/j.supflu.2009.01.003.
[37] B.C. Pérez Molina, M. Johannsen, Adsorption equilibria of benzoic acid on silica gel from supercritical carbon dioxide, J. Supercrit. Fluids 54 (2010) 237–242, https://doi.org/10.1016/j.supflu.2010.05.015.
[38] J.S. Porto, K. Tanida, Y. Sato, S. Takishima, H. Masuoka, Adsorption dynamics of benzene on activated carbon in the presence of supercritical carbon dioxide, J. Chem. Eng. Jpn. 28 (1995) 388–392, https://doi.org/10.1252/jcej.28.388.
[39] S.-H. Chou, D.S.H. Wong, C.-S. Tan, Adsorption and diffusion of benzene in activated carbon at high pressures, Ind. Eng. Chem. Res. 36 (1997) 5501–5506, https://doi.org/10.1021/ie970327h.
[40] S. LIM, S.S.H. RIZVI, Adsorption and desorption of cholesterol in continuous supercritical fluid processing of anhydrous milk fat, J. Food Sci. 61 (1996) 817–820, https://doi.org/10.1111/j.1365-2621.1996.tb12209.x.

[41] K. Sawada, M. Ueda, Evaluation of the dyeing mechanism of an acid dye on protein fibers in supercritical CO2, Dyes Pigments 63 (2004) 77–81, https://doi.org/10.1016/j.dyepig.2004.01.008.

[42] A.S. Özcan, A. Özcan, Adsorption behavior of a disperse dye on polyester in supercritical carbon dioxide, J. Supercrit. Fluids 35 (2005) 133–139, https://doi.org/10.1016/j.supflu.2004.12.006.

[43] Y. Han, Y. Yang, P. Wu, Adsorption equilibria of cis-5, 8,11,14,17-eicosapentaenoic acid ethyl ester and cis-4,7,10,13,16,19-docosahexaenoic acid ethyl ester from supercritical carbon dioxide on silica gel, J. Chem. Eng. Data 53 (2008) 16–19, https://doi.org/10.1021/je700147c.

[44] B. Su, H. Xing, Y. Han, Y. Yang, Q. Ren, P. Wu, Adsorption equilibria of cis-5,8,11,14,17-eicosapentaenoic acid ethyl ester and cis-4,7,10,13,16,19-docosahexaenoic acid ethyl ester on c18-bonded silica from supercritical carbon dioxide, J. Chem. Eng. Data 54 (2009) 2906–2913, https://doi.org/10.1021/je801003d.

[45] X. Yang, M.A. Matthews, Equilibrium and kinetics properties of p-dichlorobenzene and toluene on silica gel in dense CO2 by chromatography analysis, Chem. Eng. J. 93 (2003) 163–172, https://doi.org/10.1016/S1385-8947(02)00254-1.

[46] Y. Iwai, M. Higuchi, H. Nishioka, Y. Takahashi, Y. Arai, Adsorption of supercritical carbon dioxide + 2,6- and 2,7-dimethylnaphthalene isomers on NaY-type zeolite, Ind. Eng. Chem. Res. 42 (2003) 5261–5267, https://doi.org/10.1021/ie0302847.

[47] J. Gregorowicz, Adsorption of eicosane and 1,2-hexanediol from supercritical carbon dioxide on activated carbon and chromosorb, Fluid Phase Equilib. 238 (2005) 142–148, https://doi.org/10.1016/j.fluid.2005.09.020.

[48] C. Erkey, G. Madras, M. Orejuela, A. Akgerman, Supercritical carbon dioxide extraction of organics from soil, Environ. Sci. Technol. 27 (1993) 1225–1231, https://doi.org/10.1021/es00043a025.

[49] Y.-K. Ryu, K.-L. Kim, C.-H. Lee, Adsorption and desorption of n-hexane, methyl ethyl ketone, and toluene on an activated carbon fiber from supercritical carbon dioxide, Ind. Eng. Chem. Res. 39 (2000) 2510–2518, https://doi.org/10.1021/ie990673u.

[50] I. Smirnova, J. Mamic, W. Arlt, Adsorption of drugs on silica aerogels, Langmuir 19 (2003) 8521–8525, https://doi.org/10.1021/la0345587.

[51] M. Škerget, Ž. Knez, Supercritical fluid adsorption and desorption of lipids on various adsorbents, Acta Chim. Slov. 54 (2007) 688–692. https://dk.um.si/IzpisGradiva.php?lang=eng&id=26036.

[52] M. Sato, M. Goto, A. Kodama, T. Hirose, New fractionation process for citrus oil by pressure swing adsorption in supercritical carbon dioxide, Chem. Eng. Sci. 53 (1998) 4095–4104, https://doi.org/10.1016/S0009-2509(98)00212-7.

[53] C.S. Tan, D.C. Liou, Adsorption equilibrium of toluene from supercritical carbon dioxide on activated carbon, Ind. Eng. Chem. Res. 29 (1990) 1412–1415, https://doi.org/10.1021/ie00103a045.

[54] J. Benkhedda, J.-N. Jaubert, D. Barth, C. Zetzl, G. Brunner, Adsorption and desorption of m-xylene from supercritical carbon dioxide on activated carbon, Sep. Sci. Technol. 36 (2001) 2197–2211, https://doi.org/10.1081/SS-100105913.

[55] Á. Sastre, Á. Martín, E. Alonso, Adsorption of nickelocene and ruthenocene on mesoporous silica MCM-48 and activated carbon supports in supercritical carbon dioxide, J. Supercrit. Fluids 117 (2016) 138–146, https://doi.org/10.1016/j.supflu.2016.06.013.

[56] M. Pantić, Ž. Knez, Z. Novak, Supercritical impregnation as a feasible technique for entrapment of fat-soluble vitamins into alginate aerogels, J. Non-Cryst. Solids 432 (2016) 519–526, https://doi.org/10.1016/j.jnoncrysol.2015.11.011.

[57] I. Ushiki, H. Tsuji, S. Takishima, Y. Ito, Y. Sato, H. Inomata, Measurement and correlation of adsorption equilibria of propylene glycol monomethyl ether acetate on activated carbon in the presence of supercritical carbon dioxide, Fluid Phase Equilib. 513 (2020) 112556, https://doi.org/10.1016/j.fluid.2020.112556.

[58] X. Hu, S. Zheng, L. Zhu, A.R. Tanyi, H. Lan, Y. Hong, Y. Su, H. Wang, J. Li, Adsorption of 2-phenylethyl alcohol on silica aerogel from saturated solution in supercritical CO2, J. Supercrit. Fluids 79 (2013) 41–45, https://doi.org/10.1016/j.supflu.2013.02.015.

[59] I. Ushiki, M. Ota, Y. Sato, H. Inomata, Measurements and Dubinin–Astakhov correlation of adsorption equilibria of toluene, acetone, n-hexane, n-decane and methanol solutes in supercritical carbon dioxide on activated carbon at temperature from 313 to 353K and at pressure from 4.2 to 15.0MPa, Fluid Phase Equilib 344 (2013) 101–107, https://doi.org/10.1016/j.fluid.2013.01.014.

[60] S.E. Bozbag, L.C. Zhang, M. Aindow, C. Erkey, Carbon aerogel supported nickel nanoparticles and nanorods using supercritical deposition, J. Supercrit. Fluids 66 (2012) 265–273,-https://doi.org/10.1016/j.supflu.2012.02.027.

[61] Y. Zhang, B. Cangul, Y. Garrabos, C. Erkey, Thermodynamics and kinetics of adsorption of bis(2,2,6,6-tetramethyl-3,5-heptanedionato)(1,5-cyclooctadiene) ruthenium (II) on carbon aerogel from supercritical CO2 solution, J. Supercrit. Fluids 44 (2008) 71–77, https://doi.org/10.1016/j.supflu.2007.08.010.

[62] H. Lü, S. Liu, Y. Li, F. Meng, D. Yao, X. Mo, Adsorption equilibrium for Z-ligustilide on C18-bonded silica from supercritical carbon dioxide, Chin. J. Chem. Eng. 24 (2016) 1680–1686, https://doi.org/10.1016/j.cjche.2016.04.020.

[63] I. Ushiki, N. Takahashi, M. Koike, Y. Sato, S. Takishima, H. Inomata, Adsorption kinetics of rhodium (III) acetylacetonate onto mesoporous silica adsorbents in the presence of supercritical carbon dioxide, J. Supercrit. Fluids 135 (2018) 137–144, https://doi.org/10.1016/j.supflu.2018.01.017.

[64] M.J. Tenorio, C. Pando, J.A.R. Renuncio, J.G. Stevens, R.A. Bourne, M. Poliakoff, A. Cabañas, Adsorption of Pd(hfac)2 on mesoporous silica SBA-15 using supercritical CO2 and its role in the performance of Pd–SiO2 catalyst, J. Supercrit. Fluids 69 (2012) 21–28, https://doi.org/10.1016/j.supflu.2012.05.003.

[65] M.A.E. Cunha, R.F. Neves, J.N.S. Souza, L.F. França, M.E. Araújo, G. Brunner, N.T. Machado, Supercritical adsorption of buriti oil (Mauritia flexuosa Mart.) in γ-alumina: a methodology for the enriching of anti-oxidants, J. Supercrit. Fluids. 66 (2012) 181–191, https://doi.org/10.1016/j.supflu.2011.10.021.

[66] I. De Marco, E. Reverchon, Starch aerogel loaded with poorly water-soluble vitamins through supercritical CO2 adsorption, Chem. Eng. Res. Des. 119 (2017) 221–230, https://doi.org/10.1016/j.cherd.2017.01.024.

[67] I. Ushiki, N. Takahashi, T. Shimizu, Y. Sato, M. Ota, R.L. Smith, H. Inomata, Adsorption equilibria of rhodium acetylacetonate with MCM-41, MSU-H, and HMS silica substrates in supercritical carbon dioxide for preparing catalytic mesoporous materials, J. Supercrit. Fluids 120 (2017) 240–248, https://doi.org/10.1016/j.supflu.2016.05.032.

[68] I. Ushiki, Y. Sato, M. Ota, H. Inomata, Adsorption equilibria of volatile organic compounds on various adsorbents in supercritical carbon dioxide: measurement and analysis by the Dubinin-Astakhov equation, Fluid Phase Equilib. 420 (2016) 58–67, https://doi.org/10.1016/j.fluid.2016.02.007.

[69] G. Caputo, M. Scognamiglio, I. De Marco, Nimesulide adsorbed on silica aerogel using supercritical carbon dioxide, Chem. Eng. Res. Des. 90 (2012) 1082–1089, https://doi.org/10.1016/j.cherd.2011.11.011.

[70] I. Ushiki, M. Koike, T. Shimizu, Y. Sato, S. Takishima, H. Inomata, Measurement and modeling of adsorption equilibria of cobalt (III) acetylacetonate on MCM-41 mesoporous silica in the presence of supercritical carbon dioxide with methanol co-solvent, J. Supercrit. Fluids 140 (2018) 329–335, https://doi.org/10.1016/j.supflu.2018.06.015.
[71] S.E. Bozbag, U. Unal, M.A. Kurykin, C.J. Ayala, M. Aindow, C. Erkey, Thermodynamic control of metal loading and composition of carbon aerogel supported Pt–Cu alloy nanoparticles by supercritical deposition, J. Phys. Chem. C 117 (2013) 6777–6787, https://doi.org/10.1021/jp311641g.
[72] C.D. Saquing, D. Kang, M. Aindow, C. Erkey, Investigation of the supercritical deposition of platinum nanoparticles into carbon aerogels, Microporous Mesoporous Mater. 80 (2005) 11–23, https://doi.org/10.1016/j.micromeso.2004.11.019.
[73] I. Ushiki, K. Kikuchi, Y. Sato, Y. Ito, S. Takishima, H. Inomata, Adsorption equilibria of VOCs (n-octane, propylene glycol monomethyl ether, ethanol, and 2-propanol) on activated carbon under supercritical carbon dioxide conditions, Fluid Phase Equilib. 462 (2018) 59–64, https://doi.org/10.1016/j.fluid.2018.01.027.
[74] C. Erkey, Supercritical Fluids and Organometallic Compounds, first ed., Elsevier B.V, Amsterdam, 2011.
[75] Y. Zhang, D. Kang, M. Aindow, C. Erkey, Preparation and characterization of ruthenium/carbon aerogel nanocomposites via a supercritical fluid route, J. Phys. Chem. B 109 (2005) 2617–2624, https://doi.org/10.1021/jp0467595.
[76] X. Yang, C.T. Lira, Adsorption equilibrium of benzaldehyde and benzyl alcohol onto polymeric resin from supercritical CO2, Adsorption 23 (2017) 63–71, https://doi.org/10.1007/s10450-016-9819-4.
[77] H. Lü, G. Wang, M. Zhang, Z. Geng, M. Yang, Y. Sun, Adsorption equilibrium of citric acid from supercritical carbon dioxide/ethanol on cyano column, Chin. J. Chem. Eng. 23 (2015) 905–911, https://doi.org/10.1016/j.cjche.2013.10.002.
[78] W. Zhou, G. Anitescu, L.L. Tavlarides, Polychlorinated biphenyl (PCB) partition equilibrium between St. Lawrence River sediments and supercritical fluids, J. Supercrit. Fluids. 29 (2004) 31–42, https://doi.org/10.1016/S0896-8446(03)00012-3.
[79] C.-C. Lai, C.-S. Tan, Heat effects for toluene adsorption on activated carbon from supercritical carbon dioxide, Fluid Phase Equilib. 111 (1995) 127–141, https://doi.org/10.1016/0378-3812(95)02769-B.
[80] M. Modell, R.C. Reid, Thermodynamics and Its Applications, Prentice-Hall, Englewood Cliffs, NJ, 1983.
[81] C.A. Eckert, D.H. Ziger, K.P. Johnston, S. Kim, Solute partial molal volumes in supercritical fluids, J. Phys. Chem. 90 (1986) 2738–2746, https://doi.org/10.1021/j100403a036.
[82] F. Recasens, E. Velo, M.A. Larrayoz, J. Puiggené, Endothermic character of toluene adsorption from supercritical carbon dioxide on activated carbon at low coverage, Fluid Phase Equilib. 90 (1993) 265–287, https://doi.org/10.1016/0378-3812(93)85068-W.
[83] S.K. Jha, G. Madras, Modeling of adsorption equilibria in supercritical fluids, J. Supercrit. Fluids 32 (2004) 161–166, https://doi.org/10.1016/j.supflu.2003.12.010.
[84] J.W. King, Supercritical fluid adsorption at the gas-solid interface, in: Supercrit Fluids, American Chemical Society, Washington, DC, 1987, pp. 13–150.
[85] J.H. Chen, D.S.H. Wong, C.S. Tan, R. Subramanian, C.T. Lira, M. Orth, Adsorption and desorption of carbon dioxide onto and from activated carbon at high pressures, Ind. Eng. Chem. Res. 36 (1997) 2808–2815, https://doi.org/10.1021/ie960227w.
[86] S. Sircar, Gibbsian surface excess for gas adsorption revisited, Ind. Eng. Chem. Res. 38 (1999) 3670–3682, https://doi.org/10.1021/ie9900871.

[87] T.G. Squires, M.E. Paulaitis, No Title, ACS Publishing, Washington, DC, Chicago, IL, 1985.

[88] E. Alonso, F.J. Cantero, J. García, M.J. Cocero, Scale-up for a process of supercritical extraction with adsorption of solute onto active carbon. Application to soil remediation, J. Supercrit. Fluids. 24 (2002) 123–135, https://doi.org/10.1016/S0896-8446(02)00016-5.

[89] M.J. Cocero, J. García, Mathematical model of supercritical extraction applied to oil seed extraction by CO2+saturated alcohol—I. Desorption model, J. Supercrit. Fluids. 20 (2001) 229–243, https://doi.org/10.1016/S0896-8446(01)00068-7.

[90] R. Ishii, K. Ooi, Reaction in supercritical CO2; an intercalation of 4-phenylazoaniline between layers of montmorillonite pillared with tetramethylammonium ions, Chem. Commun. (1998) 1705–1706, https://doi.org/10.1039/A803540G.

[91] S. Haji, C. Erkey, Removal of dibenzothiophene from model diesel by adsorption on carbon aerogels for fuel cell applications, Ind. Eng. Chem. Res. 42 (2003) 6933–6937, https://doi.org/10.1021/ie030518m.

[92] S.E. Bozbag, N.S. Yasar, L.C. Zhang, M. Aindow, C. Erkey, Adsorption of Pt(cod)me2 onto organic aerogels from supercritical solutions for the synthesis of supported platinum nanoparticles, J. Supercrit. Fluids 56 (2011) 105–113, https://doi.org/10.1016/j.supflu.2010.10.045.

[93] B. Cangül, L.C. Zhang, M. Aindow, C. Erkey, Preparation of carbon black supported Pd, Pt and Pd–Pt nanoparticles using supercritical CO2 deposition, J. Supercrit. Fluids 50 (2009) 82–90, https://doi.org/10.1016/j.supflu.2009.04.001.

[94] R. Leyva-Ramos, C.J. Geankoplis, Diffusion in liquid-filled pores of activated carbon. I. Pore volume diffusion, Can. J. Chem. Eng. 72 (1994) 262–271, https://doi.org/10.1002/cjce.5450720213.

[95] T. Funazukuri, C.Y. Kong, S. Kagei, Impulse response techniques to measure binary diffusion coefficients under supercritical conditions, J. Chromatogr. A 1037 (2004) 411–429, https://doi.org/10.1016/j.chroma.2004.03.043.

Chapter 7

Synthesis of nanostructured composites of metals by supercritical deposition (SCD)

A nanostructured composite material is defined as a material that has a structure such that at least one of its phases has one or more dimensions (length, width or thickness) in the nanometer size range, usually ranging from 1 to 100 nm. Such materials include polycrystalline materials with nanometer sized crystallites, materials with surface protrusions spatially separated by distances on the order of nanometers, granular or porous materials with grain sizes in the nanometer range or nanometer sized metallic clusters dispersed within a porous or polymeric matrix (supported-metal nanoparticles). Among these materials, supported metal nanoparticles have attracted a great deal of interest and their unique properties are directly related to the specific particle morphology (size and shape), metal dispersion, concentration and the electronic properties of the metal within their host environment.

7.1 Supported nanoparticles on inorganic and carbon supports

Nanoparticles supported on inorganic or carbonaceous supports with high surface area are currently used extensively as catalysts for chemical transformations. Catalysis is a particularly important application since roughly 85%–90% of all products are made using chemical processes and catalysts are used in about 90% of these processes. Catalysts are generally metal or metal oxide nanoparticles with sizes from 1 to 10 nm with significant surface-to-volume ratios. The surface atoms are chemically more active compared to the bulk atoms because they usually have fewer adjacent coordinate atoms and unsaturated sites or more dangling bonds. At the same time, the imperfection of the particle surface induces additional electronic states in the band gap, which act as electron or hole trap centers. In a system containing only a few hundred atoms, a large fraction of these atoms is located on the surface. Because surface atoms tend to be coordinatively unsaturated, there is a large energy associated

Synthesis of Nanostructured Materials in Near and/or Supercritical Fluids
https://doi.org/10.1016/B978-0-444-64089-5.00001-9

with this surface. The smaller the nanocrystal, the larger is the contribution of the surface energy to the overall energy of the system. Because of their high surface energy, nanoparticles tend to form larger particles by agglomeration, which diminishes the catalytic activity. Therefore, supports with high surface areas are often used to serve as carriers and stabilizers of catalytically active components (i.e., nanoparticles). The supports not only prevent the agglomeration but also hinder the oxidation of nanoparticles and consequently the intrinsic properties and morphologies of nanoparticles are conserved. Supports have also a crucial role in the transport of the reactants and the products to and away from the catalytic nanoparticles. Moreover, in some cases they may participate in the catalytic reaction as well.

SCD is an alternative and promising way to deposit metal nanoparticles onto surfaces of porous solid supports. Schematic representation of the SCD process is given in Fig. 7.1.

The first step in the process is the preparation of a solution of metal complex in a supercritical fluid (SCF). The resulting solution is contacted with a porous substrate which leads to the adsorption of the metal complex on the surface of the porous substrate. Subsequently, the metal complex can be converted to its metal form and with subsequent growth of atoms leading to supported metallic nanoparticles as shown in Fig. 7.1. The conversion of the adsorbed metal complex into metallic nanoparticles can be carried out:

1. Before depressurization in the SCF phase by injecting a reducing agent, such as hydrogen or an alcohol into the solution or by thermal treatment.

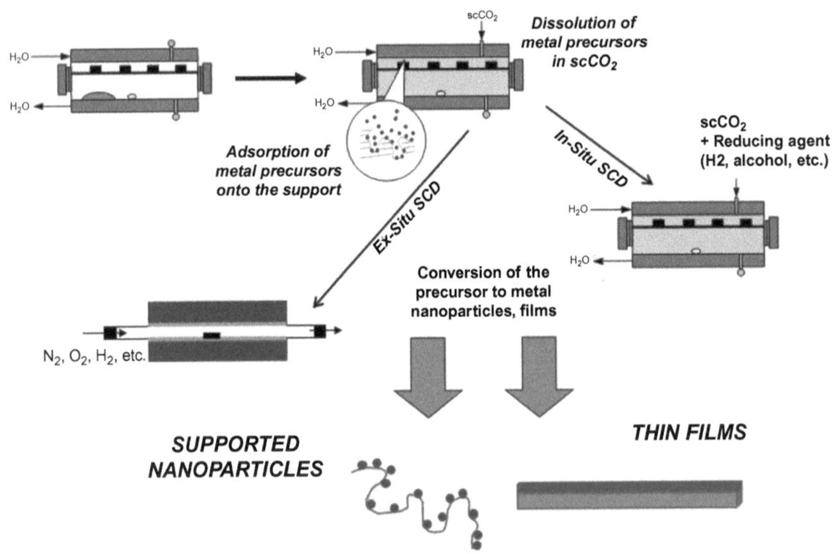

FIG. 7.1 SCD method for the preparation of supported nanoparticles and films using $scCO_2$.

2. After depressurization of the solution. In this case, the sample is subjected to heat in an inert atmosphere where the metal complex decomposes or the sample is subjected to hydrogen.

The metal complexes that have been used so far have been chelate complexes and organometallics. The solubility behavior of these complexes was discussed in Chapter 5 and the adsorption of these complexes on the surfaces of porous supports were discussed in Chapter 6. A summary of the studies carried out in the literature on preparation of supported nanoparticles on various inorganic and carbonaceous supports is given in Table 7.1.

The experimental setup located in my laboratory for SCD is given in Fig. 7.2. A 57-mL custom-made stainless-steel high-pressure vessel is used for the deposition process. It is equipped with two sapphire windows both of which have a diameter of 2.5 cm, a vent line, a rupture disc assembly, a pressure transducer, and a thermowell. The vessel is sealed with poly(ether ether ketone) O-rings inside and outside the sapphire windows.

Highly adsorbent supports may need to be dried in a vacuum oven to remove adsorbed moisture and other impurities in the support prior to SCD. The support must be weighed after drying and before placement in the vessel to be able to gravimetrically determine the amount of uptake after deposition. Especially in cases where support material and the metal complex are both in powder form, placement of the support material into a pouch is beneficial to be able to separate the metal complex and the deposited support for the determination of metal complex uptake after SCD gravimetrically.

Before each experiment, the metal complex and the support material are placed together inside the vessel along with a stirring rod. A corrosion resistant stainless-steel (Grade 316) screen can be placed along the center of the vessel to avoid contact between the stirring rod and the pouch containing the support material. The vessel is then sealed with O-rings and placed on a magnetic stirrer. Since the presence of air inside the vessel may interfere with the deposition process, the air inside the vessel is displaced by flowing CO_2 at ambient conditions through the vessel prior to deposition.

A pressure transducer and a thermocouple are connected to the vessel to monitor pressure and temperature. The vessel temperature is controlled by means of a circulating heater/cooler. Depending on the deposition temperature, various fluids may be used as the heating medium. Water can be used for deposition processes up to 90°C. For temperatures up to 150°C, silicon oil can be used. Usage of the silicon oil is not recommended for processes with temperatures less than 80°C since it cannot be properly circulated due to its high viscosity at low temperatures. For temperatures higher than 150°C, special heat transfer fluids may be used.

When the temperature reaches the deposition condition, CO_2 is pumped into the vessel via a syringe pump. The metal complex starts to dissolve in $scCO_2$ and dissolved complex adsorbs onto the substrate. After allowing

TABLE 7.1 Summary of the studies in the literature on preparation of supported nanoparticles on high surface area substrates by supercritical deposition.

Metal	Metal complex	Morphology	Substrate	Solvent medium	Deposition temperature (°C)	Deposition pressure (MPa)	Conversion method	Particle size range (nm)	Ref.
Au	$HAuCl_4$	Nps	MCM-48	$scCO_2$ + H_2O	70	10–17	1	2–28	[1]
	$HAuCl_4 \cdot 3H_2O$	Nps	CNT	$scCO_2$ + MeOH	50	10	1	10	[2]
	$HAuCl_4 \cdot 3H_2O$	Nps	Graphene	$scCO_2$ + MeOH	50	10	1	10	[2]
	Au(acac)me_2	Nps	CB	$scCO_2$	80	15.5	1	1.2–4.8	[3]
	Au(acac)me_2	Nps	Al_2O_3	$scCO_2$	80	15.5	1	5.2–22.2	[3]
	Au(acac)me_2	Nps	TiO_2	$scCO_2$	80	15.5	1	11.9–34.9	[3]
Ag	Ag(cod)(hfac)	Nps	Vycor glass	$scCO_2$	50	20	3	4–5	[4]
	Ag(cod)(hfac)	Nps	Synthetic opal	$scCO_2$	50	20	3	4–5	[4]
	Ag(NO_3)	Nps	SWCNT	$scCO_2$ + EtOH	60–65	10–12	1	4–6	[5][a]
	Ag(NO_3)	Nps	CNF	$scCO_2$ + EtOH	60–65	10–12	1	8	[5][a]
	Ag(acac)	Nps	AC	scH_2O	400	30	–	1–80	[6]
	Ag(cod)(hfac)	Nps	Plasma treated Si	$scCO_2$	60, 120, 180	13	1	18–50	[7]
	Ag(cod)(hfac)	Nps		$scCO_2$	60, 120, 180	13	1	20–70	[7]
	Ag(NO_3)	Nps, Nrs			30–80	8–24	3	3–14	[8]
	Ag(NO_3)	Nps, Nrs			30–80	8–24	3	~5.2	[8]

	Ag(cod)(hfac)	Nps	Plasma treated glass	$scCO_2$ + Alcohols	36	24.7	3	2–50	[9]
	Ag(cod)(hfac)	Nps	SBA-15	$scCO_2$ + Alcohols	55	9.5	–	–	[10]
	Ag(cod)(hfac)	Nps	KIT-6	$scCO_2$	55	20	3	4–8	[11]
			ECNF	$scCO_2$ + EtOH					
			CNT	$scCO_2$					
			SA						
Co	$CoCp_2$	Nps	MCM-41	$scCO_2$	70	11	2	~15	[12]
	$CoCp_2$	Nps	Al-MCM-41	$scCO_2$	70	11	2	~20	[12]
	$CoCp_2$	Nps	MCM-41	$scCO_2$	200	14	2	20–30	[13]
	$CoCp_2$	Nps	MCM-41	$scCO_2$	70	11	2	~20	[14]
	$Co(acac)_3$	Nps	RFA	$scCO_2$	70	20	3	–	[15]
Cu	$Cu(hfac)_2{\cdot}H_2O$	Nps	Silica	$scCO_2$ + Alcohols	100–150	20–22	1	3–22	[16]
	$Cu(NO_3)_2{\cdot}5H_2O$	Nps, Nrs	SBA-15	$scCO_2$ + EtOH	50	22	3	4–5	[17]
	CuDI6	Nps	CA	$scCO_2$	35	10.6	3	15	[18]
	CuDI6	Nps	RFA	$scCO_2$	35	10.6	3	15	[18]
	CuDI6	Nps	SA	$scCO_2$	35	10.6	3	15	[18]
	$CuCl_2{\cdot}2H_2O$	Nps	GO	$scCO_2$	80	18	–	5–10	[19][a]
	$Cu(NO_3)_2{\cdot}3H_2O$	Nps	SBA-15	$scCO_2$ + Alcohols	50	14–26	3	20–80	[20]

Continued

TABLE 7.1 Summary of the studies in the literature on preparation of supported nanoparticles on high surface area substrates by supercritical deposition—cont'd

Metal	Metal complex	Morphology	Substrate	Solvent medium	Deposition temperature (°C)	Deposition pressure (MPa)	Conversion method	Particle size range (nm)	Ref.
Fe_3O_4	$Fe(NO_3)_3$	Nps	3DHPC	$scCO_2$ + EtOH	200	20	3	11–15	[21][a]
Fe_2O_3	Fe $(NO_3)_3{\cdot}9H_2O$	Nps	SiO_2	scH_2O	500	40–42	–	~15	[22]
Fe_2O_3	Fe $(NO_3)_3{\cdot}9H_2O$	Nps	SiO_2	$scCO_2$ + EtOH	60	12	2	2–14	[23][a]
Ir	Ir(hfac)(cod)	Nps	Si wafer	$scCO_2$	45–184	10.3	1	~8–40	[24]
Ni	$Ni(cp)_2$	Nps	MWCNT	$scCO_2$	70	17	1	10–60	[25]
	$Ni(acac)_2$	Nps, Nrs	CA	$scCO_2$	60	30	3	5–13	[26]
	Ni $(NO_3)_2{\cdot}6H_2O$	Nps	SBA-15	$scCO_2$ + EtOH	35	9	3	–	[27][a]
	$NiCl_2{\cdot}6H_2O$	Nws	SBA-15	$scCO_2$ + EtOH	35	9	3	–	[27][a]
	Ni $(hfac)_2.2H_2O$	Nps	SBA-15	$scCO_2$ + EtOH	80	13.5	3	–	[27][a]
	Ni $(hfac)_2.2H_2O$	Nps	C14	$scCO_2$ + EtOH	80	13.5	1	2	[27][a]
	$Ni(Cp)_2$	Nps	SBA-15	$scCO_2$	60	9.5–11	1	1–30	[27]
	$Ni(Cp)_2$	Nps	SBA-15	$scCO_2$	60	11	1	<2	[27]
	$Ni(Cp)_2$	Nps	C14	$scCO_2$	60	11	1	10–25	[27]
	$Ni(Cp)_2$	Nps	SBA-15	$scCO_2$ + EtOH	60	11	1	1–2	[27][a]
	$Ni(acac)_2$	Nps	C14	$scCO_2$ + EtOH	80	13.5	1	2	[27][a]
	$Ni(Cp)_2$	Nps	rGO	scCO[illegible]	60	10	3	8.5–15	[28]

Pb	$Pb(acac)_2$	Nps	MWCNT	$scCO_2$ + MeOH	200	10	1	5–10	[29]
Pd	Pd $(hfac)_2 \cdot xH_2O$	Nps	MWCNT	$scCO_2$	–	–	1	5–10	[30, 31]
	$Pd(hfpd)_2$	Nps	SA	$scCO_2$	40	27.6	3	<6, 15–20	[32]
	$Pd(hfac)_2$	Nps	CMP	$scCO_2$	40	5.5–27.6	3	1–3	[33]
	$Pd(hfac)_2$	Nps	SiO_2 AlO	$scCO_2$	50–80	8.5	3	2–10	[34]
	$Pd(acac)_2$	Nps	CB	$scCO_2$	40	20	1	3–100	[35]
	$Pd(acac)_2$	Nps	Papam modified sg	$scCO_2$ + THF	40	10	1, 3	3–5	[36]
	$Pd(hfac)_2$	Nps	SBA-15	$scCO_2$	40	8.5	1, 3	8–9	[37]
	$Pd(hfac)_2$	Nps	CNT	$scCO_2$ + EtOH	80	15	1	5	[38]
	$Pd(hfac)_2$	Nps	CNT		50	20	–	<5	[39]
	$Pd(hfac)_2$	Nps	SBA-15	$scCO_2$ + MeOH	40–80	8.5–14	3	5–8	[40]
	$Pd(hfac)_2$	Nps	ZTC	$scCO_2$	40	8	1	2–5	[41]
	$Pd_2(dba)_3$	Nps	SBA-15	$scCO_2$	84	41.4	3	~55	[42]
	$Pd(hfac)_2$	Nps	CNT	$scCO_2$	55	9.5	–	1–21	[10]
	$Pd(hfac)_2$	Nps	Graphene	$scCO_2$ + EtOH	50	18	3	1.3–7.7	[43]
	$Pd(hfac)_2$	Nps	rGO		50	18	3	1.7–23.8	[43]
	$Pd(hfac)_2$	Nps	CNT	$scCO_2$	50	18	3	3.2–11.1	[43]
	$Pd(hfac)_2$	Nps	CB	$scCO_2$	50	18	3	2.8–9.1	[43]
	$Pd(OAc)_2$	Nps	SiO_2	$scCO_2$	60	20	3	~3	[44]
	$Pd(OAc)_2$	Nps	TiO_2	$scCO_2$	60	20	3	~3	[44]
	$Pd(DMG)_2$	Nps	Al_2O_3	$scCO_2$ + Acetone	90	27.6	1	1–17	[45]
	$Pd(PTQD)_2$	Nps	Al_2O_3		90	27.6	1	0.5–8	[45]
					300	20	3	9–26	

Continued

TABLE 7.1 Summary of the studies in the literature on preparation of supported nanoparticles on high surface area substrates by supercritical deposition—cont'd

Metal	Metal complex	Morphology	Substrate	Solvent medium	Deposition temperature (°C)	Deposition pressure (MPa)	Conversion method	Particle size range (nm)	Ref.
	Pd(acac)$_2$	Nps	TiO_2	$scCO_2$ + Acetone	80	17.2	1	16–17	[46]
	Pd(phen)Cl_2	Nps	SBA-15	$scCO_2$	80	20.6	1	~10	[47]
	(Pd(Pyr)$_2$)Cl_2	Nps	MWCNT	$scCO_2$	80	17.2	1	4–11	[48]
	(Pd(Pyr)$_2$)Cl_2		MWCNT	$scCO_2$ + Alcohols					[49]
				$scCO_2$					
				$scCO_2$					
				$scCO_2$					
Pt	Pt(cod)me$_2$	Nps	CA	$scCO_2$	80	27.6	1	4.7 & 1000	[50]
	Pt(cod)me$_2$	Nps	CA	$scCO_2$	80	18.6	2	3.7 & 4.7	[50]
	Pt(cod)me$_2$	Nps	CA	$scCO_2$	80	27.6	3	2.3	[50]
		Nps	CA	$scCO_2$	80	27.6	3	2.7	[50]
	Pt(cod)me$_2$	Nps	CA	$scCO_2$	80	27.6	3	3.6	[50]
	Pt(cod)me$_2$	Nps	CA	$scCO_2$	80	27.6	3	17.1	[50]
	Pt(cod)me$_2$	Nps	CA	$scCO_2$	80	27.6	3	1	[51]
	Pt(cod)me$_2$	Nps	CA	$scCO_2$	80	27.6	3	2.5	[51]
	Pt(cod)me$_2$	Nps	CA	$scCO_2$	80	27.6	3	3	[51]

	Pt(cod)me$_2$	Nps	CA	scCO$_2$	80	27.6	3	3	[51]
	Pt(cod)me$_2$	Nps	SA	scCO$_2$	80	27.6	3	3.9	[52]
	Pt(cod)me$_2$	Nps	CB	scCO$_2$	80	27.6	3	2	[52]
	Pt(cod)me$_2$	Nps	Alumina	scCO$_2$	80	27.6	3	2.7	[52]
	Pt(cod)me$_2$	Nps	Alumina	scCO$_2$	80	27.6	3	1.2	[52]
	Pt(cod)me$_2$	Nps	SiO_2	scCO$_2$	80	27.6	3	6.4	[52]
	Pt(cod)me$_2$	Nps	SiO2	scCO$_2$	80	27.6	3	4.6	[52]
	Pt(cod)me$_2$	Nps	CB	scCO$_2$	80	20	1	2–8	[35]
	Pt(cod)me$_2$	Nps	CB	scCO$_2$	70	13.6–24.2	2, 3	1.2–2	[53]
	Pt(cod)me$_2$	Nps	CB	scCO$_2$	40	8.3	3	1–5	[54]
	Pt(cod)me$_2$	Nps	SBA-15	scCO$_2$	60	10–15	1	2–3	[55]
	Pt(cod)me$_2$	Nps	γ-Alumina	scCO$_2$	80	27.6	3	1–2.7	[56]
	Pt(cod)me$_2$	Nps	RFA	scCO$_2$	35	10.7	3	2–3.3	[57]
	Pt(cod)me$_2$	Nps	SnO_2	scCO$_2$	80–100	15.5–25	1	2–13	[58]
	Pt(cod)me$_2$	Nps	Papam modified sg	scCO$_2$ + Acetone	40	10	1, 3	3–4	[36]
	Pt(acac)$_2$	Nps			40	10	1, 3	0.8–1.6	[59]
	Pt(acac)$_2$	Nps	Papam modified sw	scCO$_2$ + Acetone	120	25	3	1–5	[60]
	Pt(cod)me$_2$	Nps	CB	scCO$_2$	80	27.6	3	3.5	[61]
	Pt(cod)me$_2$	Nps	C	scCO$_2$	80	27.6	3	2.3–2.6	[61]
	Pt(cod)me$_2$	Nps	SWCNT	scCO$_2$	80	27.6	3	2.3–2.6	[61]
	Pt(cod)me$_2$	Nps	MWCNT	scCO$_2$	80	27.6	3	–	[61]
	Pt(cod)me$_2$	Nps	GO	scCO$_2$	80	27.6	3	–	[61]
	Pt(cod)me$_2$	Nps	WO_3-ZrO_2	scCO$_2$	80	27.6	3	43	[61]
	Pt(cod)me$_2$	Nps	TiO_2	scCO$_2$	80	27.6	3	8.1	[61]
	Pt(cod)me$_2$	Nps	Al_2O_3	scCO$_2$	100–120	20–15	3	3– 11	[62]
	Pt(cod)me$_2$	Nps			120–200	10	1	<5	[63]

Continued

TABLE 7.1 Summary of the studies in the literature on preparation of supported nanoparticles on high surface area substrates by supercritical deposition—cont'd

Metal	Metal complex	Morphology	Substrate	Solvent medium	Deposition temperature (°C)	Deposition pressure (MPa)	Conversion method	Particle size range (nm)	Ref.
	Pt(hfac)$_2$	Nps	Al_2O_3	$scCO_2$	200	12	1	2–6	[64]
	Pt(acac)$_2$	Nps	MWCNT	$scCO_2$	70	24.5	3	1.1–5.1	[65]
	Pt(cod)me$_2$	Nps	Graphene	$scCO_2$	70	24.5	3	2.1–17.5	[65]
	Pt(cod)me$_2$	Nps	Graphene	$scCO_2$	70	24.5	3	1.2–5.9	[65]
	Pt(cod)me$_2$	Nps	MWCNT	$scCO_2$	80	15.5	1	1.3–6.9	[66]
	Pt(cod)me$_2$	Nps	C	$scCO_2$	80	15.5	1, 3	1.9–6.5	[66]
	Pt(cod)me(n-C_3F_7)	Nps	TiO_2	$scCO_2$	80	15.5	1, 3	1.9–5.1	[66]
		Nps	TiO_2	$scCO_2$	80	15.5	1	2.4–5.8	[66]
	Pt(cod)me(i-C_3F_7)	Nps	TiO_2	$scCO_2$	80	15.5	1	2.4–4.8	[66]
	Pt(cod)me(n-C_4F_9)	Nps	TiO_2	$scCO_2$	80	15.5	1	2.2–4.9	[66]
		Nps	TiO_2	$scCO_2$	80	15.5	1, 3	2.1–3.7	[66]
	Pt(cod)me(n-C_6F_{13})	Nps	TiO_2	$scCO_2$	50	13.2	2, 3	1–7.5	[67]
		Nps	MCM-41	$scCO_2$	60	20	3	2–3	[68][a]
	Pt(cod)me(n-C_8F_{17})	Nps	C	$scCO_2$	60	12	3	1–2	[69]
		Nps	MOF	$scCO_2$ + EtOH	60	14	3	1–6	[70]
	Pt(cod)me$_2$	Nps	Graphene		200	24	1	3–4, 8–9	[71]
	Pt(cod)me$_2$	Nps	CeO_x/Al_2O_3	$scCO_2$	35	12	3	1.2–2.6	[72]
	H_2PtCl_6	Nps	MWCNT	$scCO_2$	80	27.6	3	1–9	[73]

	Pt(cod)me$_2$	Nps	Graphene	scCO$_2$ + MeOH	35	10.7	3	1–7	[74]
	Pt(acac)$_2$	Nps	CA	scCO$_2$	60	13	3	3.5–5.5	[28]
	Pt(acac)$_2$	Nps	GA	scCO$_2$	35	12	3	1.5–4	[75]
	Pt(cod)me$_2$		rGO	scCO$_2$					
	Pt(cod)me$_2$		rGO/CB	scCO$_2$					
	Pt(cod)me$_2$			scCO$_2$					
	Pt(cod)me$_2$			scCO$_2$					
	Pt(cod)me$_2$								
Ru	Ru(acac)$_3$·xH_2O	Nps	MWCNT	scCO$_2$	–	–	1	~1	[30]
	RuCl$_3$·H_2O	Nps	CNT	scH$_2$O	–	–	2	>5	[76]
	RuCl$_3$·H_2O	Nps	Amorphous carbon	scH$_2$O	–	–	2	~5	[76]
	Ru(cod)(tmhd)$_2$	Nps	CA	scCO$_2$	80	27.6	3	1.7	[77]
	Ru(cod)(tmhd)$_2$	Nps	CA	scCO$_2$	80	27.6	3	2.4	[77]
	Ru(cod)(tmhd)$_2$	Nps	CA	scCO$_2$	80	27.6	3	3.8	[77]
	Ru(cod)(tmhd)$_2$	Nps	CA	scCO$_2$	80	27.6	3	1.8	[77]
	Ru(cod)(tmhd)$_2$	Nps	CA	scCO$_2$	80	27.6	3	2.8	[77]
	Ru(cod)(tmhd)$_2$	Nps	SA	scCO$_2$	80	27.6	3	2.6	[78]
	Ru(acac)$_2$	Nps	CA	scCO$_2$	80	27.6	3	2.5	[77]
	Ru(acac)$_2$	Nps	CA	scCO$_2$	80	27.6	3	4	[77]
		Nps	MCM-41	scCO$_2$	150	20	1	1.6–8.2	[79]
		Nps	MOF	scCO$_2$ + MeOH	200	9.7	3	2	[80][a]
		Nps	C	scCO$_2$ + EtOH	45	10	3	1–6	[81][a]
		Nps	C		45	10	3	1–5	[81]
		Nps	GLC		–	–	–	2–5	[82]

Continued

TABLE 7.1 Summary of the studies in the literature on preparation of supported nanoparticles on high surface area substrates by supercritical deposition—cont'd

Metal	Metal complex	Morphology	Substrate	Solvent medium	Deposition temperature (°C)	Deposition pressure (MPa)	Conversion method	Particle size range (nm)	Ref.
	Ru(cod)(tmhd)$_2$	Nps	rGO	$scCO_2$ + EtOH	80	13.5	3	2.5–4.5	[28]
	$RuCl_3 \cdot H_2O$			scH_2O					
	$RuCl_3 \cdot H_2O$			$scCO_2$					
	Ru(acac)$_3$								
	$RuCl_3$								
	Ru(cod)(tmhd)$_2$								
Rh	Rh(acac)$_2 \cdot xH_2O$	Nps	MWCNT	$scCO_2$	–	–	1	3–5	[30]
	$Rh_2(OAc)_4$	Nps	SBA-15	$scCO_2$ + THF	75	16	1	6	[83]
	Rh(acac)$_3$	Nps	MCM-41	$scCO_2$	60	15	3	~2.4	[84]
	Rh(acac)$_3$	Nps	MSU-H	$scCO_2$	60	15	3	~4.4	[84]
	Rh(acac)$_3$	Nps	HMS	$scCO_2$	60	15	3	~12.9	[84]
V_2O_5	V(acac)$_3$	Nps	MWCNT	$scCO_2$	50–70	13.8–20.7	3	<20	[85]

1: In situ chemical conversion. 2: In situ thermal conversion. 3: Ex situ conversion.
[a]*Initially impregnation in alcohol, then precipitation in SCF.*

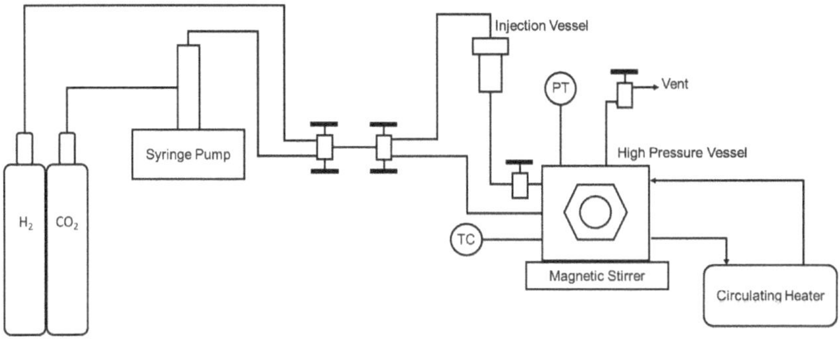

FIG. 7.2 Schematic representation of the experimental setup for SCD of supported nanoparticles.

sufficient time for adsorption, the metal complex is then converted to its metal form either by in situ or ex situ conversion methods.

During in situ thermal conversion of the metal complex, the temperature of the vessel is increased to the desired value by increasing the temperature of the heating medium within the circulating heater. In isovolumetric processes, due to the rise in temperature, pressure within the vessel also increases. The maximum pressure that the vessel can endure should not be exceeded during in situ thermal conversion process. After allowing a certain time for the metal complex to be thermally converted to its metal form, the vessel is depressurized.

For in situ chemical conversion of the metal complex, a certain ratio of $scCO_2$ and the reducing agent, generally H_2, are mixed within the 30 mL injection vessel. Then, $scCO_2$ and reducing agent mixture is injected into the vessel. Calculation steps of the amount and pressure of H_2 in the injection vessel are given below:

1. Amount and pressure of hydrogen needed is calculated from the stoichiometry of the reaction between hydrogen and the metal complex. If an excess amount of reducing agent is going to be used, it is also taken into account in this calculation.
2. Since the fraction of H_2 is small compared to CO_2 in the injection vessel, H_2 can be assumed to have no contribution to the density in the injection vessel.
3. CO_2 density and amount in the injection vessel is determined from pressure, temperature, and volume of the vessel before injection.
4. After the valve between the injection and deposition vessels is opened, pressures will equilibrate. A final pressure should be calculated by assuming a final temperature for the whole system.
5. The amount and density of CO_2 in the injection vessel after injection is calculated by using final pressure and injection vessel temperature.
6. The amount of CO_2 transferred to the deposition vessel from the injection vessel is calculated from the difference between the initial and final CO_2 amount in the injection vessel.

7. Fraction of H_2 transferred to deposition vessel is calculated by dividing the amount of H_2 transferred (calculated in step 1) to the summation of amount of H_2 and CO_2 transferred.
8. Since the mole fraction of H_2 in the injection vessel does not change during the injection, amount of H_2 needed to be supplied to the injection vessel can be calculated using this fraction.
9. Pressure of H_2 in the injection vessel is calculated using ideal gas law.

H_2 is delivered to the injection vessel up to the calculated pressure (step 9). Afterwards, final injection pressure is achieved by delivering CO_2 into the injection vessel. The valve between the injection and deposition vessels is then opened and pressures equilibrate. Pressure inside the deposition vessel should match the final pressure calculated in step 4. Then, the reduction reaction is allowed to proceed for a certain time. After conversion of the metal complex is complete, the vessel is depressurized.

Depressurization of the vessel should be conducted slowly, around 100 psi/-min (~0.7 MPa/min). If the support material used in SCD is an aerogel, rapid changes in pressure may result in pore collapse. For polymer supports, fast depressurization may result in foaming. Since CO_2 is expanded to the atmosphere during depressurization using a valve, the valve may cool considerably which may result in freezing of the CO_2 and blockage of the valve

Conversion of the metal complexes can be conducted ex situ, i.e., after depressurizing the vessel. Ex situ conversion of the metal complex can be thermal or chemical at atmospheric pressure. In thermal ex situ conversion, metal complex-adsorbed supports can be placed in a porcelain combustion boat for easier handling. Process tubes that are made of inert and thermally stable materials should be used. Quartz process tubes can be used in processes with temperatures up to 1100°C. Combustion boat is positioned at the center of the quartz process tube that is placed in a tube furnace. Thermal conversion is carried out under flow of an inert gas, generally N_2. After conversion, the system should be cooled down to room temperature under flowing N_2 and in a controlled environment. Cooling down the temperature too rapidly may result in cracks in the quartz process tube and porcelain combustion boat.

During ex situ chemical conversion of the metal complex, the same setup in thermal conversion can be used but the flowing gas is changed to a reactive gas such as hydrogen or a mixture of hydrogen with nitrogen that enables the conversion of the adsorbed metal complex to its metal form.

The supports investigated so far are numerous and include silica, alumina, silica aerogel (SA), carbon aerogel (CA), carbon black (CB) and carbon nanotubes (CNTs). The most widely investigated metal has been platinum due its excellent catalytic properties in a wide variety of industrially important reactions. Figs. 7.3–7.5 contain images of Pt nanoparticles prepared by SCD on three chemically different types of commonly used supports which are silica, alumina and CB. In all of these samples, (1,5-cyclooctodiene) dimethyl

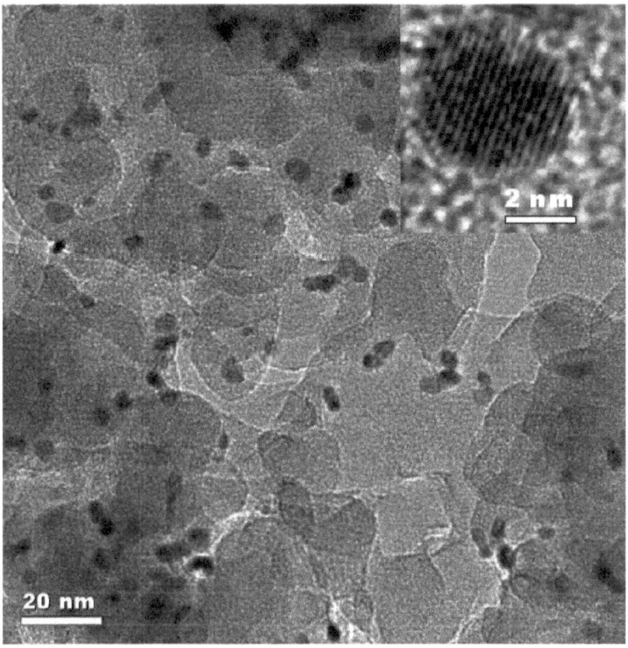

FIG. 7.3 TEM image of SiO_2 supported Pt nanoparticles prepared by SCD [78]. *(Reprinted from Y. Zhang, D. Kang, C. Saquing, M. Aindow, C. Erkey, Supported platinum nanoparticles by supercritical deposition, Ind. Eng. Chem. Res. 44 (2005) 4161–4164, https://doi.org/10.1021/ie050345w, Copyright © 2006, with permission from Elsevier.)*

FIG. 7.4 TEM image of Al_2O_3 supported Pt nanoparticles prepared by SCD [52].

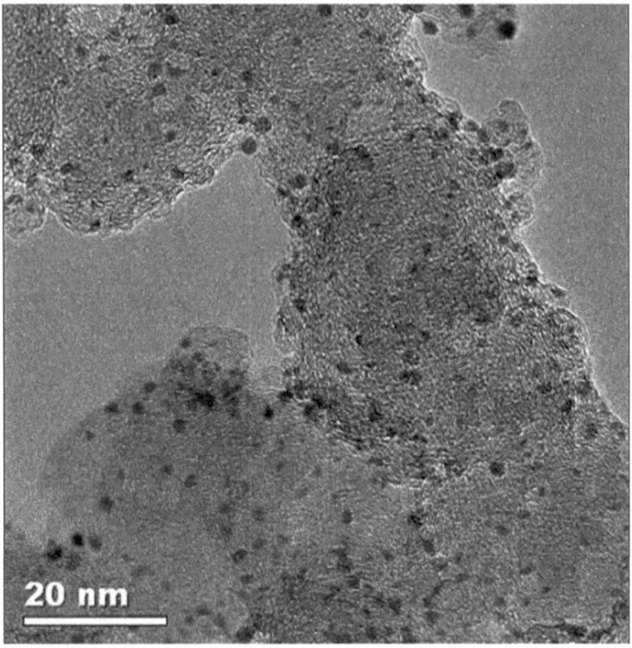

FIG. 7.5 TEM image of CB supported Pt nanoparticles prepared by SCD. *(Reprinted from A. Bayrakçeken, A. Smirnova, U. Kitkamthorn, M. Aindow, L. Türker, I. Eroğlu, C. Erkey, Pt-based electrocatalysts for polymer electrolyte membrane fuel cells prepared by supercritical deposition technique, J. Power Sources 179 (2008) 532–540, https://doi.org/10.1016/j.jpowsour.2007.12.086, Copyright © 2008, with permission from Elsevier.)*

platinum(II) ($Pt(cod)(me)_2$) was used as the metal complex and was adsorbed at 80°C and 20 MPa. The adsorbed Pt complex was converted to Pt nanoparticles in an inert nitrogen atmosphere at 300°C. Even though the metal contents of the samples are slightly different, it is important to review the general features in these images. First of all, the nanoparticles are quite uniformly distributed on the surface. The average particle size of Pt nanoparticles on all three supports is quite small.

Fig. 7.3 shows the HRTEM data obtained from one of the Pt/SiO_2 samples and reveals that slightly ellipsoidal Pt nanoparticles are dispersed either on the surface of a silica crystallite or sandwiched between crystallites [52]. No coalescence of the metal nanoparticles is observed. The character of these nanoparticles was confirmed using EDXS. The spectrum contains strong characteristic Si, O, and Pt X-ray peaks as expected, together with additional peak corresponding to Cu. This latter peak is an artifact that arises from the Cu grid on which the HRTEM sample is supported. No lattice fringes are observed in images from most of the nanoparticles, presumably because of their small sizes and random orientations. Such fringes are, however, observed occasionally, and one example is shown in the Fig. 7.3 inset image. It shows a single nanoparticle with 2.4-Å lattice fringes corresponding to the {111} planes of the face-centered cubic metallic Pt. No evidence for grain boundaries or lattice defects is observed in such images, i.e., the individual Pt nanoparticles are defect-free single crystals. The images in Figs. 7.4 and 7.5 also reveal uniformly dispersed metal nanoparticles with a mean particle size of 1.5 and 2.0 nm, respectively.

In both cases, a narrow particle size distribution is observed. Platinum nanoparticles on silica support are larger and have broader size distributions than Pt nanoparticles on CB and Al_2O_3 supports at similar metal contents. These observations indicate that particle formation and growth is primarily governed by the strength of the interaction between the metal complex and the support. The large particle sizes and the broad distributions reported in the literature for silver- (10–100 nm) and palladium- (within two discrete size ranges of <6 and 15–20 nm) supported nanocomposites using volatile fluorinated precursors [(1,1,1,5,5,5-hexafluoropentane-2,4-dione)(1,1,4,7,10,10-hexamethyltriethylene) silver, (1,1,1,5,5,5-hexafluoropentane- 2,4-dione)(tetraethylene glycol dimethyl ether) silver, and 1,1,1,5,5,5-hexafluoro-2,4-pentanedione palladium (II)] support this hypothesis. In those studies, particles larger than 100 nm were observed.

Oztuna et al. prepared graphene aerogel (GA) supported Pt electrocatalysts via SCD. They investigated the effect of metal complex conversion temperature on average Pt nanoparticle size [74]. Adsorption was conducted at 35°C and 10.7 MPa for 24 h using $Pt(cod)me_2$ as Pt complex followed by ex situ conversion under flowing N_2 at various temperatures (200–800°C). Fig. 7.6 shows an

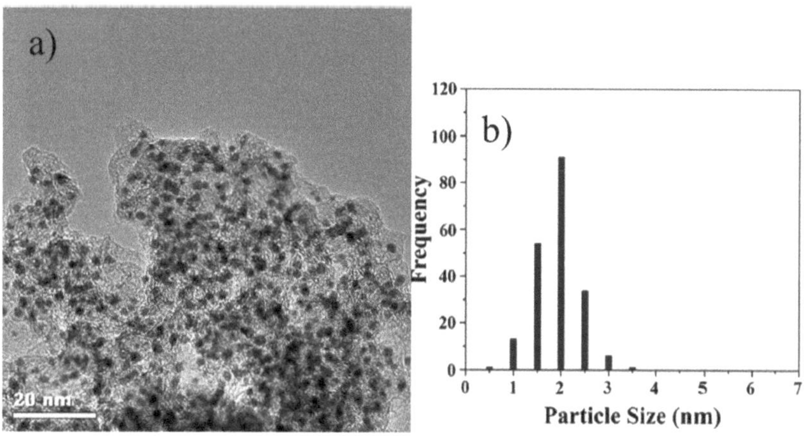

FIG. 7.6 (a) TEM image of GA supported Pt nanoparticles prepared by SCD and (b) the particle size distribution of the image given in (a). *(Reprinted from F.E. Sarac Oztuna, S.B. Barim, S.E. Bozbag, H. Yu, M. Aindow, U. Unal, C. Erkey, Graphene aerogel supported Pt electrocatalysts for oxygen reduction reaction by supercritical deposition, Electrochim. Acta 250 (2017) 174–184, https://doi.org/10.1016/j.electacta.2017.08.067, Copyright (c) 2017, with permission from Elsevier.)*

exemplary transmission electron microscopy (TEM) image and the associated particle size distribution for Pt nanoparticles supported on GA. These images show a mean particle size of 1.9 nm with a remarkably narrow particle size distribution considering a relatively high Pt loading of 32.9 wt% for the sample. XRD showed that at 200°C metal complex conversion was not complete unlike in the case of Pt/CAs [73] probably due to different surface chemistries of GA and CA.

On the other hand, average Pt nanoparticle sizes obtained from TEM images were almost identical in Pt/GAs as in Pt/CAs at the corresponding conversion temperatures and similar Pt loadings indicating that precise control of nanoparticle size is possible with SCD. Moreover, XPS C1s spectra of the electrocatalysts and blank GA showed that during thermal treatment, Pt complex conversion and nanoparticle growth occurred simultaneously with the reduction of the functional groups on GAs which caused more defect sites that Pt nanoparticles can anchor.

Since the TEM images are taken from a very small section of the sample, the uniform distribution and small sizes observed may not reflect the character of the whole sample. A comparison of particle size from TEM images and from XRD data is necessary to make sure that the particle size is uniform throughout the whole sample, since the sample sizes required for XRD is on the order of a few milligrams. Fig. 7.7 shows an XRD pattern of a CA supported Ru nanoparticles prepared by SCD and a TEM image of this sample. The average particle size in the images is around 2.5 nm compares very well with the particle size of 2.8 nm calculated using the Ru (1010) peak which indicates uniform deposition.

If the conversion of the adsorbed metal complex to metallic nanoparticles is carried out after depressurization, the metal loading in a particular sample depends on the amount of metal complex adsorbed on the substrate. Therefore,

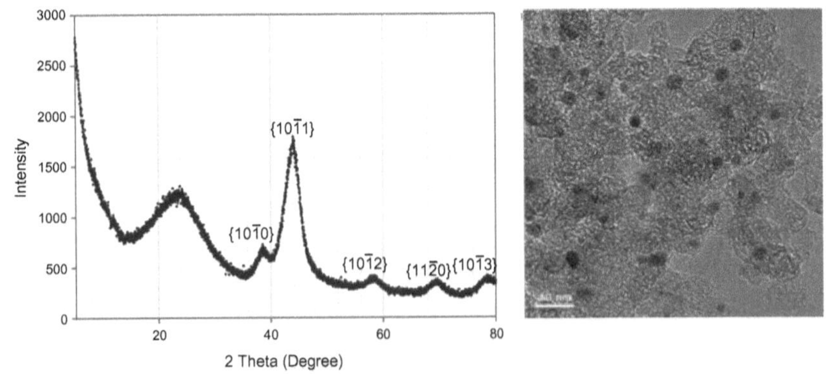

FIG. 7.7 XRD data and TEM image from a CA supported Ru nanoparticles prepared by SCD [78].

it is possible to control the metal loading by controlling the adsorbed amount of the metal complex on the support. As explained in Chapter 4, this depends on the adsorption isotherm of the metal complex between the substrate and the SCF phases. This control can be demonstrated in Fig. 7.8 where three CA supported Pt nanoparticle samples are prepared by changing the amount of platinum complex used in the supercritical adsorption step. This results in a change of the equilibrium concentration of the metal complex in the fluid phase which causes a change in the equilibrium uptake of the metal complex by the CA sample. The increasing concentration of Pt nanoparticles can be clearly seen in Fig. 7.8.

In the above argument, it was assumed that the entire metal complex is converted to its metal form. The conversion of the metal complex to metallic nanoparticles after depressurization depends on three factors, which are:

1. temperature
2. medium of decomposition (inert gas, hydrogen, etc.)
3. the support

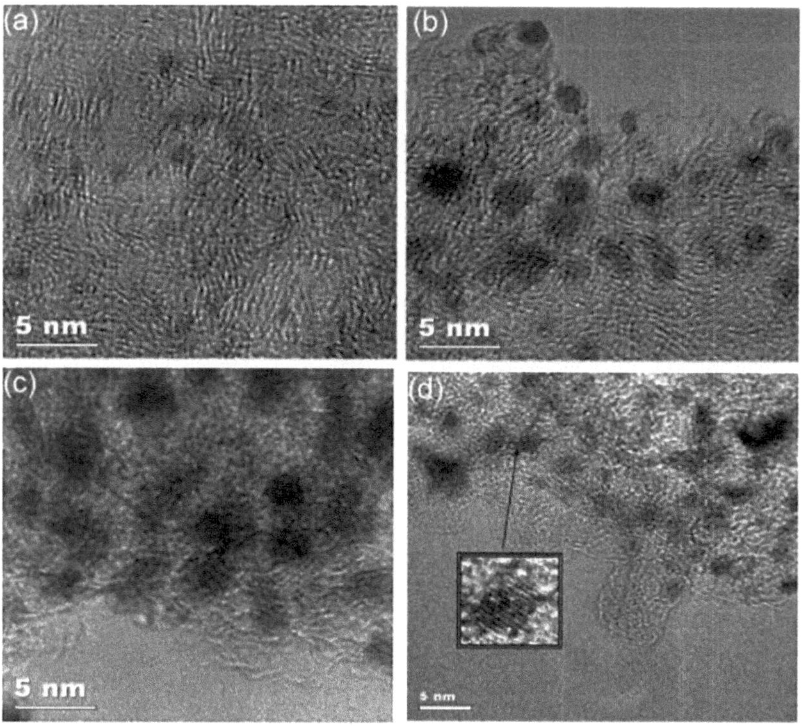

FIG. 7.8 TEM images of CA supported Pt nanoparticles with different Pt content [51].

In case of metal complexes four things can happen to the adsorbed metal complex during the conversion step, namely:

1. it may convert to metal completely
2. it may volatilize from the surface completely
3. some of the complex may convert to metal and some of it may volatilize from the surface
4. the complex may convert to other metallic complexes than pure metal.

For example, all of the Pt(cod)(me)$_2$ adsorbed on the surface of carbonaceous and organic supports can be converted cleanly to metal. However, it is not generally easy to confirm this and requires some additional characterization. One way to determine the total amount of metal is acid digestion where the sample is treated with acid. The resulting solution can then be analyzed for its metal content by Atomic Absorption Spectroscopy. If the experimental metal content agrees with the metal content calculated from the adsorbed metal complex amount (by assuming complete conversion of the metal complex to metal), one can assume complete conversion.

Thermal gravimetric analysis (TGA) is a very useful tool in determining the suitability of a metal complex for SCD. The TGA of the pure metal complex provides information about the required minimum temperature to perform the thermal reduction of the adsorbed metal complex. Furthermore, the TGA of the metal complex-loaded substrates performed under N_2 gas atmosphere simulates the thermal reduction and provides a measure of the mass loss dynamics and extent of ligand removal. A comparison of the TGA of the pure metal complex and metal complex-loaded substrates also corroborate the strength of interaction observed previously between the metal complex and the substrate.

Fig. 7.9 presents the TGA thermograms obtained in the temperature range 25–600°C. The weight loss occurred between 100°C and 200°C for both pure and impregnated samples, however, the weight loss with pure Pt(cod)me$_2$ was substantially higher than the expected weight loss of 41% if all of the ligands in Pt(cod)me$_2$ were eliminated (hereafter referred to as the theoretical weight loss). The higher weight loss is attributed to the volatilization of Pt(cod)me$_2$ that either precedes or accompanies the decomposition process. When the flow rate of the N_2 gas was increased or the temperature ramp rate was decreased, the weight loss increased which supports the view that Pt(cod)me$_2$ was volatilizing.

On the other hand, in the Pt(cod)me$_2$-impregnated CA, the weight loss (39%–40%) was close to the theoretical value (well within the instrumental precision) at all gas flow rates. This shows that the metal complex does not volatilize upon heating when it is in the adsorbed state in the CA matrix, a further indication of the strong interaction between the metal complex and the CA matrix. The adsorbed metal complex is converted to Pt and the

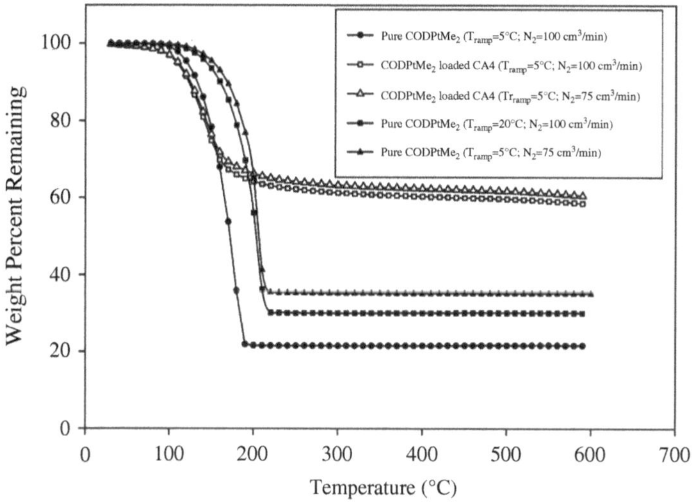

FIG. 7.9 TGA of Pt(cod)(me)$_2$ [51].

decomposition products are removed completely. This finding was further substantiated with FTIR and acid digestion experiments. FTIR analysis of the Pt/CA composite showed the absence of spectral peaks that correspond to functional groups associated with the ligands. The platinum content of one of the Pt/CA composites was determined by acid digestion. In this experiment, the sample (6.1 wt% Pt content as determined gravimetrically) was digested with a mixture of hydrochloric and nitric acid and the resulting solution was analyzed by inductively coupled plasma (ICP) Spectroscopy for its platinum content. The analysis yielded a value of 5.9 wt% for the composites' platinum content, which is in good agreement with the content determined gravimetrically. The difference is within 5%, proving that practically all of the ligand is eliminated.

TGA of pure ruthenium complexes of both bis(2,2,6,6-tetramethyl-3,5-heptanedionato) (1,5-cyclooctadiene) ruthenium (II) (Ru(cod)(tmhd)$_2$) and Ruthenium acetylacetonate (Ru(acac)$_3$), and the Ru(cod)(tmhd)$_2$-impregnated CA-21 were carried out over the range 0–600°C. The results are shown in Fig. 7.10.

The weight of pure Ru(cod)(tmhd)$_2$ and Ru(acac)$_3$ samples started to decrease at around 200°C and there was hardly any residue as the temperature reached 260°C. This behavior indicates that the Ru complexes do not decompose to elemental ruthenium, but instead that they vaporize. The weight of the Ru(cod)(tmhd)$_2$-impregnated CA sample (28.5 wt%) also started to decrease at around 200°C, but the rate of decrease with increasing temperature was smaller. Most of the weight change occurred in the temperature range 200–340°C and

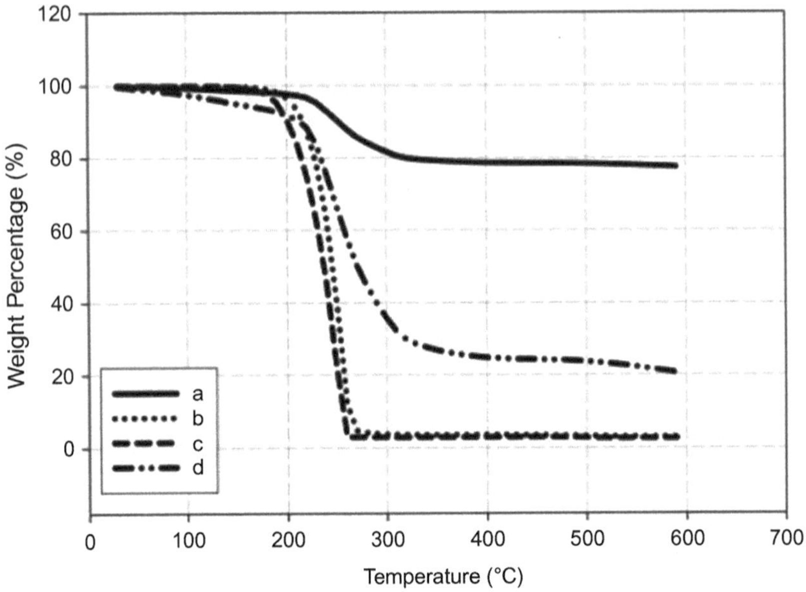

FIG. 7.10 TGA of Ru precursors: (a) carbon aerogel impregnated with Ru(cod)(tmhd)$_2$; (b) Ru (acac)$_3$; (c) Ru(cod)(tmhd)$_2$; (d) Ru(cod)(tmhd)$_2$ impregnated carbon aerogel (modified curve (a) by subtracting the amount of carbon aerogel). The heating rate is 5°C/min [77].

the weight of the sample was reduced by 22.6% when the temperature reached 600°C. The decrease in the weight of the sample was approximately equal to the weight of ligands in adsorbed Ru(cod)(tmhd)$_2$. This is also illustrated by curve d) in Fig. 7.10, which was derived from curve a) by subtracting the amount of CA in the sample, thus isolating adsorbed Ru(cod)(tmhd)$_2$. The residue was 20 wt% which compared well with 17.4 wt% of elemental Ru in Ru(cod)(tmhd)$_2$. The differences in TGA between the adsorbed and pure Ru(cod)(tmhd)$_2$ showed that decomposition of the Ru complex must be occurring in its adsorbed state without volatilization.

A variety of metal complexes, which have sufficient solubility in the SCF and a high amount of the desired metal, were shown to be eligible for SCD. Type and amount of metal complexes with their reduction methods and conditions affect particle size and particle size distribution of the supported nanoparticles. Wolff et al. and Aggarwal et al. synthesized a range of cyclooctadiene-stabilized Pt complexes with different perfluoro-alkane chains, namely Pt(cod)me(n-C_3F_7), Pt(cod)me(i-C_3F_7), Pt(cod)me(n-C_4F_9), Pt(cod)me(n-C_6F_{13}), Pt(cod)me(n-C_8F_{17}) [66, 86]. These metal complexes were used for the preparation of Pt/Al_2O_3, Pt/MCM-41, and Pt/TiO_2 nanocomposites via SCD. Dissolution and adsorption of the metal complexes were conducted at 15.5 MPa and

80°C. Afterwards, metal complex reduction was conducted using H_2 either at 15.5 MPa or 0.4 MPa. Average particle size for Pt/TiO_2 ranged between 2.9 and 5.3 nm while it was between 2.3 and 2.8 nm for Pt/Al_2O_3. Among the Pt complexes, Pt(cod)me(i-C_3F_7) resulted in the smallest particle sizes while the largest particle sizes were obtained by Pt(cod)me_2. Substitution of the methyl end group by n-C_nF_{n+1} end groups did not change particle size and particle size distribution while substitution by i-C_3H_7 resulted in opposite behavior compared to Pt(cod)me_2. The highest metal loading, 67.2 mg/g, was obtained when Pt(cod)me_2 was used while Pt(cod)me(i-C_3F_7) resulted in the nanocomposites with the lowest metal loading, 27.1 mg/g. The pressure at which the metal complexes were reduced was shown to be insignificant when the support was MCM-41, while 15.5 MPa resulted in smaller nanoparticles when Al_2O_3 was the support and 0.4 MPa resulted in smaller nanoparticles when TiO_2 was the support, which points out to the importance of support materials and metal-support interactions in terms of particle size and particle size distribution.

Reduction temperature is another significant factor influencing the particle size. When Ru(cod)$(tmhd)_2$ reduction temperature increased from 400 to 1000°C, the resulting particle size increased from 1.7 to 3.8 nm [77]. The platinum particles are more sensitive to the reduction temperature and a critical reduction temperature exists for the platinum particle growth. When Pt(cod) me_2 reduction temperature increased from 200 to 600°C, the resulting particle size increased from 2.3 to 3.6 nm [50]. When Pt(cod)me_2 was reduced at 1050°C, the obtained particle size was 17.1 nm. The increase in the reduction temperature provides more thermal energy to the individual ruthenium or platinum particles, and hence increases their mobility, allowing them to coalesce and form larger entities. This also indicates that platinum atom possesses higher mobility than ruthenium at the same temperature.

The conversion to the metal can also be carried out in the supercritical phase after the adsorption step. This can generally be accomplished by heating the solution to a higher temperature than the adsorption temperature. The adsorbed metal complex may decompose to metal and form metallic nanoparticles when heated. However, depending on the metal complex, the temperature may not be sufficiently high. It may be necessary to inject a reducing agent such as hydrogen to the solution which can react with the adsorbed metal complex at the surface and enable the formation of metal atoms which than coalesce and grow to form metallic nanoparticles.

Highly dispersed platinum particles supported on CAs with high metal content could be prepared using SCD followed by in situ thermal reduction [50]. The supercritical impregnation was carried out at 18.6 MPa for 24 h and then the temperature was increased to 135°C for thermal reduction, which also caused the pressure to increase to 31.0 MPa. The reduction in $scCO_2$ was carried out for 6 h. Representative images of both the exterior and interior

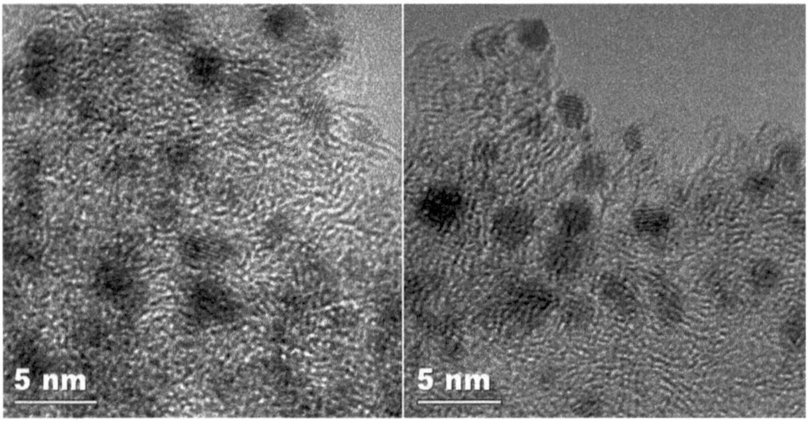

FIG. 7.11 TEM images of CA supported Pt nanoparticles prepared by thermal reduction in $scCO_2$ [51].

microstructures of the composites obtained using this method are shown in Fig. 7.11. It appears that the particle sizes are 3.7 ± 1.5 nm for the interior and 4.7 ± 1.5 nm for the exterior. In these experiments almost all of the platinum complex, Pt(cod)me_2 in the $scCO_2$ phase was adsorbed and subsequently converted to Pt in the CA substrate. For example, Pt/CA composites prepared with this method resulted in a loading of 60 and 73 wt% Pt. For both of the composites, the impregnated Pt accounted for 97% of the total Pt in the system.

Besides $scCO_2$, supercritical water (scH_2O) can also function as a processing medium for the preparation of supported nanoparticles. Fig. 7.12 shows the SEM and TEM images of Ru/CNT nanocomposites, which were fabricated by Sun et al. by treating CNTs and $RuCl_3{\cdot}H_2O$ in scH_2O at 400°C or 450°C [76].

The average particle diameter was 5 nm when the weight ratio of $RuCl_3{\cdot}H_2O$ to CNTs was 1:1. As shown in Fig. 7.12B and D, increasing the weight ratio, the amount of ruthenium nanoparticles on the CNT walls increased and nanoparticle clusters with size up to dozens of nanometers were observed on the external walls of the CNTs. According to Fig. 7.12E and F, the ruthenium particles prepared at 450°C were larger than particles obtained at 400°C. The method could also be used to impregnate Ru nanoparticles onto amorphous carbon. The interaction between the particle and the CNT wall was found to be very strong since the particles could not be removed by thorough washing, severe agitation and prolonged sonication. The Ru-CNTs nanocomposites displayed good catalytic activity for benzene hydrogenation. The product was cyclohexane and the conversion was 100% with turnover frequencies reaching 1430 moles H_2 per mole Ru per hour. It is worth mentioning that this is the only work

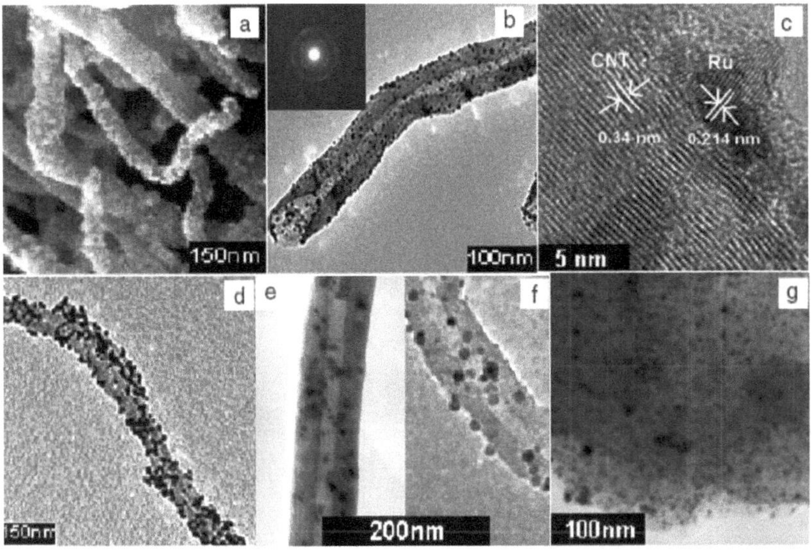

FIG. 7.12 (a) SEM image of Ru-CNT composites (the initial weight of $RuCl_3{\cdot}H_2O$ to CNT was 1:1); (b) High magnification TEM image of Ru-CNT composites; (c) HRTEM image of Ru-CNT composites; (d) TEM image of Ru-CNT composites (the initial weight of $RuCl_3{\cdot}H_2O$ to CNT was 2:1); (e) and (f) TEM images of Ru-CNT composites synthesized at 400°C and 450°C, respectively (the initial weight of $RuCl_3{\cdot}H_2O$ to CNT was 1:2); (g) TEM image of amorphous carbon decorated with Ru nanoparticles.

using scH_2O instead of $scCO_2$ as the processing medium for SCD. The solubility of inorganic salts in scH_2O is low, although they can be used for SCD. In comparison to the deposition in $scCO_2$, it is necessary to remove the water in the resulting composite after deposition in scH_2O.

Using SCD, Ye et al. was able to prepare highly dispersed nanoparticles for catalytic applications [30]. Nanoparticles of palladium, rhodium and ruthenium were deposited onto functionalized multiwalled carbon nanotubes (MWCNTs) through the reduction of adsorbed metal-β-diketone complexes by hydrogen in $scCO_2$. The deposition of palladium, rhodium and ruthenium was carried out at desired pressure and at 80°C, 250°C, and 250°C, respectively. The composites were characterized by TEM, energy dispersive X-ray spectroscopy (EDS) and X-ray photoelectron spectroscopy (XPS) analyses. As shown in Fig. 7.13, pure and highly dispersed nanoparticles, with a narrow size distribution were deposited on the surface of the MWCNTs. The particle sizes for rhodium, palladium, and ruthenium were 3–5, 5–10, and around 1 nm, respectively. Preliminary experiments demonstrated that Pd nanoparticles supported on MWCNTs were effective catalysts for hydrogenation of olefins in carbon dioxide. The Pd nanoparticle-MWCNT composite also showed a high electrocatalytic activity in oxygen reduction indicative of its potential for fuel cell application.

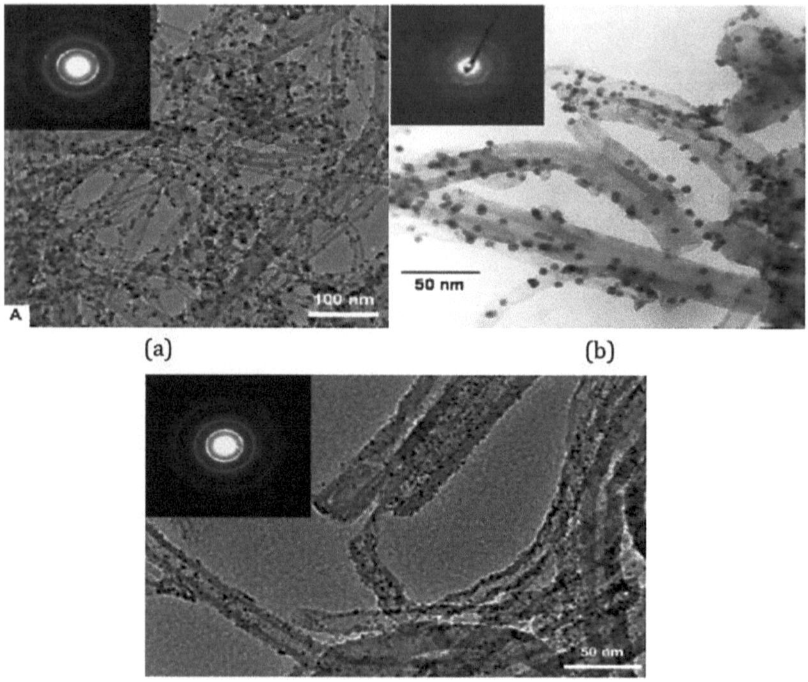

(a) (b)

(c)

FIG. 7.13 TEM image of MWCNTs decorated with (a) Rh; (b) Pd; (c) Ru nanoparticles [76].

It was found that metal films and highly dispersed nanoparticles could be produced simultaneously on the surface and within the matrix of porous monolithic materials using SCD. Saquinq et al. deposited platinum on CA monoliths with a pore diameter of 22 nm (CA-22) by impregnation of the monoliths using Pt(cod)me$_2$ at 80°C and 27.6 MPa for 24 h [50]. Subsequently, H_2 (200% excess) was charged into the vessel and the reduction was allowed to proceed for 6 h at 80°C and 27.6 MPa. As shown in Fig. 7.14, the chemical reduction in scCO_2 resulted in a radically different Pt/CA composite. The external surface of the CA monolith was coated with a Pt film and the thickness of the Pt coating was at least 100 μm as measured from Fig. 7.14a. A closer examination of the surface morphology (e.g., Figs. 7.13c and 7.14b) revealed a continuous film with irregular protrusions up to 2 μm in diameter, which are typical for disordered columnar growth and equiaxed particles $\approx$1 μm in diameter lying on the surface. The particles at the interior of the monolith were, however, small and well dispersed with an average particle size of 4.7 ± 1.8 nm.

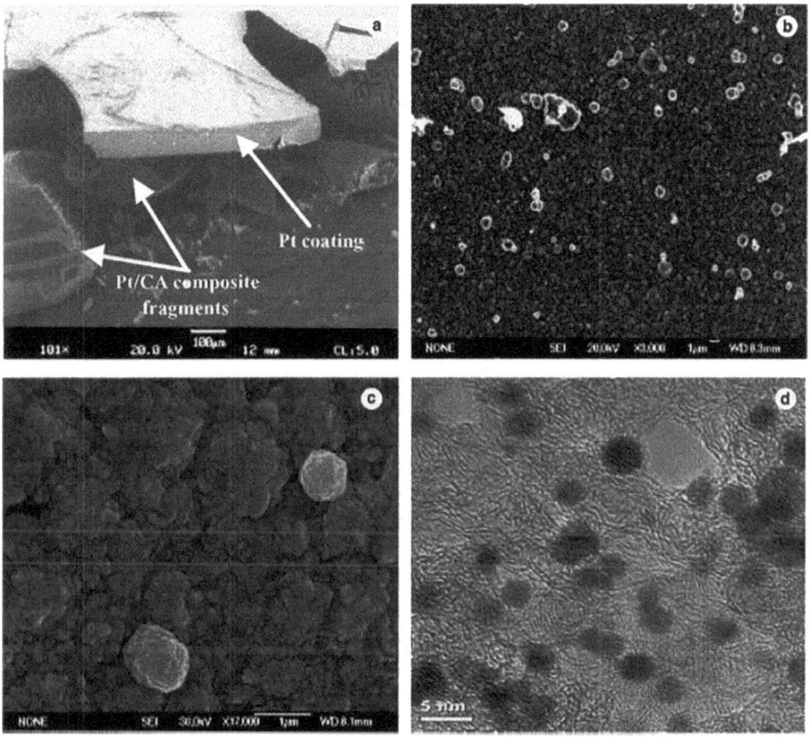

FIG. 7.14 CA supported films and nanoparticles prepared by SCD [50].

7.2 Polymer supported nanoparticles

scCO_2 displays high permeation rate in many polymers and the exposure to scCO_2 results in various extents of swelling. This is particularly advantageous for the synthesis or processing of polymer nanocomposites as well as for impregnating a wide variety of chemicals into various polymers. Moreover, the degree of CO_2 sorption/swelling in polymers, diffusion rates within the substrate, and the partitioning of solutes between the SCF and the swollen polymer can be tuned by density mediated adjustments of solvent strength via changes in temperature and pressure. Fig. 7.15 illustrates the effects of pressure and temperature on the solubility of CO_2 in polyethylene terephthalate (PET).

The weight fraction of CO_2 inside the polymer increases appreciably with increasing pressure. The effect of temperature on CO_2 sorption is more complex; until a certain percent of CO_2 sorbs inside the polymer, the sorption of CO_2 increases with decreasing temperature at a particular pressure then increases with increasing temperature. The effect of temperature on

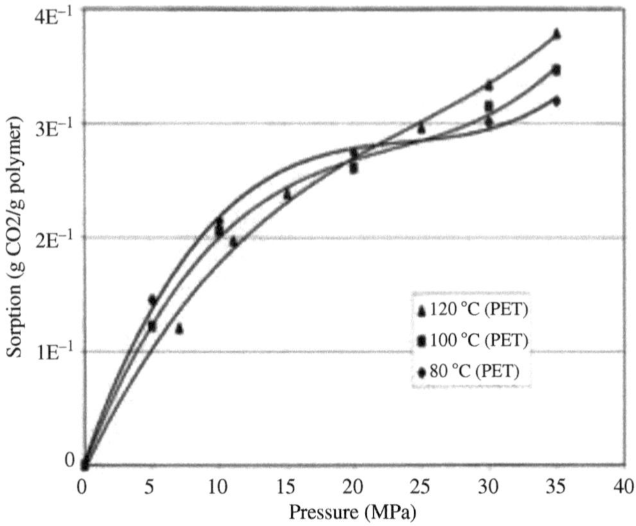

FIG. 7.15 The effects of pressure and temperature on the solubility of CO_2 in PET [87].

CO_2 sorption also exhibits a cross-over due to reaching critical concentration where the glass transition occurs. On the other hand, high pressure CO_2 can also plasticize polymers as it sorbs in them. This phenomenon is the so-called glass transition depression and it is depicted in Fig. 7.16.

In 1995, a SCF was first successfully employed by Watkins et al. for the preparation of polymer supported platinum nanoparticles [88]. Pt(cod)me$_2$ was impregnated into poly(4-methylpent-1-ene) (PMP) and poly(tetrafluoroethylene) (PTFE) assisted by scCO_2 at 80°C and 15.5 MPa for 4 h. The Pt complex was reduced by (a) hydrogen (24 h) subsequent to depressurization at 60°C and 7.0 MPa of H_2, (b) hydrogen (4 h) in CO_2 at 80°C and 15.5 MPa induced by the addition of a ~10-fold excess of H_2 to the reaction vessel prior to depressurization, and (c) thermolysis (20 h) in CO_2 induced by raising the temperature to 140°C prior to decompression (final pressure ~265 bar). Fig. 7.17 shows the TEM images of Pt/PMP synthesized by methods (a) and (b), which indicate the presence of discrete platinum clusters in PMP having a maximum diameter of approximately 15 nm for method (a) and 50 nm for method (b), respectively.

For the Pt/PMP composites synthesized by the method (b), SEM analysis of the composite cross section revealed a sharp concentration gradient between a platinum-enriched surface layer and the bulk of the substrate which contained a uniform concentration of the Pt particles smaller than 50 nm. A high concentration of larger platinum crystals (50–100 nm) was observed. TEM analysis of thermal reduction of Pt(cod)me$_2$ in CO_2-swollen PMP at 140°C revealed

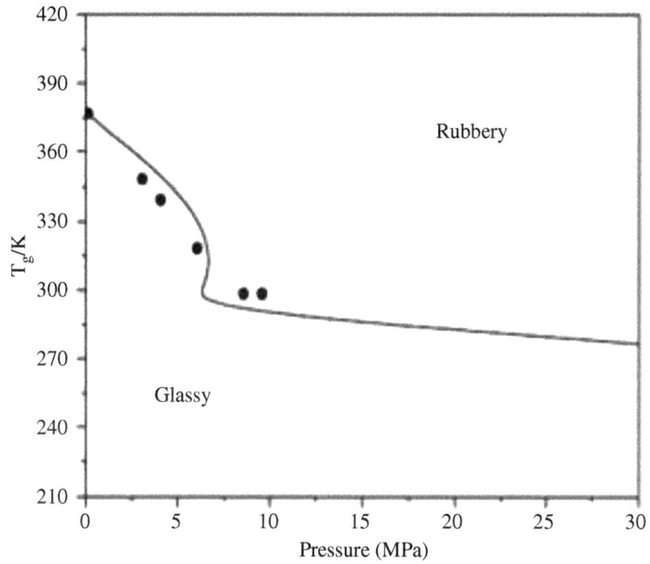

FIG. 7.16 Depression of glass transition of polymers [87].

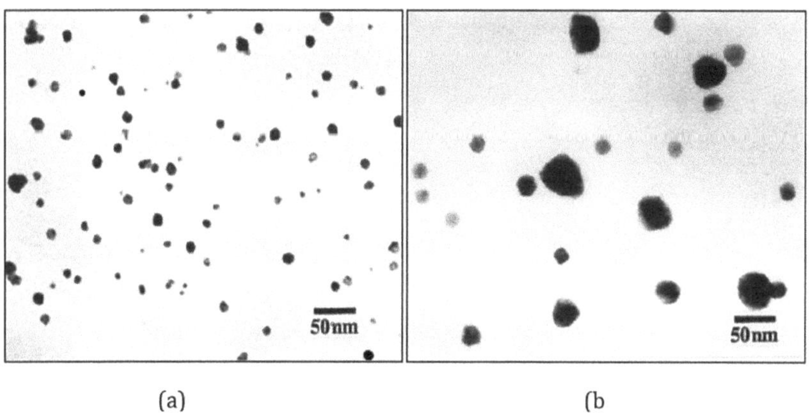

FIG. 7.17 Polymer supported Pt nanoparticles prepared by SCD [88].

feathery platinum aggregates comprised of smaller primary particles. The aggregates, which were distributed throughout the substrate, had a maximum diameter of approximately 50 nm.

Quite a few more studies have been conducted since then and a summary of the studies in the literature on SCD of nanoparticles into polymers is given in Table 7.2. Metallic nanoparticles of gold, silver, platinum and palladium have

TABLE 7.2 Summary of the studies in the literature on preparation of polymer supported nanoparticles by SCD.

Metal	Metal complex	Morphology	Substrate	Solvent medium	Deposition temperature (°C)	Deposition pressure (MPa)	Conversion method	Particle size range (nm)	Ref.
Au	Au(acac) me_2	Nps	Polyamide	$scCO_2$	40	27.58	3	3.2–18.7	[89]
			Polypropylene					~23.1	
			PTFE					~4.4	
Ag	Ag(hfpd) (tegme)	Nps	PSDB	$scCO_2$	40	27.5	3	10–100	[90]
	Ag(cod) (hfac)	Nps	BTDA-ODA Polymide	$scCO_2$	110	34.5	3	–	[91]
	Ag(cod) (hfac)	Nps	Cotton fiber	$scCO_2$	40	21	3	–	[92]
	Ag(cod) (hfac)	Nps	FAP	$scCO_2$	50	20	4	3–8	[93]
	Ag(cod) (hfac)	Nps	Polycarbonate	$scCO_2$	40	9–10.3	3	2–10	[94]
	Ag(acac)	Nps	Resin CR39	$scCO_2$ + Acetone	90	25	2, 3	3–8	[95]
	Ag(acac)	Nps	A – alumina	scH_2O	400	30	–	1–150	[96]
Cu	$Cu(HFDI)_2$	Nps	Polyarylate	$scCO_2$	45	8	3	10–40	[97]
	Cu $(hfac)_2H_2O$	Nps	PTFE	$scCO_2$	40–160	12–20	2	<20	[98]

Pd	Pd(hfac)$_2$	Nps	Aramid fiber	scCO$_2$	100–150	10–25	3	–	[99]
	Pd(hfac)$_2$	Nps	Nafion 117	scCO$_2$	80	13.6	1	5–10	[100]
	Pd(acac)$_2$	Nps	Polyimide film	scCO$_2$	40–200	19.6	3	4.6	[101]
	PdCl$_2$	Nps	PZS nanospheres	scCO$_2$	40	13	–	10–35	[102][a]
	Pd(hfac)$_2$	Nps	ABS/PEEA	scCO$_2$	80	10	2	–	[103]
	Pd(hfac)$_2$	Nps	PPI/HMDI	scCO$_2$	45	15	3	–	[104]
	Pd(hfac)$_2$	Nps	MPF	scCO$_2$	45	15	3	–	[104]
Pt	Pt(cod)me$_2$	Nps	PMP, PTGE	scCO$_2$	80	15.5	1	<100	[88]
							2	<15	[52]
	Pt(cod)me$_2$	Nps	Nafion 112	scCO$_2$	80	27.6	3	<50	[101]
	Pt(acac)$_2$	Nps	Polyimide film	scCO$_2$	40–200	19.6	3	2.5	
							3	12	
Rh	Rh(acac)$_3$	Nps	PPI/HMDI	scCO$_2$	45	15	3	1–2.5	[104]
	Rh(acac)$_3$	Nps	MPF	scCO$_2$	45	15	3	1–30	[104]
Ru	Ru(cod)(tmhd)$_2$	Nps	PDMS	scCO$_2$	40	10.34	3	~2	[105]
ZnO	Zn(TTA)$_2$	Nps	PET	scCO$_2$	80	20	3	50–80	[106]

Conversion methods: 1: In situ chemical conversion, 2: In situ thermal conversion, 3: Ex situ conversion.
[a]*Initially impregnation in alcohol, then precipitation in SCF.*

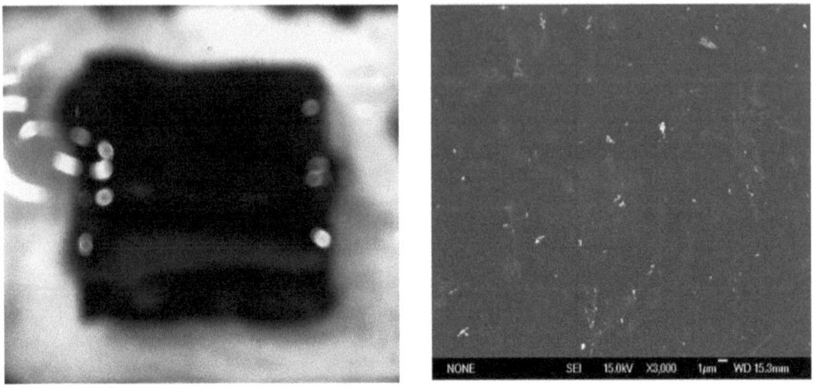

FIG. 7.18 Photographs of the Pd/Nafion®-2 membrane prepare by SCD [100].

been incorporated into a wide variety of polymers. Among the various metals, Ag has been the most widely investigated due to its antimicrobial activity.

The Pd/Nafion membranes with 1.19 and 2.15 mg/cm^2 Pd loadings, named as Pd/Nafion-1 and Pd/Nafion-2, respectively, were characterized for their morphology and evaluated in direct methanol fuel cells (DMFCs). Fig. 7.18a is a photograph of the Pd/Nafion-2 membrane. The square region in the middle of the membrane (2.5 × 2.5 cm) which was exposed to the Pd complex changed from clear to nontransparent black due to Pd impregnation. The area that was masked with Teflon film was still transparent indicating little Pd was incorporated into this region. Fig. 7.18b shows a SEM image of the Pd/Nafion-2 membrane surface.

The membrane had a uniform surface morphology and no cracks were observed. The Pd/Nafion-2 membrane surface was also studied by X-ray diffraction. As shown in Fig. 7.19, the diffraction peak at $2\Theta = 12–22°$ was assigned to the combination of the amorphous ($2\Theta = 16°$) and crystalline ($2\Theta = 17.5°$) scattering from the polyfluorocarbon chains of Nafion. The corresponding peak position from 40° to 85° represents the diffraction of specific crystal surfaces of the palladium particle, indicating that Pd metal was incorporated in the membrane.

Fig. 7.20a also shows the TEM image of the cross section of the Pd/Nafion-2 membrane. A thin (0.2–0.3 μm) dense band of agglomerated Pd particles can be observed near the membrane surface and a significant number of isolated Pd particles can be seen deeper into the membrane. The number density of particles decreases as does the particle size with increasing distance from the surface. This is illustrated by a higher resolution image at 1 μm from the surface, as shown in the squared area in Fig. 7.20. The Pd particle size estimated from Fig. 7.20b is about 8 nm in diameter.

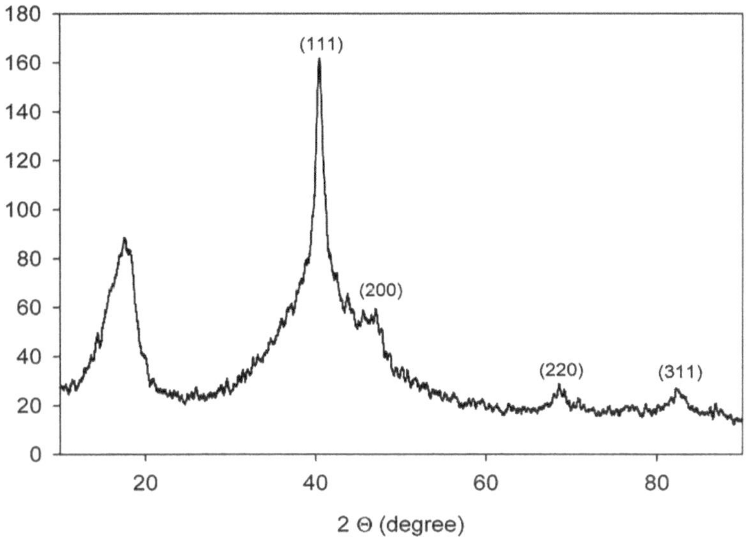

FIG. 7.19 XRD of Pd/Nafion®-2 [100].

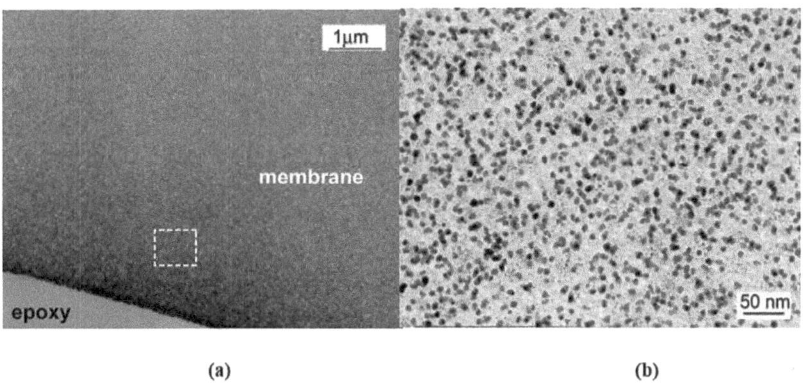

FIG. 7.20 SEM and TEM Images of Pd/Nafion®-2 membrane prepared by SCD [100].

A composite of Nafion with Pt was also prepared. To investigate the effects of the high pressure conditions used for impregnation and the elevated temperatures used for thermal reduction on the chemical structure of Nafion, composite films of platinum/Nafion 112 were prepared using SCD [107]. The changes in the chemical structure of Nafion 112 were monitored by FTIR. The IR spectra of pure Nafion 112 and Pt(cod)me_2 is shown in Fig. 7.21.

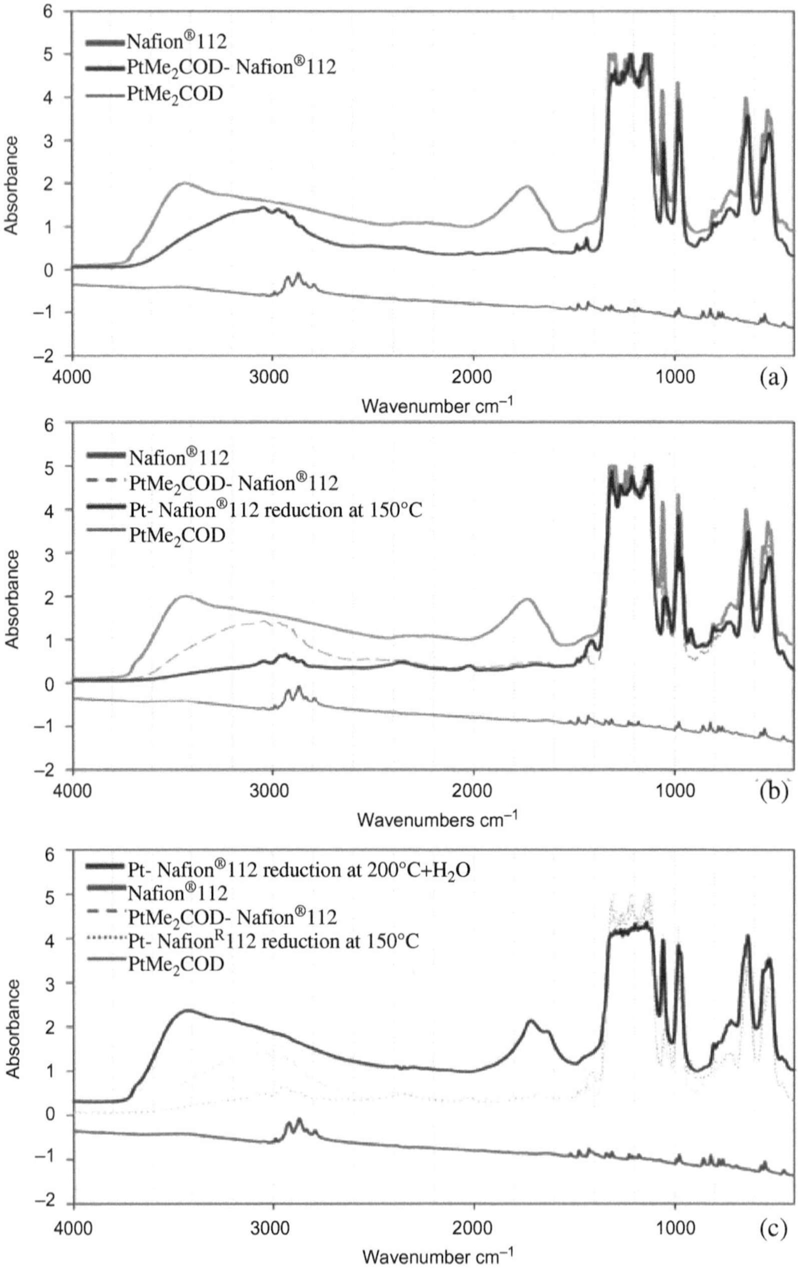

FIG. 7.21 Effects of impregnation and thermal treatment on chemical structure. (a) The IR spectrum of Nafion impregnated with $PtMe_2$(cod) (compared with pure Nafion™ 112 and $PtMe_2$(cod). (b) The IR spectrum of Nafion™ impregnated with $PtMe_2$(cod) and subjected to thermal treatment at 150°C for 2 h. (c) The IR spectrum of Nafion™ impregnated with $PtMe_2$(cod) and subjected to heat treatment at 200°C for another 2 h and saturated with water [107].

The detailed assignment of peaks in the IR spectra of Nafion 112 was given by Gruger et al. [108]. The two plateaus between 1600 and 3800 cm^{-1} were assigned to water or acidic water (H_3O^+). The peak at 1058 cm^{-1} was assigned to sulfonic groups, which is one of the most important functional groups in Nafion. The IR spectrum of Pt(cod)me_2 impregnated Nafion 112 after $scCO_2$ deposition is also shown in Fig. 7.21a. Based on this spectrum, although one of the plateaus assigned to water disappeared, the main structure of Nafion 112 was intact. The new peaks which appeared around 1430, 1480, 2850, and 3050 cm^{-1} were assigned to Pt(cod)me_2. The Pt(cod)me_2 impregnated into Nafion 112 was subjected to thermal treatment at 150°C for 2 h and the IR spectrum of the resulting Pt-Nafion composite is shown in Fig. 7.21b. The plateaus assigned to water disappeared completely, but the main structure of Nafion 112 was almost intact. However, the peaks assigned to Pt(cod) me_2 were still apparent indicating that Pt(cod)me_2 did not decompose completely at 150°C during the 2 h. Further thermal reduction was carried out at 200°C for another 2 h. The resulting Pt-Nafion film was immersed in water for 24 h. After wiping the water off from the surface of the film by tissues, the IR spectrum of this film was taken and shown in Fig. 7.21c. The structure of the Nafion is slightly changed. The plateaus assigned to water appeared again, which means after Pt(cod)me_2 impregnation and thermal reduction, the Nafion film still had the ability to absorb water. The peaks assigned to Pt(cod)me_2 disappeared completely, which shows that treatment at 200°C for 2 h is sufficient to decompose Pt(cod)me_2 completely. The structure of the Pt-Nafion nanocomposites was investigated further using TEM and the images are shown in Fig. 7.22.

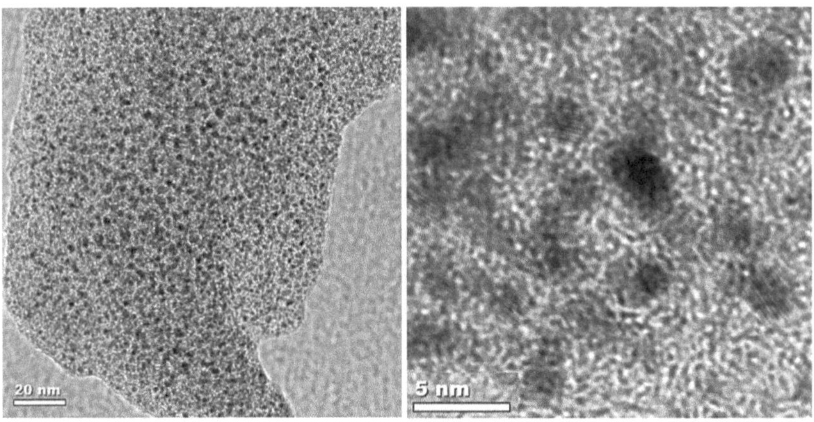

(a) (b)

FIG. 7.22 TEM image of Nafion supported Pt nanoparticles prepared by SCD [87].

Fig. 7.22A shows the overview of the nanocomposite's structure. The platinum particles appear as uniformly distributed dark features against the background contrast from the Nafion support. Individual platinum crystals with lattice fringes can be seen under high resolution (Fig. 7.22B). The average crystallite size is 2.5 nm and the distribution is very narrow (1.7–3.2nm).

Along similar lines, Kim et al. investigated SCD of palladium particles onto Nafion 117 membranes [109]. The membranes were impregnated with $Pt(acac)_2$ at 20 MPa and 80°C for 4 h in $scCO_2$ followed by chemical reduction using various concentrations of $NaBH_4$ solution. The electron probe micro analysis (EPMA) in which the cross-section of Pd/Nafion membranes was characterized for the Pd concentration are presented in Fig. 7.23.

The results indicate that the Pd/Nafion composite membranes showed different morphologies depending on the concentration of the reducing agent ($NaBH_4$). They obtained Pd particles of 4–5 nm in 0.5 mM solution of $NaBH_4$,

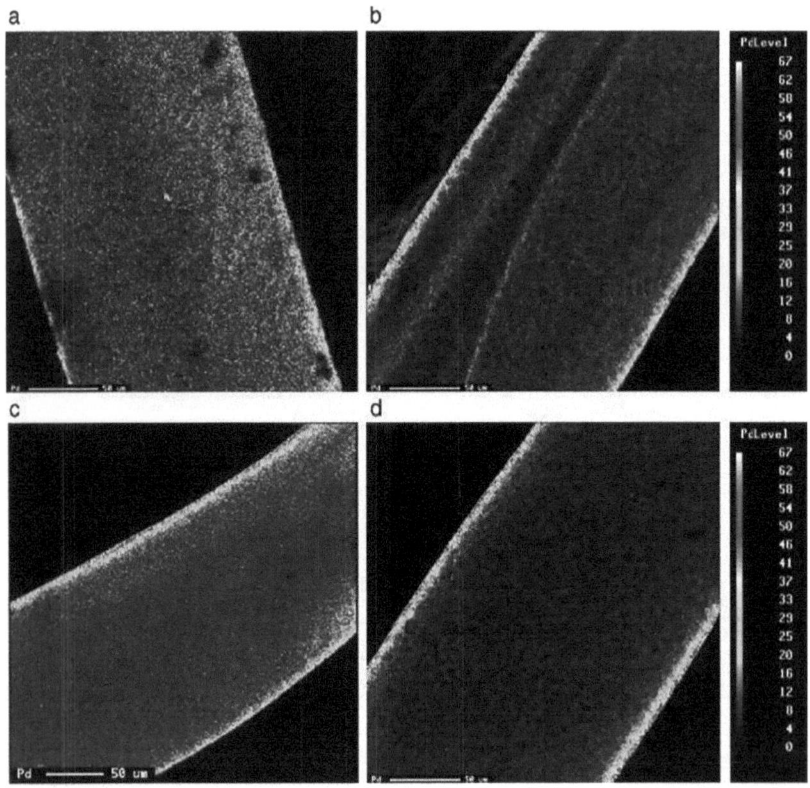

FIG. 7.23 Electron probe micro analysis (EPMA) of the cross-section of Pd/Nafion® membranes prepared by SCD [109].

whereas for 10 and 100 mM solutions, the average diameter of the particles was 20–70 nm. They observed that the ion conductivity of Pd/Nafion reduced in 0.5 mM (0.0833 S cm^{-1}) $NaBH_4$ solution was higher than that of the Nafion 117 (0.0740 S cm^{-1}). However, the increase of the reducing agent concentration led to a decrease in the ion conductivity to a level even less than that of the Nafion 117 which was attributed to the increase of the Pd particle size. It was suggested that the ability of the particles with diameters of 20–70 nm to conduct protons was less than 4–5 nm particles. Additionally, the permeability of the composite membranes decreased with the increasing $NaBH_4$ concentration. They also tested the composite membranes in a DMFC with 2 M methanol at 80°C with air in the cathode. As a result of the competition phenomenon between the ionic conductivity (0.0765 S cm^{-1}) and permeation rate (4.38×10^{-7} cm^2 s^{-1}), the composite membranes prepared using 2 mM reducing agent had the best cell performance and 44% higher current density than that of Nafion 117.

A more recently used metal complex conversion technique is laser irradiation [93]. The method consists of irradiation of the metal complex after being adsorbed on substrates in $scCO_2$. Ag nanoparticles were synthesized in this fashion using laser irradiation (λ = 532 nm). The average particle size was in the range of 3–8 nm with a narrow size distribution. A significant achievement of this study was the formation of periodic layered structures of nanoparticles residing horizontally on the polymer surface as shown in Fig. 7.24.

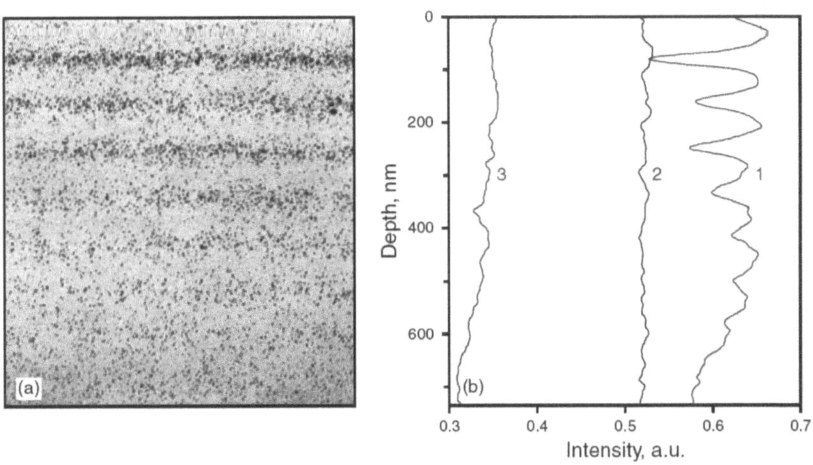

FIG. 7.24 TEM images of fluorinated acrylic polymer supported Ag nanoparticles prepared via SCD (decomposed by laser irradiation at laser dose of ~200 J/cm^2); (b) – averaged profiles of TEM images for different precursor conversion cases: laser (1), UV-Hg lamp (2), and thermal processing (3) [4].

The interlayer distance was as short as 90 nm and at least 5–6 periodic layers each made up of Ag nanoparticles are clearly seen in Fig. 7.24A and B. The authors report that at lower laser dozes the concentration of Ag particles decreased, and the period of structures grew (120–180 nm). At higher laser doses, concentration of generated Ag particles increased, however periodic layered structure in polymer/Ag(hfac)cod ((1,5-cyclooctadiene)(hexafluoroacetylacetonato) silver(I)) film became visually less pronounced. Regarding the mechanism of the observed effect of the laser formation of short-period nano-sized Ag particle assemblies, we note that these structures are formed under coherent laser radiation only. In other words, interference and formation of standing waves in polymer film is most likely to be responsible for such effect. In this case, both photochemical decomposition of Ag complex and the effect of optical tweezers can be responsible for the preferential assembling of Ag nanoparticles from reduced Ag atoms in antinodes of standing waves, thus causing formation of periodic layered structures.

Mauricio et al. synthesized ZnO/PET by SCD [106]. The ZnO precursor zinc bis(2-thenoyltrifluoroacetonate) ($Zn(TTA)_2$) was synthesized using zinc acetate dihydrate ($Zn(CH_3COO)_2 \cdot 2H_2O$) and thenoyltrifluoroacetone (HTTA). Deposition of $Zn(TTA)_2$ onto PET was carried out in a high-pressure vessel with a 25.0 mL internal volume. SCD was carried out at 85°C and 20 MPa. After impregnation, the vessel was depressurized and $Zn(TTA)_2$/PET was converted to ZnO/PET using hydrazine as the reducing agent. SEM micrographs, shown in Fig. 7.25 indicated that the surface is covered with spherical ZnO nanoparticles with particle size ranging from 50 to 80 nm.

Cu/PTFE composites were prepared by SCD with impregnation temperature and pressure varying between 40–160°C and 12–20 MPa, respectively [98]. $Cu(hfac)_2 \cdot H_2O$ was used as Cu complex. Reduction of the Cu complex was conducted in situ by increasing the vessel temperature and pressure to 230°C and 25 MPa, respectively. Effects of deposition temperature and pressure were investigated. Impregnated amount of the Cu complex increased with increasing deposition temperature from 40 to 120°C, and decreased above 120°C. Impregnated amount increased with increasing pressure up to 16 MPa and plateaued at higher pressures. SEM images showed that Cu nanoparticles penetrated deeper into the Cu/PTFE composite when they were synthesized at 180°C compared to that at 160°C. TEM images showed that Cu nanoparticles were dispersed on PTFE homogeneously with Cu nanoparticle sizes not larger than 20 nm.

Ge et al. synthesized Ru nanoparticles dispersed in polydimethylsiloxane (PDMS) films using SCD [105]. ($Ru(cod)(tmhd)_2$) was used as Ru complex in $scCO_2$ at 40°C and 10.34 MPa. Metal complex conversion was conducted ex situ under flowing N_2 at 250°C and ambient pressure. TEM images, shown in Fig. 7.26, revealed that Ru nanoparticles were dispersed throughout PDMS with diameters of up to 2 nm.

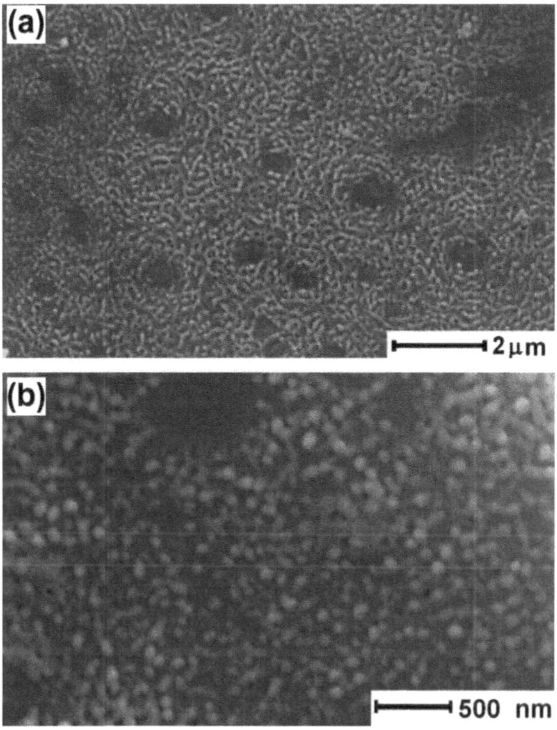

FIG. 7.25 SEM images of (a) ZnO/PET surface and (b) surface spherical particles [106].

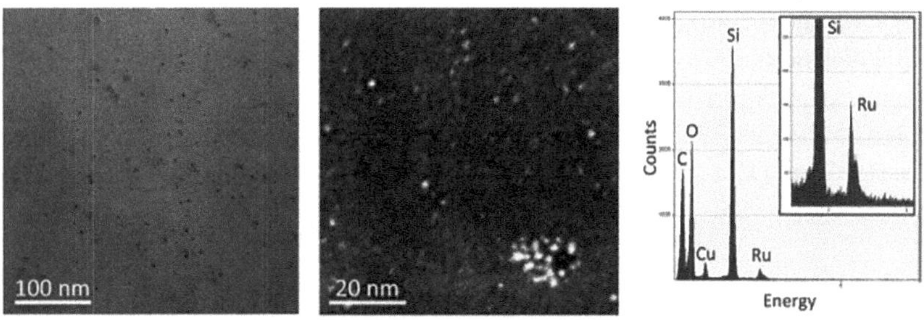

(a) (b) (c)

FIG. 7.26 TEM and SEM data from a cryo-ultramicrotomed section through a Ru/PDMS composite (a) BF-TEM, (b) HAADF-STEM, and (c) STEM EDXS [105].

Parenago et al. prepared composites with Rh and Pd impregnated on mesoporous phenol-formaldehyde (MPF) resin and poly(propylene imine) dendrimer cross linked with 1,6-hexamethylendiisocyanate (PPI/HMDI) using SCD [104]. Rh and Pd complexes were $Rh(acac)_3$ and $Pd(hfac)_2$, respectively. Impregnation was conducted at 45°C and 15 MPa. Reduction of the precursors were conducted with H_2 at 5 MPa and 50°C. Particle size distribution of Rh nanoparticles of Rh/PPI-HMDI composite was narrow and between 0.5 and 3 nm while that of Rh/MPF was relatively wide and between 1 and 50 nm, with an accumulation between 1 and 5 nm.

Polymer supported nanoparticles can also be prepared via postgrafting processing in $scCO_2$. In this process, metal complexes are dissolved in a monomer phase prior to free radical suspension polymerization or bulk polymerization. The metal complex molecules may be trapped and be distributed uniformly inside the polymer matrix during the polymerization reaction. The metal complex can then be converted to nanoparticles using conventional methods or at supercritical conditions as previously described. The exposure of the polymer/metal nanoparticle composite to $scCO_2$ results in the CO_2 sorption-induced swelling and the plasticization of the polymer. The quick depressurization of the system results in the generation of a porous polymer/nanoparticle system. In this fashion, improved surface properties can be obtained. It is also possible to subject a polymer/metal complex mixture to the $scCO_2/H_2$ mixture to end up with polymer/metal nanoparticle composites. $scCO_2$ can contact with the already immersed metal precursor and with the presence of H_2, the conversion reaction for nanoparticle production occurs. The metal amount can be controlled through the metal complex concentration within the prepolymer solution. On the other hand, the nanoparticle size depends on $scCO_2/H_2$ concentration in the polymer which can be tuned by varying the density of CO_2, and thus its solubility in the polymer. However, one needs to consider the possible extraction of metal complex during $scCO_2$ processing and take into account the extraction kinetics and thermodynamics. Nevertheless, polymer supported Ag nanoparticles with average size of 15 nm were synthesized in this fashion. Fig. 7.27 shows the photographs and TEM of such materials. By comparing Fig. 7.27A and B one can infer the transformation of the metal complex. One can also observe a slight increase in porosity due to foaming upon exposure to CO_2 as described previously. The TEM images demonstrate good dispersion of Ag nanoparticles on PMMA.

7.3 Supported bimetallic and multimetallic nanoparticles

Multimetallic nanoparticles are composed of two or more different metals. They are of both academic and industrial interest as their properties are often different from either of pure nanoparticles.

Similar to bulk alloys, bimetallic nanoparticles exhibit a very wide range of combinations and compositions. Bimetallic nanoparticles (A_mB_n) can be generated with more or less controlled amount (m + n) and composition (m/n) with main mixing patterns given in Fig. 7.28 [110].

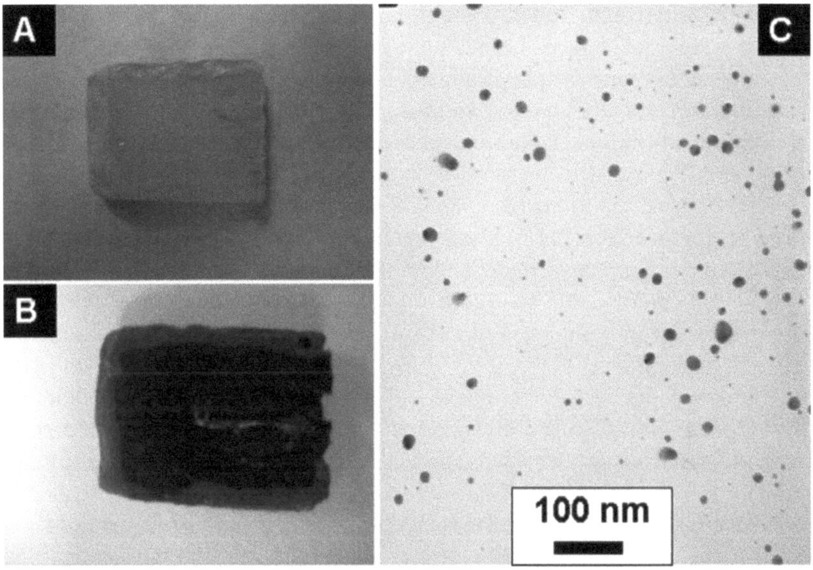

FIG. 7.27 (A) Digital photographs showing the PMMA block containing Ag precursor complex prepared by bulk polymerization; (B) the same PMMA block following the H_2-assisted reduction in $scCO_2$; (C) and the TEM image of the same Ag nanocomposite product [87].

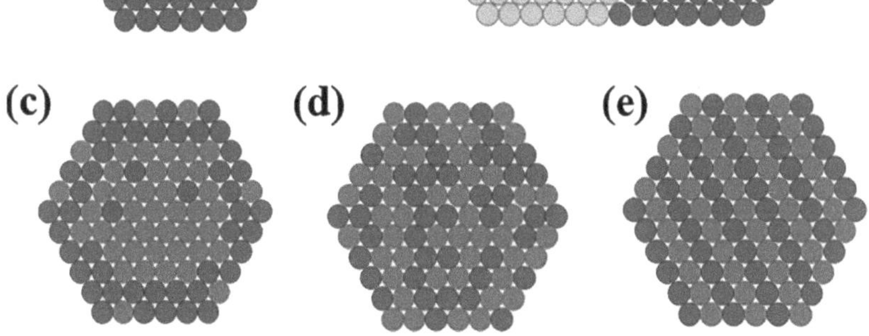

FIG. 7.28 Schematic representation of the examples of some possible configurations for bimetallic NPs: (a) core-shell, (b) segregated/Janus particle, (c) surface segregated, (d) disordered alloy, (e) ordered alloy. The illustrations display cross sections of the NPs showing an hypothetical face-centered cubic (FCC) (111) plane [110].

The formation and growth of single component nanoparticles usually consist of a number of steps including nucleation, diffusion and growth. All these steps are usually governed by kinetic phenomena [111]. The degree of segregation/mixing and atomic ordering in A_mB_n bimetallic nanoparticles depends on a variety of factors including relative strengths of A-A, B-B, and A-B bonds, relative surface energies of bulk A and B, relative atomic sizes, strength of binding to surface ligands (or surfactants) and specific electronic/magnetic effects [112]. The structure of the nanoparticle and the degree of A-B segregation may also depend on the kinetics of nucleation and growth for which the method and conditions of nanoparticle generation (type of particle source, metal complex, support, surfactant, temperature, pressure, etc.) becomes important.

The knowledge of the phase behavior of the alloys is important. Although the phase diagrams of bulk alloy systems have been extensively reported, their extension to the nanometer sized regime is not trivial since the perturbations in both morphology and electron density due to finite sizes at the nanoscale can give rise to nonbulk like properties [113]. For instance, pairs of elements such as FeAg [112] or AuPt [114] have large miscibility gaps in bulk crystalline state but can mix as nanoparticles. PtRu bimetallic nanoparticles with 1.5 nm diameter and with a Pt:Ru ratio of 0.2 display an FCC closed-packed crystal structure whereas bulk alloy of the same species with the same composition exhibit a hexagonal-closed packed (HCP) structure [115]. These observations point out that differences may exist between the structural properties predicted by the bulk phase diagrams and those displayed by metal nanoparticles.

In catalysis, supported bimetallic nanoparticles are used not only to enhance the performance but also for economical reasons [116]. The activity of bimetallic nanoparticles depends on the metal concentration, morphology, and atomic ordering. In alloy nanoparticles, the intrinsic catalytic activity adjustment originates from two cumulative phenomena called the ligand and the strain effect. The addition of another metal causing alteration in the electron density of the system (ligand effect) and the metal–metal bond length change (strain effect) may lead to improved catalytic activity [117, 118]. The incorporation of a second metal into the structure may also prevent the poisoning of the catalyst and prolong its aging behavior [119, 120]. Since catalytic reactions occur only at catalytic sites residing on the surface of nanoparticles, a significant proportion of the atoms are effectively wasted.This is particularly important for expensive catalytic metals such as Pd and Pt [121]. In order to reduce costs, there is, therefore, considerable interest in synthesizing A-B core-shell NP catalysts, where A is a relatively inexpensive metal (e.g., Co, Ni, Cu), which is (generally) less catalytically active residing at the core, and B is a more expensive, more catalytically active metal (typically Pd or Pt) residing at the shell. Synergetic electronic effects toward catalytic reactions can also be significant in alloy or core-shell NPs via the introduction of a change in the d-band center of the metal via the addition of second metal [122, 123]. Good examples of these phenomena are highest intrinsic activity of Pt_3Ni [124] and dealloyed $PtCu_3$ [125] as compared to pure Pt toward the ORR.

Bimetallic nanoparticles, in particular, core-shell nanoparticles, have very special optical properties that can be tuned with structural and compositional modification. The relative modification of core and shell radii produces spectral shifts in a wide range that is in some cases as important as the morphological changes in a single component NP along with some possible applications in surface enhanced Raman spectroscopy (SERS) [126].

Two approaches naturally arise in order to synthesize bimetallic nanoparticles using SCD which are simultaneous and sequential deposition. Fig. 7.29 shows three major steps of SCD which are dissolution of a metal complex in a SCF followed by adsorption of the metal complex onto the surface of the support and conversion of the metal complex to its metal form by various routes.

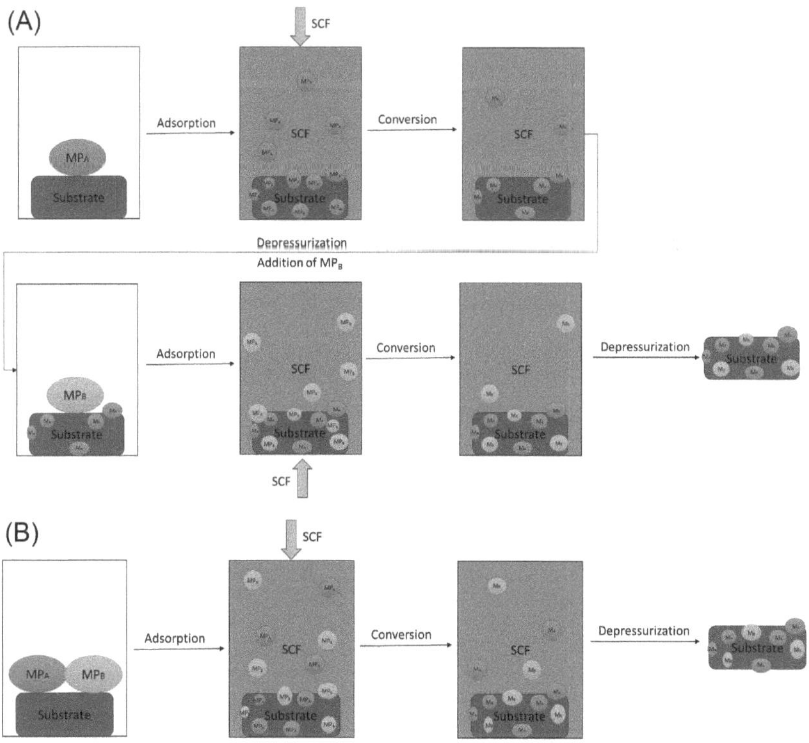

FIG. 7.29 Sequential (a) and simultaneous (b) SCD with in situ metal precursor conversion. M: Metal, MP: Metal precursor, subscripts A and B denote two metals [110].

In simultaneous SCD, one starts with two metal complexes by dissolving them simultaneously in $scCO_2$ and then carries out the same subsequent steps as in single component SCD, whereas in sequential SCD the single component SCD procedure is applied twice. The summary of studies in the literature on SCD of bimetallic nanoparticles is given in Table 7.3.

PtRu nanoparticles supported on CNTs were prepared by Lin et al. with particle sizes around 5–10 nm [143]. The metal complexes $Pt(acac)_2$ and $Ru(acac)_2$ were simultaneously adsorbed on CNTs in $scCO_2$ at 200°C and 8 MPa. A small amount of methanol was added to deposition vessel. The metal

TABLE 7.3 Supported bimetallic and trimetallic nanoparticles prepared using $scCO_2$.

Metals	Substrate	SCD	Metal/Metal molar ratio	Metal (wt%)	Particle size	Ref.
Pt-Ru	MWCNT	1	48-52	21.30[a]	2.8 ± 0.8	[127]
Pt-Cu			62-38	23.13[a]	6.6 ± 3.5	
Pt-Au			37-63	4.28[a]	9.3 ± 3.7	
Pt-Pd			50-50	31.77[a]	9.2 ± 3.3	
Pt-Ni			66-34	21.75[a]	5.7 ± 2.2	
Pt-Ru	MWCNT	1	43-57	–	5	[128]
Pt-Ru	CNT	1	50-50	–	~2	[129]
	Ketjenblack		50-50	–	~2	[129]
Pt-Pd	CB	1	92-100	16[b]	4.3 ± 2.7	[35]
Pt-Pd	CB	2	2-1	18-6	3.2	[130]
			1-2	15–10	3.9	
			1-4	8-7	5.1	
Pt-Pd	Papam modified silicon wafer	1	1-1 to 10-1	–	1.25 ± 0.25 – 8 ± 2	[131]
Fe-Ni	Silica gel		1-1		8 ± 2	[36]

TABLE 7.3 Supported bimetallic and trimetallic nanoparticles prepared using $scCO_2$—cont'd

Metals	Substrate	SCD	Metal/ Metal molar ratio	Metal (wt%)	Particle size	Ref.
Fe-Ni	Papam modified silicon wafer	1	1-1	–	$1 \pm 0.2 - 9 \pm 2$	[131]
Mg-Ni	Mg	1	2-1	0.75^c	2–35	[132]
Mg-Cu			1-2	0.40^c	15–60	
Ni-Mo	Al_2O_3	1				[133]
Pt-Ir-Co	CA	1	59-34-7	20.8^b	0.5–5.5	[134]
Au-Ag	Al_2O_3, TiO2, CB	1	70-30	4^b	4–28	[3]
Pt-Fe	Graphene	1	40-60	11^a	~1.6	[135]
Rh-Pt	SBA-15	1	81-19	3.8^b	~6.1	[136]
Pt-Fe	GCM	1	40:60	12.2^a	~2	[137]
Pt-Fe-Co	GCM	1	36:60:4	11^a	~2	
Pt-Cu	CA	2	1-1, 1-3	20^a, 15^a	1.8–4.5	[138]
Pt-Pd	C	1	2-1	3^b	1–12	[139]
Pt-Ni	Graphene	2	40-60	31^a	3–6	[140]
Pt-Fe	Graphene	2	70-30	23^a	2.5–5.5	
Pt-Cu	Graphene	2	71-29	24^a	3.5–6	
Pt-Cu	CA	1	1-3	23^a	2–7	[141]
		2	1-3	23.2^a	2–8	
Pd-Ru	SiO_2	1	1-10 to 10-1	-	~2.9	[142]
Cu-Ru		1	1-10 to 10-1		–	
Pt-Ru	CNT	1	45-55	4.1^a	5–10	[143]

1: Simultaneous; 2: Sequential; AC: activated carbon; a: Pt wt%; b: total metal wt%; c: Ni or Cu to Mg ratio.

complex conversion was conducted at 16 MPa and took 15 min with in situ chemical conversion using H_2 as the reducing agent. Yen et al. synthesized bimetallic Pt-Ru, Pt-Cu, Pt-Au, Pt-Pd, and Pt-Ni nanoparticles supported on MWCNTs via simultaneous SCD with $scCO_2$ at 10 MPa [127]. Deposition temperature was 70°C for Pt-Au and 200°C for the others. Conversion of metal complexes were conducted by in situ chemical reduction with H_2 at 25 MPa. Resulting electrocatalysts had average nanoparticle sizes around 2.8–9.3 nm.

Bayrakçeken et al. prepared PtPd/CB via sequential SCD where Pt(cod)me$_2$ was dissolved in $scCO_2$ and adsorbed onto CB at 80°C and 20 MPa [130]. After equilibrium was reached, H_2 in CO_2 ($P_{H2} = 0.8$ MPa, $P_{Total} = 27.6$ MPa) was added to the vessel for chemical conversion of the Pt complex. The vessel was depressurized after 5 h, and Pt/CB was collected. Afterwards, this procedure was repeated with Pd(acac)$_2$. Instead of bare CB, Pt/CB was added to the vessel as the substrate for deposition of Pd(acac)$_2$. By this way PtPd/CB were prepared with Pt:Pd atomic ratios ranging between 2:1 and 1:4. Average nanoparticle sizes on these catalysts were between 1 and 10 nm. These bimetallic catalysts were also compared with Pt/CB and previously reported Pd/CB [35]. The Pd particle sizes in Pd/CB ranged between 3 and 100 nm. Images also revealed that the larger Pd nanoparticles (>20 nm) have an irregular morphology suggesting that these may form by the agglomeration of smaller nanoparticles migrating across the support surface instead of by coarsening due to the atomic diffusion of metallic Pd. An increase in Pt metal loading did not influence the average Pt particle size for Pt/CB; whereas the addition of Pd resulted in an increase in the PtPd bimetallic particle size for the PtPd/CB catalysts. When TEM images for Pt/CB and PtPd/CB were compared, it was seen that the particle size distributions were broader for PtPd/CB than those for Pt/CB, with the diameters for most of the particles lying between 2 and 10 nm. EDXS analysis on bimetallic particles showed that the smaller particles (diameters between 1 and 4 nm) were Pt-rich while the larger particles (diameters greater than 5 nm) were Pd-rich. TEM images indicate that as Pd loading increased, the tendency of the Pd particles to agglomerate and form larger particles also increased. The average size of the large particles decreased from 15 to 5 nm as the Pt:Pd ratio changed from 1:4 to 1:2.

Barim et al. prepared PtCu/CA via sequential SCD with different adsorption orders (either Pt or Cu precursor first) and Pt:Cu mole ratios [138]. Pt(cod)me$_2$ and Cu(II) trifluoroacetylacetonate (Cu(tfa)$_2$) were used as metal complexes. SCD runs were conducted at 10.7 MPa and 35°C. The conversion of Cu(tfa)$_2$ was carried out at 200°C under flowing H_2, and the conversion of Pt(cod)me$_2$ was carried out at 200°C at atmospheric pressure under flowing N_2 for 2 h. PtCu/CA with Pt:Cu molar ratio of 1:1 and 1:3 were prepared with Pt loading of ~19 and ~14 wt%, respectively. The samples prepared in each batch were divided into three lots, which were then annealed at three different temperatures, 600°C, 800°C, and 950°C. For the Cu-complex-first adsorbed

PtCu/CA with Pt:Cu atomic ratio 1:1 annealed at 600°C, pure fcc Cu peaks could clearly be observed. Furthermore, the Cu peaks were very sharp whereas the PtCu peaks were broader, indicating that the PtCu alloy nanoparticles were significantly smaller than the separate Cu nanoparticles. Conversely, when the same sample was prepared by adsorption of Pt complex first, it did not exhibit any separate Cu peaks; instead, only PtCu fcc alloy peaks were observed in the XRD data. Regardless of adsorption order, PtCu/CA with Pt:Cu atomic ratio of 1:3 also exhibit fcc PtCu alloy peaks for all three annealing temperatures. Moreover, the samples annealed at 600°C and 800°C had unalloyed fcc Cu peaks; but the unalloyed Cu reflection peaks disappeared after annealing at 950°C. The intensities of the Cu peaks were higher when Cu complex was firstly adsorbed. Degree of PtCu alloying and PtCu alloy particle size both increased at higher annealing temperatures. Morphological properties of the prepared samples were investigated by TEM imaging and EDXS analysis. Fig. 7.30 shows EDXS mapping of PtCu/CA samples with Pt:Cu molar ratio of 1:3 annealed at 800°C with different adsorption orders of the complexes (a) Cu complex first, (b) Pt complex first.

Both PtCu/CA with Pt:Cu atomic ratio of 1:3 showed that Pt-Cu nanoparticles had variable compositions; i.e., some particles were richer in Pt while others were richer in Cu. As the annealing temperature increased, PtCu nanoparticle compositions became more uniform, indicating a greater degree of alloying. Increasing annealing temperature also increased the average particle size from 1.8 to 4.5 nm and resulted in the elimination of separate Cu nanoparticles on the CA surface.

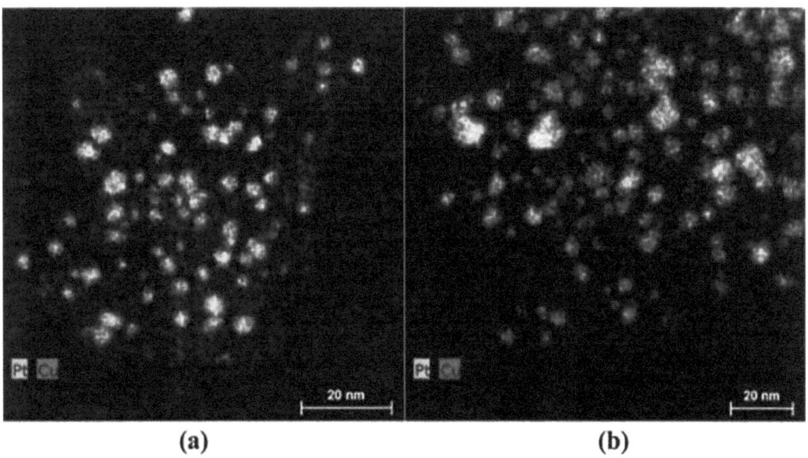

FIG. 7.30 EDXS mapping of PtCu/CA samples with Pt:Cu molar ratio of 1:3 annealed at 800°C prepared by sequential SCD with different adsorption orders of the precursors (a) Cu precursor first, (b) Pt precursor first [138].

Later, Barim et al. compared PtCu/CA prepared by simultaneous and sequential SCDs [141]. SCDs were conducted at 17.2 MPa and 80°C during dissolution and adsorption. Targeted atomic ratio of Pt:Cu was 1:3. In sequential SCD, Pt(cod)me_2 complex conversion was carried out in situ at 135°C, while $Cu(tfa)_2$ precursor conversion was conducted by H_2 injection to the vessel. For the precursor conversion during simultaneous SCD, firstly, the temperature of the vessel was raised to 135°C and kept at this temperature for 6 h. Then, the temperature of the vessel was cooled down to 80°C and H_2 injection was performed as in sequential SCD. The Pt:Cu molar ratios prior to annealing were determined from the shift in the {111} plane angle using Vegard's Law. Prior to annealing, the Pt:Cu molar ratios of the electrocatalysts prepared via the simultaneous approach were closer to the target Pt:Cu molar ratio of 1:3. For the sequentially prepared samples, the Pt:Cu ratio was much lower, indicating the presence of unalloyed Cu or of different compositions of alloyed PtCu nanoparticles compared to those prepared via the simultaneous approach. This is to be expected since in the sequential approach, the Pt nanoparticles were formed prior to the adsorption of $Cu(tfa)_2$, making it more difficult for the Cu atoms to be inserted into the FCC Pt lattice, since the conversion temperature was not high enough to enable such mobility. Conversely, in the simultaneous approach, both Pt(cod)me_2 and $Cu(tfa)_2$ adsorption and nucleation of Pt and Cu atoms occur at the same time. TEM images showed very homogenous distributions of metal nanoparticles in all samples before and after annealing with average PtCu particle size ranging between 3.2 and 4.4 nm. The particle size distributions of the samples before annealing were significantly narrower for both electrocatalysts and became broader after annealing, due to Ostwald ripening during the high-temperature annealing process. However, particle growth was not excessive, and the PtCu nanoparticles were still distributed homogenously on the CA support. Furthermore, the sequential approach resulted in a slightly broader particle size distribution than the simultaneous approach. The EDXS maps from the PtCu/CA sample with simultaneous SCD showed dispersed Pt and Cu with better Pt dispersion before annealing. Furthermore, there were large unalloyed Cu nanoparticles, however, the frequency of these large nanoparticles was quite low. Cu was mainly present in dispersed form on the CA support. EDXS maps of the sequentially prepared PtCu/CA before annealing showed dispersed Pt and Cu, however the output signals were very low. For this sample, PtCu composition before annealing was more uniform on CA when compared to the simultaneous approach. Moreover, EDXS maps of sequentially prepared PtCu/CA before annealing did not exhibit any large unalloyed Cu nanoparticles, which is in agreement with the XRD data. Since Cu was highly dispersed on the CA support, no Cu peaks were observed in the XRD pattern. On the other hand, the EDXS maps of PtCu/CA prepared via simultaneous SCD show better-resolved metal nanoparticles with much more uniform compositions of Pt and Cu after annealing. This indicates that during the high-temperature annealing the Cu atoms became more mobile and diffused into the Pt lattice which is in agreement with the shift in {111} plane angle in XRD patterns of samples before and after annealing. Moreover, the EDXS data from simultaneously prepared PtCu/CA showed large unalloyed Cu nanoparticles after annealing. They showed very uniform composition of PtCu alloy nanoparticles, indicating that the annealing process is sufficient for both alloying and nanoparticle growth. The EDXS maps also show large unalloyed Cu nanoparticles, which agrees with the XRD pattern.

Zhou et al. prepared PtFe nanoparticles supported on 3-D honeycomb structured graphene (HSG) with a Pt:Fe molar ratio of 2:3 and a Pt loading of 11 wt %. [135]. $Pt(hfac)_2$ and $Fe(acac)_2$ were used as the metal complexes. During SCD, tetrahydrofuran (THF) was injected into the vessel to modify the polarity of CO_2 and enable dissolution of the precursors. SCD was conducted at 60°C and 30 MPa. The reduction of the precursors was carried out using a mixture of borane-THF in scCO2 at 80°C for 30 min followed by depressurization at 50°C and washing the sample with a continuous feed of $scCO_2$. Fcc disordered PtFe alloy formation was observed according to XRD results. Particle sizes ranged between 1.4 and 2.5 nm. PtFe and PtFeCo supported on graphene cellular monolith (GCM) aerogels were also prepared by Zhou et al. via SCD at 15 MPa and 60°C [137]. During SCD, GCM alcogel was used and the alcogel drying and the adsorption of the precursors was simultaneously conducted. The alcohol contained in the pores of the alcogel also served as an entrainer for the dissolution of $Pt(hfac)_2$, $Co(hfac)_2{\cdot}H_2O$ and $Fe(acac)_2$.

Kolla and Smirnova prepared PtIrCo/CA with Pt:Ir:Co metal ratio of 59:34:7 and total metal loading of 20.8 wt% via SCD [134]. $Pt(cod)me_2$, Ir(hfac), and $Co(hfac)_2$ were used as the metal complexes. SCD was conducted at 82°C and 20 MPa. Metal complex conversion was carried out after depressurization with flowing H_2/Ar at 600°C or 900°C. Resulting PtIrCo alloy nanoparticles had an average particle size of 1.5 and PSD was between 0.5 and 5.5 nm.

7.4 Metallic films

SCD is also an attractive alternative to the current techniques used for reactive deposition of metallic films. Even though chemical vapor deposition provides high purity films, it suffers from drawbacks such as the low vapor pressure of the metal complexes, mass transfer limited kinetics and generation of thermo-mechanical stresses. Aqueous plating techniques suffer from drawbacks such as generation of large volumes of waste water and poor control of the deposition process. SCFs with the ability to solubilize metal complexes at high mass transfer rates, zero surface tension and complete miscibility with reductant gases such as hydrogen, are promising media for reactive metal deposition. A summary of the studies in the literature on SCD of films si given in Table 7.4.

TABLE 7.4 Summary of the studies in the literature on preparation of films by SCD.

Metal	Metal complex	Substrate	Solvent medium	Deposition temperature (°C)	Deposition pressure (MPa)	Conversion method	Ref.
Ag	Ag(cod)(hfac)	Poly (etherether ketone)	$scCO_2$	80–110	15.1	3	[154]
	Ag(cod)(hfac)	Polyimide	$scCO_2$	150	34.5	3	[155]
Au	Au(acac)me_2	SiO_2	$scCO_2$	60–125	15	1	[156]
		TiN	$scCO_2$	60–125	15	1	[156]
CeO_2	Ce$(tmhd)_4$	YSZ	$scCO_2$ + H_2O	250–300	10–15	3	[157]
Co	Co$(cp)_2$	Silicon wafer	$scCO_2$	285–320	22–26	1	[158]
		TaN coated Si wafer	$scCO_2$				
		TiN coated Si wafer	$scCO_2$				
Cu	Cu$(tmod)_2$	TiN coated Si wafer	$scCO_2$	220–270	10–13.8	1	[153]
	Cu$(tmhd)_2$	TaN coated Si	$scCO_2$ + Alcohols	270–300	20–23	1	[159]
	Cu$(hfac)_2$	Ta coated Si	$scCO_2$	70–120	15	1	[160]
	Cu$(hfac)_2$	Si	$scCO_2$	70–120	15	1	[160]
	Cu$(hfac)_2$	Pd seeded Si	$scCO_2$	200	20	1	[161]
	Cu$(hfac)_2$	Si wafer	$scCO_2$	250–400	10–15	1	[162]
	Cu$(hfac)_2$			180–400	10–15	1	[163]

	Cu(hfac)$_2$	Fe	scCO_2	180–400	10–15	1	[163]
	Cu(hfac)$_2$	Cu	scCO_2	180–400	10–15	1	[163]
	Cu(hfac)$_2$	WN	scCO_2	180–400	10–15	1	[163]
	Cu(hfac)$_2$	Etched Si	scCO_2	180–400	10–15	1	[163]
	Cu(hfac)$_2$	Au coated Si	scCO_2	180–320	8	1	[164]
	Cu(hfac)(2-butyne)	TiN coated Si	scCO_2	225	20	1	[165]
	Cu(hfac)(COD)	Si	scCO_2	225	20	1	[165]
	Cu(hfac)(VTMS)	Si	scCO_2	225	20	1	[165]
	Cu(hfac)(2-methyl-1-hexene-3)	Si	scCO_2	225	20	1	[165]
		Si					
	Cu(tmhd)$_2$		scCO_2	170–200	11	1	[166]
		Ru					
Ni	Ni(cp)$_2$	Si	scCO_2	60	14–18	1	[161]
	Ni(cp)$_2$	Silicon wafer	scCO_2	175–200	19–23	1	[158]
	Ni(cp)$_2$	TaN coated Si wafer	scCO_2	175–200	19–23	1	[158]
	Ni(cp)$_2$		scCO_2	175–200	19–23	1	[158]
		TiN coated Si wafer					
Pd	Pd(hfac)$_2$	Porous alumina	scCO_2	60	13.8	1	[167]
	PdCp(π-C_4H_7)	Porous alumina	scCO_2	60	13.8	1	[167]

Continued

TABLE 7.4 Summary of the studies in the literature on preparation of films by SCD—cont'd

Metal	Metal complex	Substrate	Solvent medium	Deposition temperature (°C)	Deposition pressure (MPa)	Conversion method	Ref.
Ru	$Ru_3(CO)_{12}$	Ta coated Si	$scCO_2$	175–300	20–25	1	[168]
	$Ru(cod)(tmhd)_2$	Ta coated Si	$scCO_2$	175–300	20–25	1	[168]
	$Ru(cod)(tmhd)_2$	Si wafer	$scCO_2$	260	14–17	1	[169]

Conversion methods: 1: In situ chemical conversion, 2: In situ thermal conversion, 3: Ex situ conversion.
Abbreviations: Metal complexes—dba, dibenzylideneacetone; DI6, 1,1,1,3,5,5,6,6,6-nonafluorohexane-2,4-diiminate
hfpd, 1,1,1,5,5,5-hexafluoropentane-2,4-dione; hmte, hexamethyltriethylene; Me, methyl; tegme, tetraethylene glycol dimethyl ether; cp, cyclopentadienyl. VTMS, vinyltrimethylsilane; tmod, 2,2,7-trimethyloctane-3,5-dionato; TTA, 2-thenoyltrifluoroacetonate; DMG, dimethylglioxime; PTQD, phenanthrenequinonedioxime; phen, phenanthroline, OAc, acetate.
Supports—3DHPC, three-dimensional hierarchical porous carbon networks; ABS/PEEA, acrylonitrile–butadiene–styrene/poly(ether-ester-amide); Al_2O_3, Alumina; BTDA–ODA, 3,3*,4,4*-benzophenone tetracarboxylic acid dianhydride–4,4*-oxydianiline; CA-4, CA with pore diameter of 4 nm;; ECNF, electrospun carbon nanofibre; GO, graphene oxide; Nafion, Nafion 112 film; PSDB, poly(styrene-divinylbenzene); SiO_2, silica. SWCNT, Single walled carbon nanotube; FAP, Fluorinated acrylic polymer; CNF, Carbon fiber; CMP, Conjugated microporous polymer; Papam, Polyamidoamine; TEOS, Tetraethyl orthosilicate; RFA, resorcinol-formaldehyde aerogel; sg, Silica gel; sw, Silicon wafer; YSZ, yttrium stabilized zirconia; ZTC, zeolite templated carbon, rGO, reduced graphene oxide, GLC, graphene-like layered carbon.

As previously mentioned, micro- or nanosized conformal films with high aspect ratios can be prepared via SCD. For making films using SCD, there are usually two different heating configurations within the reactors which are hot-wall or cold-wall. Hot-wall reactors are essentially isothermal furnaces and offer relatively easy temperature control. In cold-wall reactors, the substrate is heated directly either by induction or by radiant heating while the rest of the reactor remains cooler. Since many deposition reactions are endothermic, the coating of the reactor walls is thus prevented. These reactors can be operated in batch or continuous flow modes and as a consequence of the higher metal complex concentrations in CO_2, the formed particles further grow on the substrate to form continuous metallic films.

For designing a cold-wall reactor, one must consider the style of the heated stage, types of heating sources, mounting of the substrate on a stage, sealing, and possible natural convection of the SCF. A couple of studies exist in literature on the detailed design of cold-wall reactor. Peng and co-workers reported deposition of metal oxide films by pyrolysis of metal diketonates in a reactor with heating stage although the reactor was not specified as cold-wall reactor [144]. The schematic diagram of the experimental setup with a 110 mL high pressure cold-wall reactor is given in Fig. 7.31. Heating of the stage was carried out by three cartridge heaters (Tempco) which were placed inside the stage. A thermocouple used to control the stage temperature was fixed to the stage.

Zhao and co-workers designed a cold-wall reactor with a different configuration [145]. They desired to reduce natural convection of $scCO_2$ by mounting the heating stage facedown. A ceramic heater was used as a stage inside the

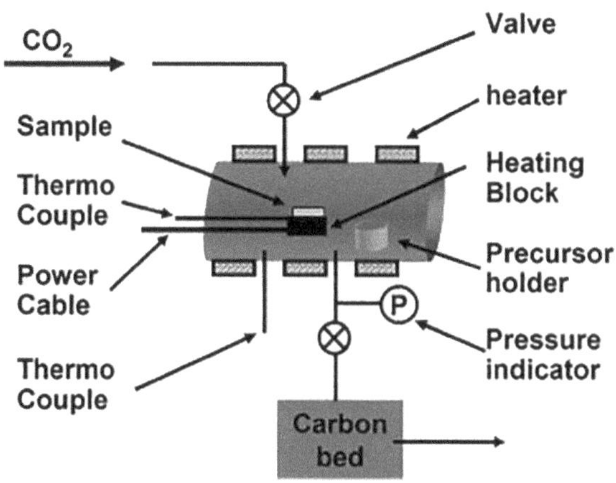

FIG. 7.31 Experimental set up of the metal oxide deposition in $scCO_2$ with cold-wall reactor [144].

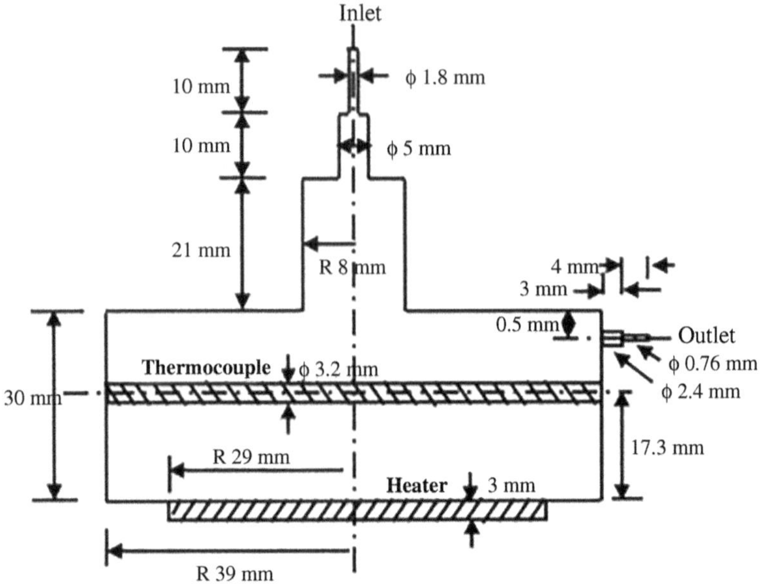

FIG. 7.32 Cold-wall reactor set-up from the cross-section [146].

15 mL stainless-steel chamber reactor. The substrate was fixed to the back side of the heater.

In a cold wall reactor, natural convection occurs due to the density difference of the SCF due to the temperature difference between the fluid adjacent to the heating stage where the temperature is high and the reactor wall. Watkins' group also studied the fluid dynamics in the cold wall reactor to understand the effect of the natural convection [146]. They performed CFD simulations with CFD Fluent software for two different cold-wall reactors similar to each other. The shematic diagram of their cold-wall reactor is given in Fig. 7.32.

A thermocouple was inserted into a thermowell from the side wall and moved to the middle of the vessel to measure the temperature from different locations along the length of the reactor. They examined the temperature differences 0.2 and 0.4 mm above the stage for two different cold-wall reactors. First one had higher clearance than the second one (51 and 12.7 mm). The initial temperature of the stage was 400 K and then it was set to 498 K. Temperature distributions were shown as Figs. 7.33 and 7.34 for the vessel with high clearance and for the vessel with low clearance, respectively.

They observed that in the reactor with a larger clearance, temperature gradient was highly uniform unlike the other reactor. Changes in the temperature

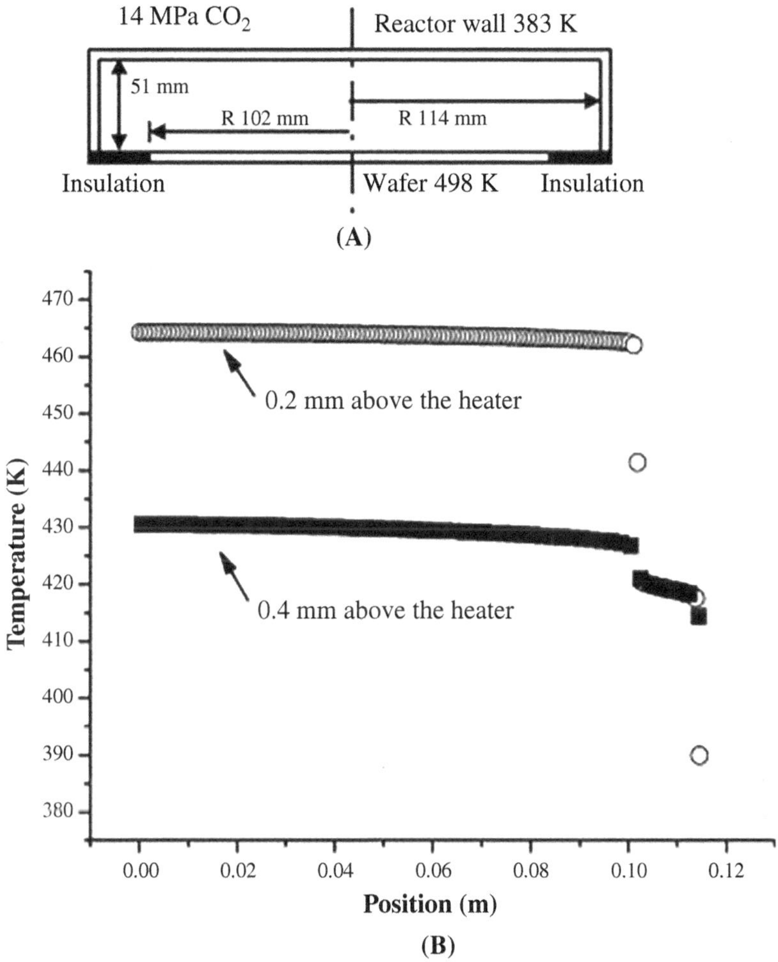

FIG. 7.33 Simulation result for a reactor with high clearance (51 mm): (A) Schematic diagram of the dimensions and boundaries; (B) Temperature distributions of 0.2 mm and 0.4 mm above the heating stage from center ($x = 0$) to wall [146].

gradient was not desirable for deposition of conformal films. Hence, they suggested that a cold-wall vessel with low clearance is not a suitable design for thin film deposition. They also simulated the temperature difference between the surface of the heater and the scCO2 in the vessel. Simulation was performed in a batch mode with CO2 at 18.5 MPa and a stage temperature of 498 K. Wall temperature was measured as 393 K which indicated a 100 K difference between the heater and the wall temperature. They reported that such a gradient was necessary during film deposition to prevent the formation of a coating on

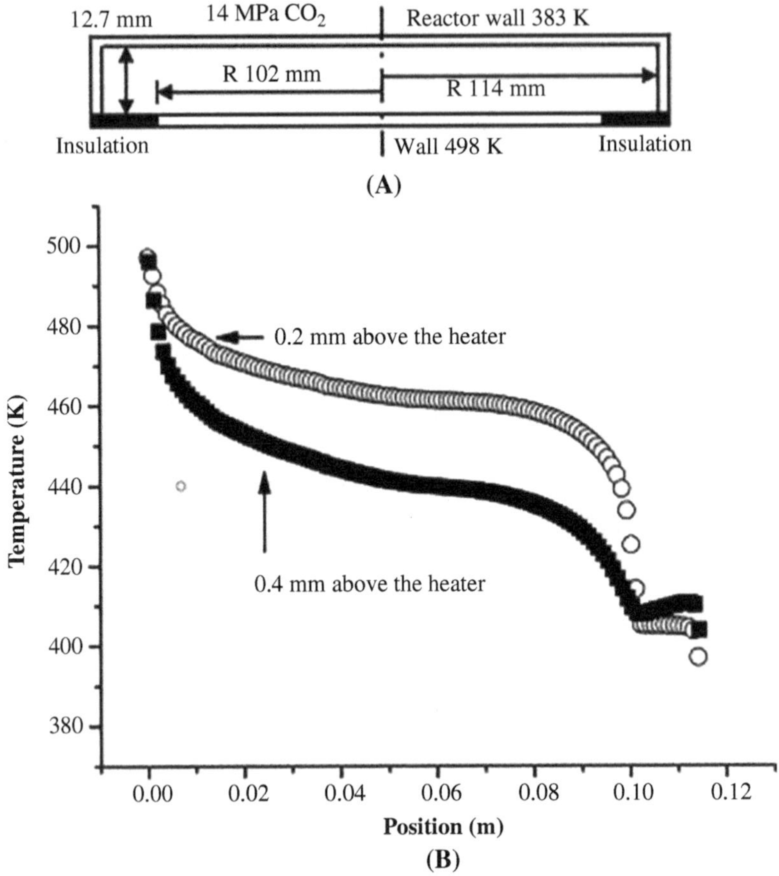

FIG. 7.34 Simulation result performed in a reactor with low clearance (12.7 mm): (A) Schematic diagram of the dimensions and boundaries; (B) Temperature distributions of 0.2 mm and 0.4 mm above the heating stage from center ($x = 0$) to wall [146].

walls of the vessel. Fig. 7.35 shows the temperature change with the distance from the heating pedestal.

Fig. 7.36 illustrates the SEM micrographs of the micro- and nanofilms synthesized using SCD. In Fig. 7.36A and B, the dark gray features are the etched silicon wafers (substrate) and the light gray features are the metals deposited within the trenches of the silicon. Fig. 7.36A represents the films deposited within trenches of Pd-seeded Si wafers from solutions of nickelocene in $scCO_2$ using a continuous-flow cold-wall reactor at 60°C and 18 MPa. The narrow seams in the 100 nm wide by 800 nm deep trenches reveal that film growth occurs uniformly from both sides of the trench. Fig. 7.36B displays the image of a Cu film deposited onto an etched Si wafer

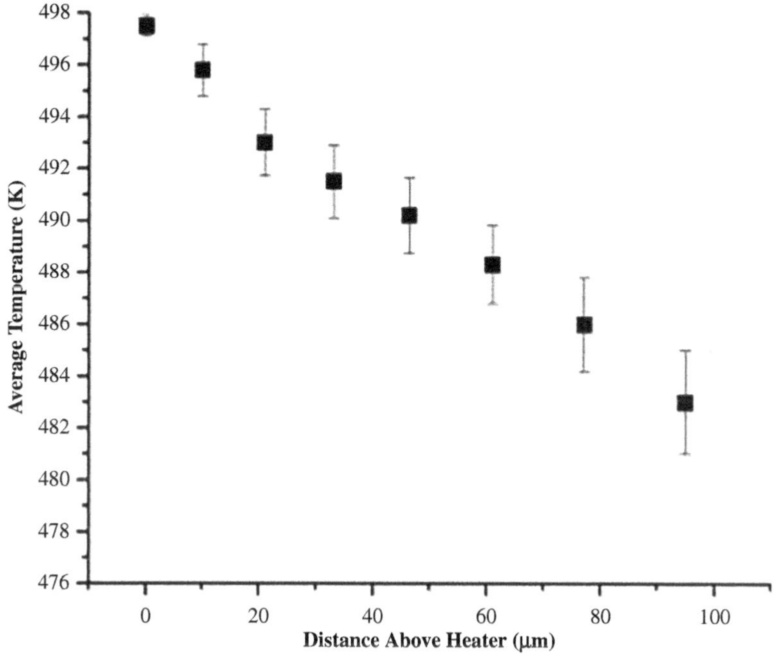

FIG. 7.35 Temperature change with the distance above the heated stage at 18.5 MPa with substrate temperature at 498 K and wall temperature at 393 K [146].

from a CO_2 solution of Cu(hfac)(butyne) at 225°C and 22 MPa. Step coverage is exceptional despite the presence of a number of trenches that are considerably narrower at the entrance (100 nm) than the midpoint. In Fig. 7.36C the SEM image shows the continuous ruthenium film deposition in a cold wall reactor on the planar silica substrates at 14.5 MPa pressure and 260°C of substrate temperature. AFM data within the inset of Fig. 7.36C indicates a mean surface roughness of 1 nm.

Cabañas et al. deposited gold films on different substrates such as metal, ceramic, and polymeric with seeding Ni, Pd, or Pt clusters via H_2 assisted SCF deposition in a cold-wall high pressure reactor [147]. They performed experiments in a batch mode with a low deposition temperature of 125°C. Steps that they followed are same as H_2-assisted Cu film deposition except the deposition time. The deposition was carried out 3–4 h. Thickness of the deposited films varied between 90 and 400 nm for 4 h deposition time depending on the substrate type and precursor loading. Examples of Au films are illustrated in Fig. 7.37 (a) on a bare Si at 125°C and 150 bar and (b) on etched Si wafer with 0.08 wt% loading 125°C and 150 bar.

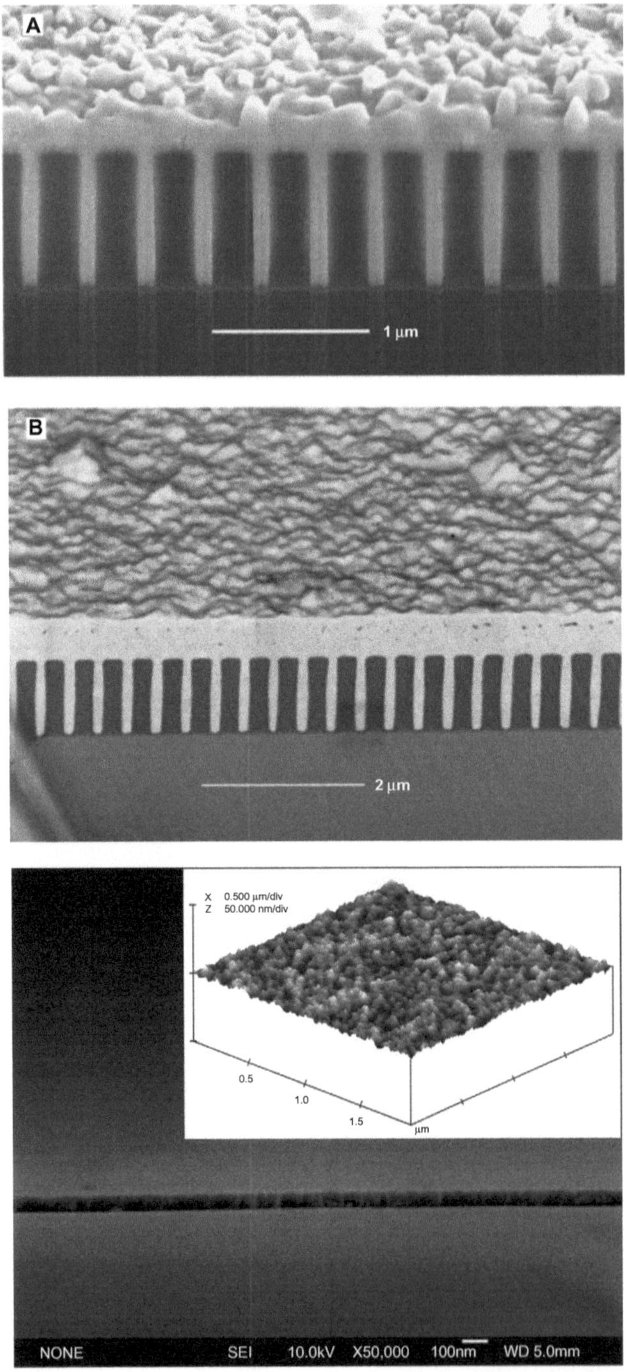

FIG. 7.36 SEM micrographs of micro and nanofilms deposited on silicon using H_2 assisted SCD; (A) Ni film deposited at 60°C and 14 MPa using a hot-wall continuous flow reactor; (B) Cu film deposited at 225°C and 22 MPa using a cold wall reactor; (C) Cross-sectional image of conformal Ru film of 77 nm thickness deposited at 269°C and 14.5 MPa. The inset in (C) is the AFM image of the film showing the mean surface roughness of 1 nm [87].

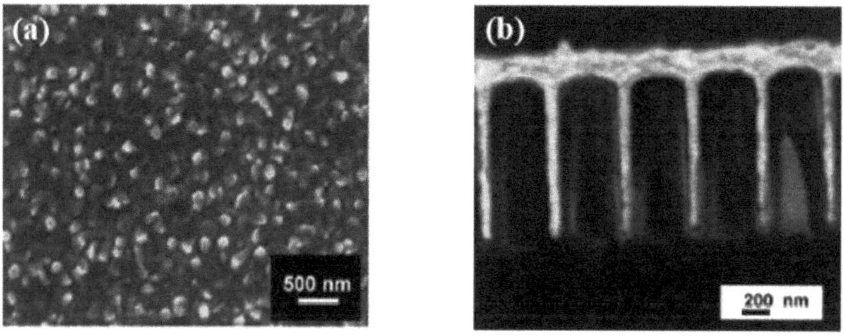

FIG. 7.37 Au film coated on Si and etched Si wafer at 150 bar and 125°C in cold-wall reactor [147].

Hunde and Watkins reported cobalt and nickel film growth on bare SiO_2 and TaN or TiN coated SiO_2 in scCO2 with the assistance of a reducing agent, H_2 using a cold-wall reactor [148]. The experiments were performed in batch mode, so that substrates were mounted on a stage and precursors (~0.2–0.5 wt%) were placed into the reactor as a first step. CO_2 was introduced (at 220–260 bar for Co deposition and 190–230 bar for Ni deposition) to the vessel after the reactor was heated to 70°C. Different from previous methods, H2 was injected after the stage was heated to specified temperature (285–320°C for Co deposition and 175–200°C for Ni deposition). While pedestal heating was running, the heating of the walls was stopped. Due to the unique nature of the cold-wall reactor, the wall temperatures of the reactor remained in a range of 90–120°C and the film growth occurred only on the substrate. Deposition times were typically 30–45 min. They observed that even a little increase in temperature leaded to a much higher growth rate of the Co films. For instance, the thickness was 150 nm at 285°C whereas it was 680 nm at 300°C at a precursor loading 0.23 wt% and hydrogen to precursor ratio of 80:1.

O'Neil and Watkins studied ruthenium film growth process with H_2 assisted SCD in a cold-wall reactor [149]. Film deposition was performed at different pressures (20–25 MPa) with substrate temperatures of 175–300°C on SiO_2 and tantalum coated Si wafer. The experimental procedure was the same as the Cu film deposition. They reported that heating rate of the pedestal changed with the set temperature of the stage. For example, it took 2 min to reach 150°C whereas 300°C was reached in 3.5 min. Films were deposited for 30 min and even shorter times were used at high temperatures. Film thickness varied from 7 to 110 nm depending on the type of the precursor and substrate, precursor loading, substrate temperature, hydrogen amount and deposition time. They also indicated that film thickness did not give enough information to attain reaction kinetics. For instance, at the same temperature, reported thickness for the 5 min. deposited film is almost half of the 30 min. deposited one. Hence, it was more likely the deposition was completed earlier than 30 min.

O'Neil and Watkins also reported metal oxide film on Si and etched Si deposition in $scCO_2$ using cold-wall reactor [150]. Typically, substrate temperature was varied between 250 and 300°C for the deposition of cerium (CeO_2), hafnium (HfO_2), titanium (TiO_2), niobium (Nb_2O_5), tantalum (Ta_2O_5), zirconium (ZrO_2), and bismuth oxides (Bi_2O_3). Few drops of H_2O were introduced in some runs as oxidizing agent. They obtained films having a thickness in the range of 21–263 nm. The batch mode experiments were conducted in 80 mL cold-wall high pressure vessel. First, substrates were fixed to the stage and known amounts of precursors (0.14–2.8 wt%) were loaded. In the use of precursors of Ce, Hf, Zr, and Bi, the vessel was purged with N_2 at 150°C otherwise it was 60°C. After pressurizing the vessel with CO_2 (11–20 MPa) or dried CO_2, it took 1 h to dissolve the precursor. In the experiments conducted with H_2O, certain amount of water (2, 5, 10, or 15 μL) was introduced to CO_2. Film growth took 30 min. Annealing films at 400°C for 1 h was enough to remove carbon, but to get rid of mixed oxide state of the films, calcination was performed at 800°C for 1 h (to obtain TiO_2 and Bi_2O_3) or 6h (for other metal oxide films). They examined the effect of moisture content of the CO_2. For instance, when dried CO_2 was introduced, film thickness of the CeO_2 was 21 nm while the film that was deposited with the as delivered CO_2 was 58 nm. They also reported the effect of the addition of water to dry CO_2 on film thickness. In the deposition of CeO_2, adding 2 μL water (concentration of 0.11 mmol in CO_2) yields 216 nm thick film and introducing 5 μL water (concentration of 0.275 mmol in CO_2) resulted in a 247 nm film thickness. Further addition of water causes reduction in the film thickness. Successfully deposited films are shown in Fig. 7.38.

Zhao and co-workers revealed a study of metal oxide film coatings in a cold-wall reactor via continuous flow SCD [151]. They examined deposition of titanium oxide (TiO_2), bismuth oxide (Bi_2O_3), bismuth titanate (BIT, $Bi_4Ti_3O_{12}$) films on SiO_2 coated Si wafers at various stage temperatures (250–350°C). Titanium di(i-propoxide)bis(2,2,6,6-tetramethyl-3,5-heptanedionato) ($Ti(O\text{-}i\text{-}Pr)_2(tmhd)_2$) and triphenylbismuth (Bi(ph)3) were Ti source and Bi source, respectively. Unlike the previous deposition method, they used different vessels to dissolve precursors. The illustration of the experimental set-up is given in Fig. 7.39.

First, excess amounts of metal complexes were loaded to the vessels. Then, all the vessels were heated to 50°C and filled with CO_2 at 10 MPa. Saturated precursor solutions were diluted with pure CO2 streams at different flow rates before deliverys to the cold-wall reactor. Pedestal heater in the cold-wall vessel was heated to desired temperature. They observed particle generation inside the

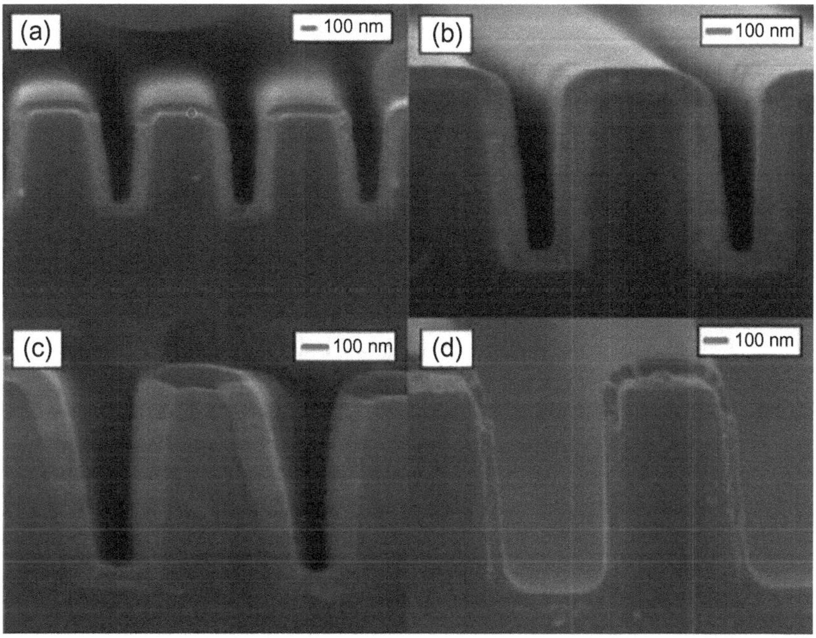

FIG. 7.38 SEM images of the metal oxides films on etched Si wafer (a) HfO_2; (b) CeO_2; (c) ZrO_2; (d) TiO_2 [150].

fluid when the flow rate was higher than 150 mL/min and residence time less than 10 s. For the film growth of Bi_2O_3, Bi complex was dissolved in $scCO_2$ in the presence of oxygen (0.05 mol/L) since Bi precursor does not contain any oxygen. However, this procedure was not applied in the deposition of BIT, because Ti and Bi complexes were delivered together to the cold-wall reactor and Ti complex contains oxygen. Ti:Bi ratio was controlled by adjusting the flow rates from the corresponding vessels. They reported that the suitable deposition temperature was 300°C for TiO_2 and Bi_2O_3 films. Hence, they conducted BIT deposition at 300°C by trying different Ti:Bi ratios to find the optimum value. Fig. 7.40 shows the BIT coatings with different Ti:Bi ratios and TiO_2 film coating.

Tekeli produced WO_3 films on a conductive glass, fluorine doped tin oxide (FTO) by using SCD [152]. A cold-wall reactor was designed for deposition of the films onto heated substrates from supercritical solutions at substantially lower temperatures than the substrate temperature (Fig. 7.41).

Experimental setup for SCD of WO_3 films was given in Fig. 7.42. WO_3 films were deposited using 0.4 wt% tungsten hexacarbonyl ($W(CO)_6$) solutions in supercritical CO_2. Four types of deposition procedures were

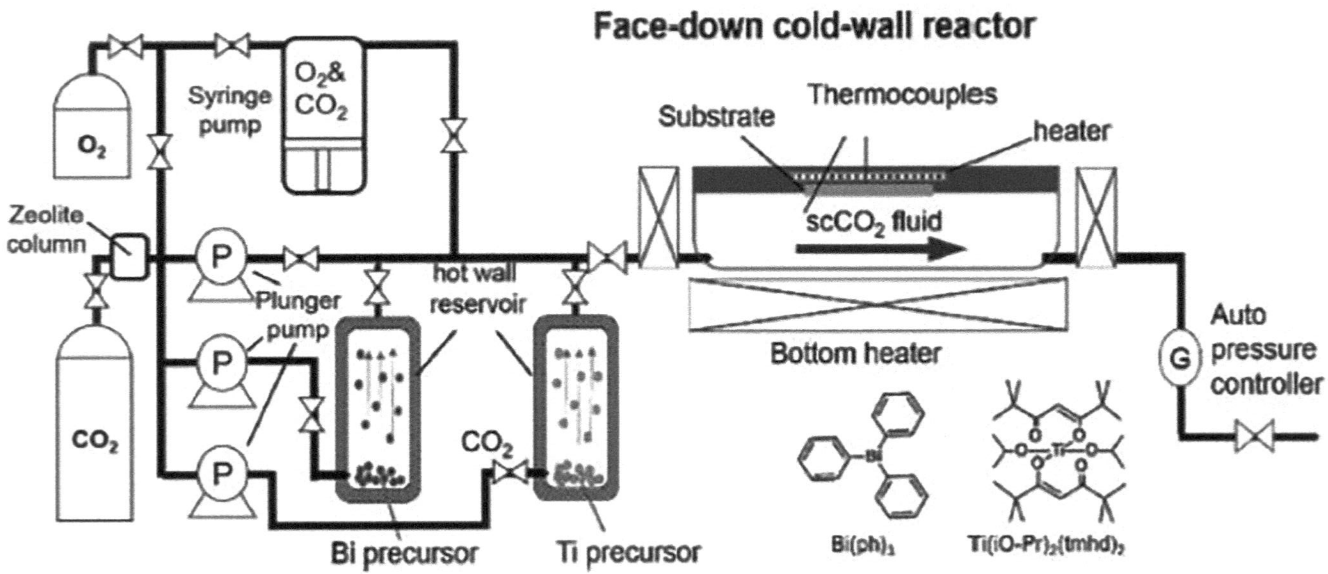

FIG. 7.39 Diagram of the continuous flow supercritical deposition experimental set-up for the deposition of bismuth titanate film [145].

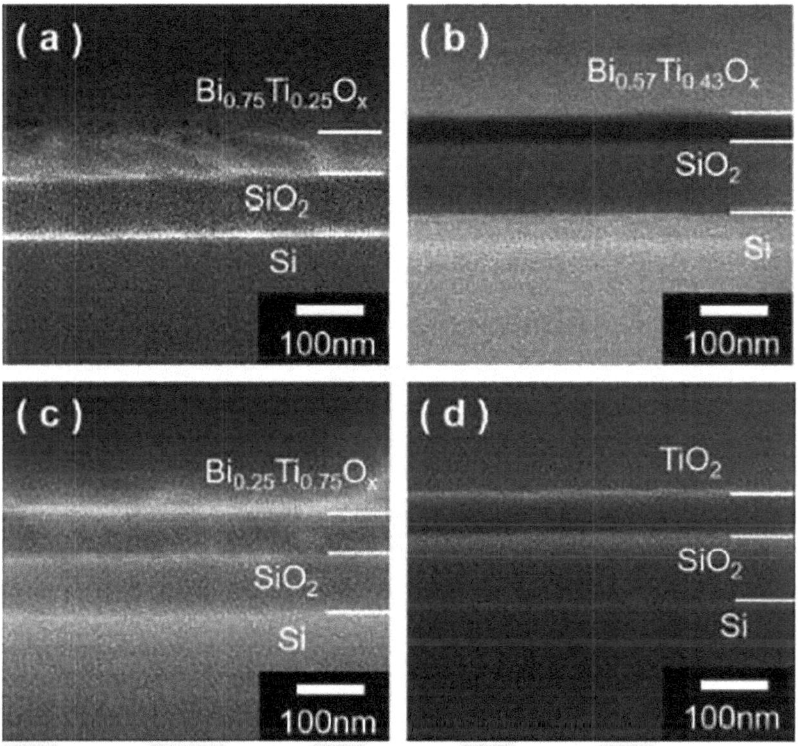

FIG. 7.40 SEM images of deposited bismuth titanate with different compositions films and titanium oxide film in cross-sectional view: (a) $Bi_{0.75}Ti_{0.25}O_x$, (b) $Bi_{0.57}Ti_{0.43}O_x$, (c) $Bi_{0.25}Ti_{0.75}O_x$, and (d) TiO_2 [151].

investigated with no additional O_2 and with the addition of O_2 in different steps of the deposition. Depositions were carried out at 8.3 MPa and a heating stage temperature of 300°C for each method. Coating of the FTO surface was confirmed via scanning electron microscopy (SEM) and energy dispersive X-ray analysis. SEM images of calcined films with additional oxygen and bare FTO are given in Fig. 7.43. SEM results indicated that the produced films were very thin (most probably less than 100 nm). The structure of the WO_3 films was determined as mixtures of monoclinic WO_3 and an unknown phase of tungsten oxide species with a different state of tungsten from the analysis results of the grazing incidence X-ray diffraction spectra.

Studies on the kinetics of particle formation and film formation are very rare. Zong and Watkins investigated the kinetics and the reaction mechanism for the formation of copper films by the hydrogen assisted reduction of

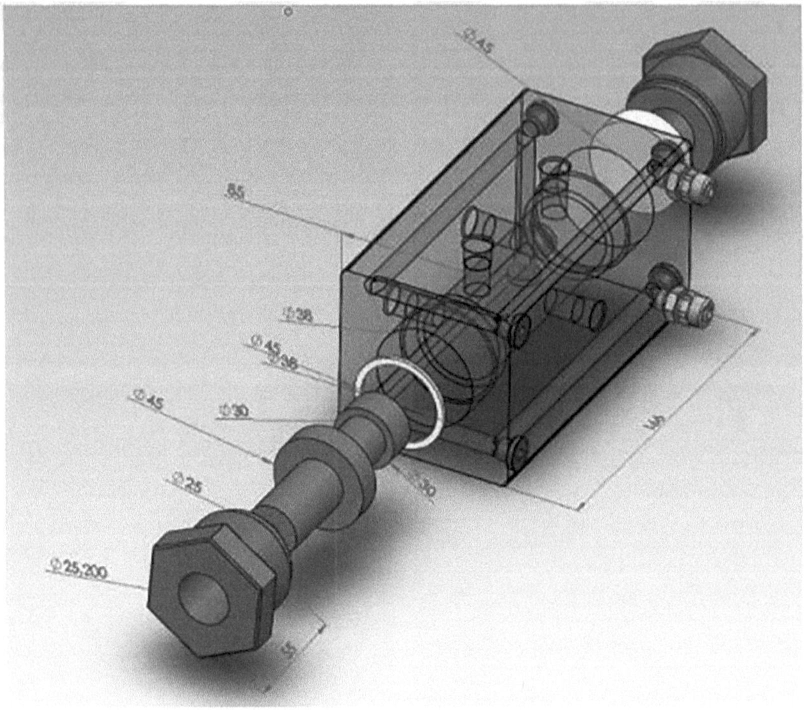

FIG. 7.41 Design of the custom-made cold-wall reactor for WO_3 thin film deposition [152].

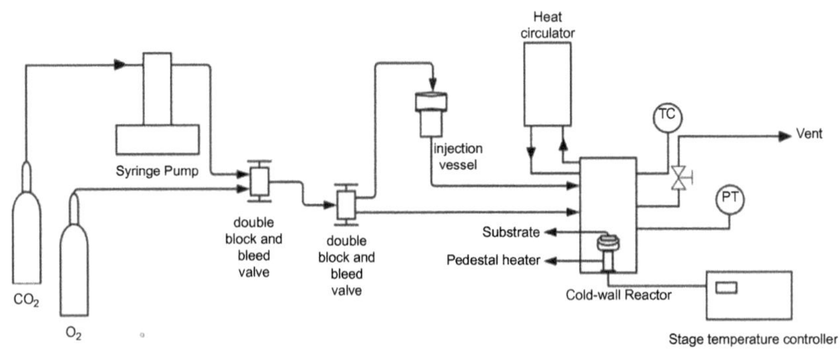

FIG. 7.42 Experimental set-up of SCD with cold-wall reactor [152].

Bis(2,2,7-trimethyloctane-3,5-dionate) Cu(II) (Cu(tmod)$_2$) in scCO_2 [153]. According to their proposed model, the film formation proceeded through the following steps:

1. metal complex and hydrogen diffuse to the surface of the substrate
2. adsorb on the surface

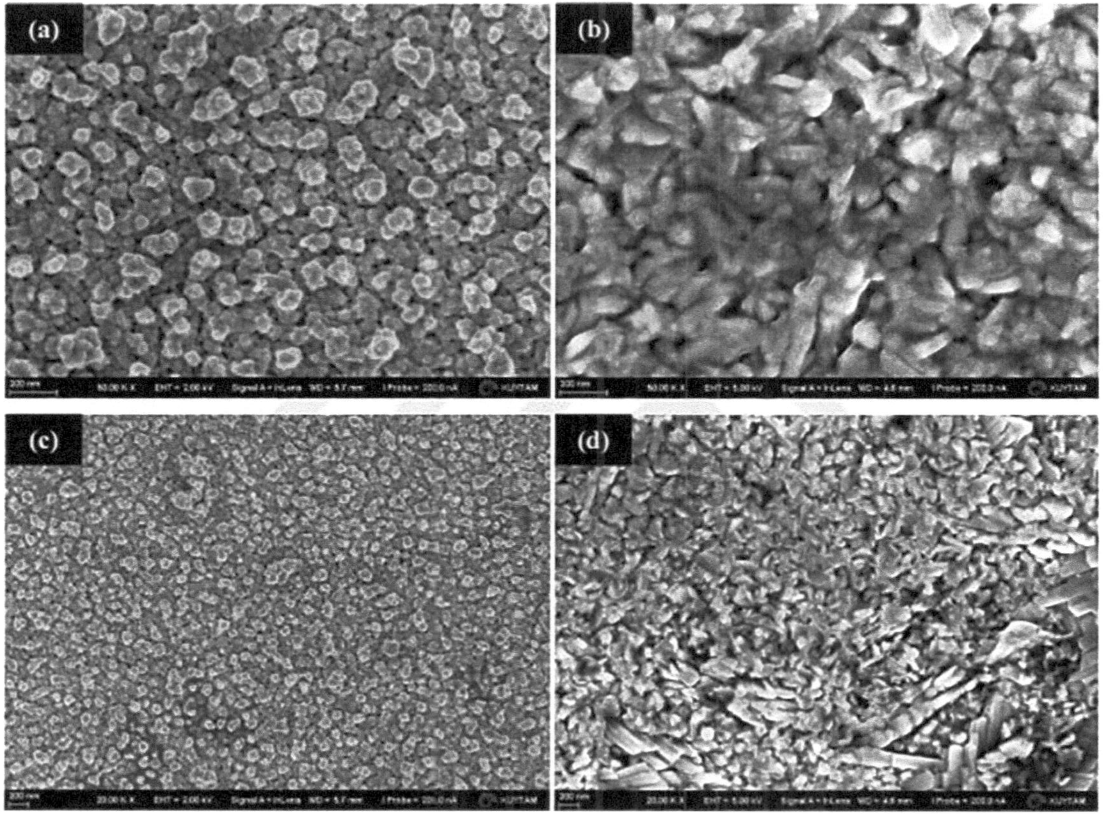

FIG. 7.43 SEM images of calcined and oxygenated WO_3 films (a, c) and bare FTO substrate (b, d) in different magnifications [152].

3. adjacent reactants undergo a surface reaction to liberate copper
4. hydrogenated ligands desorb from the surface and diffuse into the supercritical solution leaving pure metal on the surface forming conformal layers.

The rate expression developed by assuming that the surface reaction between the adsorbed ligand and the hydrogen atom to form a volatile by-product was the rate determining step predicted the effect of metal complex and by-product concentration on reaction rate very well. The model was slightly off in predicting the dependency of the reaction rate on hydrogen concentration and this was attributed to neglecting the reaction of hydrogen with copper oxides in the model.

7.5 Comparison with other methods

There are several ways to synthesize supported nanoparticles, including impregnation, deposition-precipitation, co-precipitation, sonochemical reduction, chemical vapor impregnation, sol-gel, and microemulsion using organic stabilizing agents. The control over either particle dimensions, including the particle size and distribution, or metal concentration in the composites remains the major problems for these methods. For example, the traditional impregnation, deposition-precipitation and co-precipitation methods using liquid solutions as the processing medium could cause not only an agglomeration of particles but also the collapse of the fragile supports such as aerogels due to the high surface tension of the liquid solutions. The conventional sol-gel route could be used to incorporate metal nanoparticles into the structure of various types of porous materials. However, the metal complexes in solution interfere with the polymerization chemistry resulting in materials with undesirable properties. The water-in-oil microemulsions which can be created using surfactants is another medium used for the synthesis of supported metallic nanoparticles. However, surfactants could interfere with the direct attachment of the particle to support by affecting the metal-support interactions and the removal of the surfactant could result in agglomeration of the particles due to high temperatures. Chemical vapor impregnation (CVD) is another promising route to prepare supported metal nanoparticles. In practice, however, it is often limited by the vapor pressure of the metal complex, the processing temperature requirements and mass-transfer-limited kinetics.

Ogel et al. [170] synthesized $Pt/\gamma\text{-}Al_2O_3$ catalyst for CO and NO oxidation reactions with different synthesis methods. SCD, flame spray pyrolysis (FSP), laser ablation in liquid (LAL) methods were selected as advanced synthesis methods. Catalysts synthesized by these methods were compared with conventionally synthesized (incipient wetness impregnation, IWI) catalysts. Two different precursors were used in synthesis via IWI; platinum acetyl acetonate ($Pt(acac)_2$)), and hexochloroplatinic acid ($H_2PtCl_6 \cdot H_2O$). Catalysts were named as PAA and HCP, respectively. All catalysts had a 2 wt% Pt content. Particle size and particle size distribution of the catalysts were compared for fresh catalysts via HAADF-STEM images. HCP had nanoparticles with 1–2 nm diameter, SCD and FSP resulted in nanoparticles with 2–4 nm diameter. PAA had nanoparticles with 1–2 nm diameter with a few larger particles with diameters ranging between 8 and 40 nm. LAL had broader particle size distribution between 2 and 10 nm. All catalysts were hydrothermally aged at 600°C. After aging, average particle size for PAA, HCP and LAL increased from 2.1, 1.1, and 3.4 nm to 20.1, 10.3, and 8.8 nm. Average nanoparticle size in FSP increased from 1.8 to 3.8 nm with a bimodal particle size distribution where a considerable number of small nanoparticles exist along with larger nanoparticles with diameters ranging between 5 and 15 nm. The smallest change in average particle size diameter was obtained by SCD where it increased from 2.4 to 2.8 nm. This was linked to the limited mobility of slightly larger particles (2–4 nm) compared to particles having 1 nm diameter. Catalysts were then hydrothermally aged at 700°C and 800°C. after ageing at 800°C, all catalysts suffered sintering. However, degrees of sintering were different for each catalyst. Average particle size of SCD catalyst increased to 8 nm with nanoparticles as large as 50 nm. However, nanoparticles with diameters smaller than 5 nm were still observable in HAADF-STEM images. HCP catalyst resulted in nanoparticles with 10 nm average diameter. Chlorine traces from the hexachloroplatinic acid precursor were probably still present after ageing and inhibited further particle growth in HCP catalyst. PAA resulted in the highest average particle size, 63.2 nm, with observable nanoparticle sizes ranging between 10 and 140 nm. Average particle size in LAL increased to 42.6 with very broad particle size distribution Average particle size of FSP was 7.7 nm with bimodal size distribution where 1 5 nm nanoparticles exist with nanoparticles with diameters ranging between 10 and 40 nm. Particle size and particle size distribution obtained in fresh catalyst was shown to be an important factor for sintering of the catalysts at higher temperatures. Synthesis of catalysts with slightly larger particles (2–4 nm) with SCD resulted in resistance to sintering at higher temperatures. Deposition with a broader particle size distribution, as in LAL, resulted in massive and rapid sintering of nanoparticles during hydrothermal ageing.

To compare catalyst performances for NO and CO oxidations, T_{20} and T_{50} (temperature at 20% and 50% conversion, respectively) were reported for each ageing condition (Figs. 7.44A and 7.44B).

For the fresh catalysts, CO oxidation performances increased as HCP $<$ LAL $<$ PAA $<$ SCD $<$ FSP, which was in line with previously reported optimum Pt nanoparticle size of 2–3 nm [171] for CO oxidation due to easier reducibility of these particles compared to smaller ones. After ageing at 600°C, performance changed LAL $>$ FSP $\geq$ SCD $\gg$ HCP $>$ PAA. T_{50} increased for all catalysts after ageing at 800°C but the most severe increase was obtained in LAL due to the significant increase in nanoparticle size. Long term stability and CO activity were obtained by FSP, SCD and HCP. Performances of freshly prepared catalysts for NO oxidation were in the following order: LAL $\gg$ FSP $\geq$ SCD $>$ PAA $\gg$ HCP. NO oxidation was previously reported to take place on

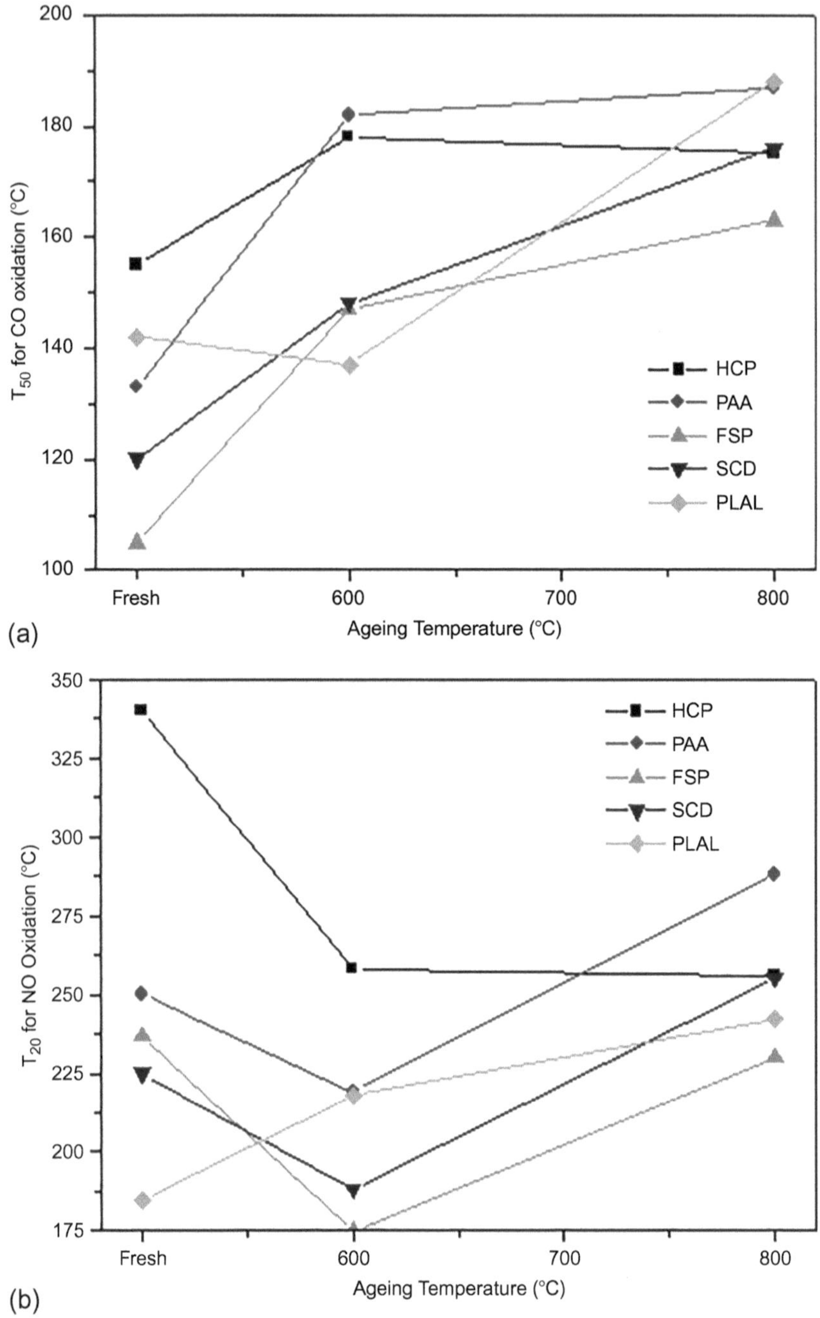

FIG. 7.44 T_{50} for CO oxidation (a) and T_{20} for NO oxidation (b) for fresh and aged Pt/Al_2O_3 prepared by different synthesis methods [170].

reduced surface Pt sites which are more abundant on larger nanoparticles [172]. Here, the catalyst with the highest average nanoparticle size (LAL), performed the best where the catalyst with the smallest average nanoparticle size resulted in the highest T_{20}. Due to the particle size effect, NO oxidation performance of the catalysts was enhanced after ageing at 600°C, except for LAL, with the most enhancements obtained with FSP and SCD. Although the average nanoparticle size did not significantly change after ageing at 600°C for SCD, increase in NO oxidation performance can be related to the occurrence of particles with nanoparticle sizes larger than 5 nm after ageing. Since average particle size and particle size distribution in FSP after ageing at 600°C were very similar to those in SCD, similar NO performances were obtained. However, severe ageing at 800°C resulted in an increase in T_{20} in all samples. In summary, this study illustrated the importance of synthesis method for catalyst preparation and activity. Particle size and particle size distribution of the catalyst are affected by the preparation method. Highly active catalysts for CO and NO oxidation were synthesized by SCD with higher endurance to hydrothermal ageing compared to different synthesis techniques such as IWI and LAL.

References

[1] M. Chatterjee, Y. Ikushima, Y. Hakuta, H. Kawanami, In situ synthesis of gold nanoparticles inside the pores of MCM-48 in supercritical carbon dioxide and its catalytic application, Adv. Synth. Catal. 348 (2006) 1580–1590, https://doi.org/10.1002/adsc.200606137.

[2] J.-W. Wu, C.-H. Wang, Y.-C. Wang, J.-K. Chang, Ionic-liquid-enhanced glucose sensing ability of non-enzymatic Au/graphene electrodes fabricated using supercritical CO_2 fluid, Biosens. Bioelectron. 46 (2013) 30–36, https://doi.org/10.1016/j.bios.2013.02.021.

[3] S. Muller, M. Turk, Production of supported gold and gold-silver nanoparticles by supercritical fluid reactive deposition: effect of substrate properties, J. Supercrit. Fluids 96 (2015) 287–297, https://doi.org/10.1016/j.supflu.2014.09.035.

[4] A.O. Rybaltovskii, Y.S. Zavorotnyi, N.V. Minaev, M.I. Samoilovich, P.S. Timashev, M.Y. Tsvetkov, V.N. Bagratashvili, Synthesis of silver nanocomposites by SCF impregnation of matrices of synthetic opal and Vycor glass by the Ag(hfac)COD precursor, Russ. J. Phys. Chem. B 3 (2009) 1106–1112, https://doi.org/10.1134/S1990793109070100.

[5] A. Niu, Y. Han, J. Wu, N. Yu, Q. Xu, Synthesis of one-dimensional carbon nanomaterials wrapped by silver nanoparticles and their antibacterial behavior, J. Phys. Chem. C 114 (2010) 12728–12735, https://doi.org/10.1021/jp104720w.

[6] O. Sawai, Y. Oshima, Deposition of silver nano-particles on activated carbon using supercritical water, J. Supercrit. Fluids 47 (2008) 240–246, https://doi.org/10.1016/j.supflu.2008.08.007.

[7] M.J. Casciato, G. Levitin, D.W. Hess, M.A. Grover, Controlling the properties of silver nanoparticles deposited on surfaces using supercritical carbon dioxide for surface-enhanced Raman spectroscopy, J. Nanopart. Res. 14 (2012) 836, https://doi.org/10.1007/s11051-012-0836-2.

[8] Q. Xu, Y. Wang, A. Wang, J. Yin, Systematical study of depositing nanoparticles and nanowires in mesoporous silica using supercritical carbon dioxide and co-solvents: morphology control, thermodynamics and kinetics of adsorption, Nanotechnology 23 (2012) 285602, https://doi.org/10.1088/0957-4484/23/28/285602.

[9] C. Lai, P. Kolla, Y. Zhao, H. Fong, A.L. Smirnova, Lignin-derived electrospun carbon nanofiber mats with supercritically deposited Ag nanoparticles for oxygen reduction reaction in alkaline fuel cells, Electrochim. Acta 130 (2014) 431–438, https://doi.org/10.1016/j.electacta.2014.03.006.

[10] E. Garrido, C. Aymonier, L. Roiban, O. Ersen, C. Labrugère, P. Gaillard, M. Lamirand-Majimel, Noble metals supported on carbon nanotubes using supercritical fluids for the preparation of composite materials: a look at the interface, J. Supercrit. Fluids 101 (2015) 110–116, https://doi.org/10.1016/j.supflu.2015.03.009.

[11] V. Arakcheev, V. Bagratashvili, A. Bekin, D. Khmelenin, N. Minaev, V. Morozov, A. Rybaltovsky, Laser assisted synthesis of silver nanoparticles in silica aerogel by supercritical deposition technique, J. Supercrit. Fluids 127 (2017) 176–181, https://doi.org/10.1016/j.supflu.2017.03.028.

[12] S.G. Aspromonte, Á. Sastre, A.V. Boix, M.J. Cocero, E. Alonso, Cobalt oxide nanoparticles on mesoporous MCM-41 and Al-MCM-41 by supercritical CO_2 deposition, Microporous Mesoporous Mater. 148 (2012) 53–61, https://doi.org/10.1016/j.micromeso.2011.07.014.

[13] S. Aspromonte, A. Sastre, A. Boix, M.J. Cocero, E. Alonso, Optimization and modelling of the supercritical CO_2 deposition of CoxOy nanoparticles in MCM41, J. Supercrit. Fluids 110 (2016) 47–55, https://doi.org/10.1016/j.supflu.2015.11.026.

[14] S.G. Aspromonte, M.D. Mizrahi, E. Alonso, J.M. Ramallo-López, A.V. Boix, Co/MCM41 catalyst in the COProx reaction prepared by supercritical CO_2 reactive deposition, Microporous Mesoporous Mater. 239 (2017) 147–157, https://doi.org/10.1016/j.micromeso.2016.10.006.

[15] S. Ünsal, M.B. Yağcı, S.E. Bozbağ, B. Deljoo, M. Aindow, C. Erkey, Supercritical deposition coupled with ammonia treatment: a new route to co-promoted N-doped carbon aerogels with high oxygen reduction reaction activity, Energy Technol. 7 (2019) 1900450, https://doi.org/10.1002/ente.201900450.

[16] S. Marre, A. Erriguible, A. Perdomo, F. Cansell, F. Marias, C. Aymonier, Kinetically controlled formation of supported nanoparticles in low temperature supercritical media for the development of advanced nanostructured materials, J. Phys. Chem. C 113 (2009) 5096–5104, https://doi.org/10.1021/jp809533n.

[17] J.-Z. Yin, Q.-Q. Xu, A.-Q. Wang, Controlled growth of copper nanoparticles and nanorods in the channels of SBA-15 by supercritical fluid deposition, Chem. Eng. Commun. 197 (2009) 627–632, https://doi.org/10.1080/00986440903249676.

[18] S.E. Bozbag, S.O. Kostenko, M.A. Kurykin, V.N. Khrustalev, A.R. Khokhlov, L. Zhang, M. Aindow, C. Erkey, Aerogel-copper nanocomposites prepared using the adsorption of a polyfluorinated complex from supercritical CO_2, J. Nanopart. Res. 14 (2012) 973, https://doi.org/10.1007/s11051-012-0973-7.

[19] Y. Meng, F. Su, Y. Chen, Synthesis of nano-Cu/graphene oxide composites by supercritical CO_2-assisted deposition as a novel material for reducing friction and wear, Chem. Eng. J. 281 (2015) 11–19, https://doi.org/10.1016/j.cej.2015.06.073.

[20] G.-Y. Qiao, Q.-Q. Xu, J.-Z. Yin, A.-Q. Wang, G. Xu, Synthesis of CuO/SBA-15 nanocomposite in ternary system of CO_2, inorganic salt and co-solvent, J. Supercrit. Fluids 128 (2017) 18–25, https://doi.org/10.1016/j.supflu.2017.05.004.

[21] L. Wang, L. Zhuo, C. Zhang, F. Zhao, Supercritical carbon dioxide assisted deposition of Fe_3O_4 nanoparticles on hierarchical porous carbon and their lithium-storage performance, Chem. A Eur. J. 20 (2014) 4308–4315, https://doi.org/10.1002/chem.201304700.

[22] M. Hosseinpour, S.J. Ahmadi, S. Fatemi, Successive co-operation of supercritical water and silica-supported iron oxide nanoparticles in upgrading of heavy petroleum residue:

suppression of coke deposition over catalyst, J. Supercrit. Fluids 100 (2015) 70–78, https://doi.org/10.1016/j.supflu.2015.02.013.

[23] E. Chamorro, M.J. Tenorio, L. Calvo, M.J. Torralvo, R. Sáez-Puche, A. Cabañas, One-step sustainable preparation of superparamagnetic iron oxide nanoparticles supported on mesoporous SiO_2, J. Supercrit. Fluids 159 (2020) 104775, https://doi.org/10.1016/j.supflu.2020.104775.

[24] M.J. Casciato, J.T. Vastola, J.C. Lu, D.W. Hess, M.A. Grover, Initial experimental design methodology incorporating expert conjecture, prior data, and engineering models for deposition of iridium nanoparticles in supercritical carbon dioxide, Ind. Eng. Chem. Res. 52 (2013) 9645–9653, https://doi.org/10.1021/ie400996g.

[25] Q. Peng, J.C. Spagnola, G.N. Parsons, Self-catalyzed hydrogenolysis of nickelocene: functional metal coating of three-dimensional nanosystems at low temperature, J. Electrochem. Soc. 155 (2008) D580, https://doi.org/10.1149/1.2946723.

[26] S.E. Bozbag, L.C. Zhang, M. Aindow, C. Erkey, Carbon aerogel supported nickel nanoparticles and nanorods using supercritical deposition, J. Supercrit. Fluids 66 (2012) 265–273,-https://doi.org/10.1016/j.supflu.2012.02.027.

[27] J. Morère, S. Royuela, G. Asensio, P. Palomino, E. Enciso, C. Pando, A. Cabañas, Deposition of Ni nanoparticles onto porous supports using supercritical CO_2: effect of the precursor and reduction methodology, Philos. Trans. R. Soc. A Math. Phys. Eng. Sci. 373 (2015) 20150014, https://doi.org/10.1098/rsta.2015.0014.

[28] J. Morère, E. Sánchez-Miguel, M.J. Tenorio, C. Pando, A. Cabañas, Supercritical fluid preparation of Pt, Ru and Ni/graphene nanocomposites and their application as selective catalysts in the partial hydrogenation of limonene, J. Supercrit. Fluids 120 (2017) 7–17, https://doi.org/10.1016/j.supflu.2016.10.007.

[29] J. García, C. Jiménez, F. Martínez, R. Camarillo, J. Rincón, Electrochemical reduction of CO_2 using Pb catalysts synthesized in supercritical medium, J. Catal. 367 (2018) 72–80, https://doi.org/10.1016/j.jcat.2018.08.017.

[30] X.-R. Ye, Y. Lin, C. Wang, M.H. Engelhard, Y. Wang, C.M. Wai, Supercritical fluid synthesis and characterization of catalytic metal nanoparticles on carbon nanotubes, J. Mater. Chem. 14 (2004) 908–913, https://doi.org/10.1039/B308124A.

[31] X.R. Ye, Y. Lin, C.M. Wai, Decorating catalytic palladium nanoparticles on carbon nanotubes in supercritical carbon dioxide, Chem. Commun. (2003) 642–643, https://doi.org/10.1039/B211350C.

[32] K.S. Morley, P. Licence, P.C. Marr, J.R. Hyde, P.D. Brown, R. Mokaya, Y. Xia, S.M. Howdle, Supercritical fluids: a route to palladium-aerogel nanocomposites, J. Mater. Chem. 14 (2004) 1212–1217, https://doi.org/10.1039/b311065f.

[33] T. Hasell, C.D. Wood, R. Clowes, J.T.A. Jones, Y.Z. Khimyak, D.J. Adams, A.I. Cooper, Palladium nanoparticle incorporation in conjugated microporous polymers by supercritical fluid processing, Chem. Mater. 22 (2010) 557–564, https://doi.org/10.1021/cm9030446.

[34] M.J. Tenorio, M.J. Torralvo, E. Enciso, C. Pando, J.A.R. Renuncio, A. Cabañas, Supercritical CO_2 as a reaction and impregnation medium in the synthesis of Pd-SiO_2 aerogel inverse opals, J. Supercrit. Fluids 49 (2009) 369–376, https://doi.org/10.1016/j.supflu.2009.03.011.

[35] B. Cangül, L.C. Zhang, M. Aindow, C. Erkey, Preparation of carbon black supported Pd, Pt and Pd-Pt nanoparticles using supercritical CO_2 deposition, J. Supercrit. Fluids 50 (2009) 82–90, https://doi.org/10.1016/j.supflu.2009.04.001.

[36] S.R. Puniredd, B.C.C. Nguan, M.P. Srinivasan, Covalent molecular assembly in supercritical carbon dioxide: formation of nanoparticles in immobilized dendrimers within a porous silica gel matrix, J. Colloid Interface Sci. 333 (2009) 679–683, https://doi.org/10.1016/j.jcis.2008.10.084.

[37] J. Morère, M.J. Tenorio, M.J. Torralvo, C. Pando, J.A.R. Renuncio, A. Cabañas, Deposition of Pd into mesoporous silica SBA-15 using supercritical carbon dioxide, J. Supercrit. Fluids 56 (2011) 213–222, https://doi.org/10.1016/j.supflu.2010.12.012.

[38] M. Majimel, S. Marre, E. Garrido, C. Aymonier, Supercritical fluid chemical deposition as an alternative process to CVD for the surface modification of materials, Chem. Vap. Depos. 17 (2011) 342–352, https://doi.org/10.1002/cvde.201106921.

[39] C.-Y. Chen, J.-K. Chang, W.-T. Tsai, C.-H. Hung, Uniform dispersion of Pd nanoparticles on carbon nanostructures using a supercritical fluid deposition technique and their catalytic performance towards hydrogen spillover, J. Mater. Chem. 21 (2011) 19063–19068, https://doi.org/10.1039/C1JM13528G.

[40] M.J. Tenorio, C. Pando, J.A.R. Renuncio, J.G. Stevens, R.A. Bourne, M. Poliakoff, A. Cabañas, Adsorption of Pd(hfac)2 on mesoporous silica SBA-15 using supercritical CO_2 and its role in the performance of Pd-SiO_2 catalyst, J. Supercrit. Fluids 69 (2012) 21–28, https://doi.org/10.1016/j.supflu.2012.05.003.

[41] E. Masika, R.A. Bourne, T.W. Chamberlain, R. Mokaya, Supercritical CO_2 mediated incorporation of pd onto templated carbons: a route to optimizing the Pd particle size and hydrogen uptake density, ACS Appl. Mater. Interfaces 5 (2013) 5639–5647, https://doi.org/10.1021/am401622w.

[42] A.J. Hunt, V.L. Budarin, J.W. Comerford, H.L. Parker, V.K. Lazarov, S.W. Breeden, D.J. Macquarrie, J.H. Clark, Deposition of palladium nanoparticles in SBA-15 templated silica using supercritical carbon dioxide, Mater. Lett. 116 (2014) 408–411, https://doi.org/10.1016/j.matlet.2013.11.075.

[43] J. Zhao, Z. Liu, H. Li, W. Hu, C. Zhao, P. Zhao, D. Shi, Development of a highly active electrocatalyst via ultrafine Pd nanoparticles dispersed on pristine graphene, Langmuir 31 (2015) 2576–2583, https://doi.org/10.1021/la5049425.

[44] K. Matsuyama, S. Tanaka, T. Kato, T. Okuyama, H. Muto, R. Miyamoto, H. Bai, Supercritical fluid-assisted immobilization of Pd nanoparticles in the mesopores of hierarchical porous SiO_2 for catalytic applications, J. Supercrit. Fluids 130 (2017) 140–146, https://doi.org/10.1016/j.supflu.2017.07.032.

[45] F. Ulusal, B. Darendeli, E. Erünal, A. Egitmen, B. Guzel, Supercritical carbondioxide deposition of γ-alumina supported Pd nanocatalysts with new precursors and using on Suzuki-Miyaura coupling reactions, J. Supercrit. Fluids 127 (2017) 111–120, https://doi.org/10.1016/j.supflu.2017.03.024.

[46] R. Camarillo, S. Tostón, F. Martínez, C. Jiménez, J. Rincón, Enhancing the photocatalytic reduction of CO_2 through engineering of catalysts with high pressure technology: Pd/TiO_2 photocatalysts, J. Supercrit. Fluids 123 (2017) 18–27, https://doi.org/10.1016/j.supflu.2016.12.010.

[47] F. Ulusal, B. Güzel, Deposition of palladium by the hydrogen assisted on SBA-15 with a new precursor using supercritical carbon dioxide, J. Supercrit. Fluids 133 (2018) 233–238, https://doi.org/10.1016/j.supflu.2017.10.023.

[48] E. Erünal, F. Ulusal, M.Y. Aslan, B. Güzel, D. Üner, Enhancement of hydrogen storage capacity of multi-walled carbon nanotubes with palladium doping prepared through supercritical CO_2 deposition method, Int. J. Hydrogen Energy 43 (2018) 10755–10764, https://doi.org/10.1016/j.ijhydene.2017.12.058.

[49] E. Erünal, F. Ulusal, Thermal effects on hydrogen storage capacity of Pd/MWCNT nanoparticles deposited at moderate pressure and temperature supercritical carbon dioxide conditions, J. Supercrit. Fluids 157 (2020) 104706, https://doi.org/10.1016/j.supflu.2019.104706.

[50] C.D. Saquing, D. Kang, M. Aindow, C. Erkey, Investigation of the supercritical deposition of platinum nanoparticles into carbon aerogels, Microporous Mesoporous Mater. 80 (2005) 11–23, https://doi.org/10.1016/j.micromeso.2004.11.019.

[51] C.D. Saquing, T.-T. Cheng, M. Aindow, C. Erkey, Preparation of platinum/carbon aerogel nanocomposites using a supercritical deposition method, J. Phys. Chem. B 108 (2004) 7716–7722, https://doi.org/10.1021/jp049535v.

[52] Y. Zhang, D. Kang, C. Saquing, M. Aindow, C. Erkey, Supported platinum nanoparticles by supercritical deposition, Ind. Eng. Chem. Res. 44 (2005) 4161–4164, https://doi.org/10.1021/ie050345w.

[53] A. Bayrakçeken, A. Smirnova, U. Kitkamthorn, M. Aindow, L. Türker, İ. Eroğlu, C. Erkey, Vulcan-supported Pt electrocatalysts for PEMFCs prepared using supercritical carbon dioxide deposition, Chem. Eng. Commun. 196 (2008) 194–203, https://doi.org/10.1080/00986440802290110.

[54] S.-Y. Ang, D.A. Walsh, Highly stable platinum electrocatalysts for oxygen reduction formed using supercritical fluid impregnation, J. Power Sources 195 (2010) 2557–2563, https://doi.org/10.1016/j.jpowsour.2009.10.097.

[55] C.S. Lin, F.L.-Y. Lam, X. Hu, W.Y. Tam, K.M. Ng, Synthesis of metallic nanostructures using chemical fluid deposition, J. Phys. Chem. C 112 (2008) 10068–10072, https://doi.org/10.1021/jp801845q.

[56] S. Haji, Y. Zhang, C. Erkey, Atmospheric hydrodesulfurization of diesel fuel using Pt/Al_2O_3 catalysts prepared by supercritical deposition for fuel cell applications, Appl. Catal. A Gen. 374 (2010) 1–10, https://doi.org/10.1016/j.apcata.2009.10.044.

[57] S.E. Bozbag, N.S. Yasar, L.C. Zhang, M. Aindow, C. Erkey, Adsorption of Pt(cod)me2 onto organic aerogels from supercritical solutions for the synthesis of supported platinum nanoparticles, J. Supercrit. Fluids 56 (2011) 105–113, https://doi.org/10.1016/j.supflu.2010.10.045.

[58] G.I. Garrido, F.C. Patcas, G. Upper, M. Türk, S. Yilmaz, B. Kraushaar-Czarnetzki, Supercritical deposition of Pt on SnO_2-coated Al_2O_3 foams: phase behaviour and catalytic performance, Appl. Catal. A Gen. 338 (2008) 58–65, https://doi.org/10.1016/j.apcata.2007.12.019.

[59] S.R. Puniredd, C.M. Yin, Y.S. Hooi, P.S. Lee, M.P. Srinivasan, Dendrimer-encapsulated Pt nanoparticles in supercritical medium: synthesis, characterization, and application to device fabrication, J. Colloid Interface Sci. 332 (2009) 505–510, https://doi.org/10.1016/j.jcis.2009.01.011.

[60] E.E. Said-Galiyev, A.Y. Nikolaev, E.E. Levin, E.K. Lavrentyeva, M.O. Gallyamov, S.N. Polyakov, G.A. Tsirlina, O.A. Petrii, A.R. Khokhlov, Structural and electrocatalytic features of Pt/C catalysts fabricated in supercritical carbon dioxide, J. Solid State Electrochem. 15 (2011) 623–633, https://doi.org/10.1007/s10008-010-1169-7.

[61] B. Kaya, S. Irmak, A. Hesenov, O. Erbatur, C. Erkey, Effect of support materials on supported platinum catalyst prepared using a supercritical fluid deposition technique and their catalytic performance for hydrogen-rich gas production from lignocellulosic biomass, Bioresour. Technol. 123 (2012) 723–726, https://doi.org/10.1016/j.biortech.2012.07.098.

[62] E.E. Said-Galiev, A.Y. Vasil'Kov, A.Y. Nikolaev, A.I. Lisitsyn, A.V. Naumkin, I.O. Volkov, S.S. Abramchuk, O.L. Lependina, A.R. Khokhlov, E.V. Shtykova, K.A. Dembo, C. Erkey, Structure of mono- and bimetallic heterogeneous catalysts based on noble metals obtained by means of fluid technology and metal-vapor synthesis, Russ. J. Phys. Chem. A 86 (2012) 1602–1608, https://doi.org/10.1134/S0036024412100214.

[63] M. Watanabe, T. Akimoto, E. Kondoh, Effects of surface modification of carbon nanotube on platinum nanoparticle deposition using supercritical carbon dioxide fluid, Phys. Status Solidi. 209 (2012) 2514–2520, https://doi.org/10.1002/pssa.201228305.

[64] J. Zhao, L. Zhang, T. Chen, H. Yu, L. Zhang, H. Xue, H. Hu, Supercritical carbon-dioxide-assisted deposition of Pt nanoparticles on graphene sheets and their application as an electrocatalyst for direct methanol fuel cells, J. Phys. Chem. C 116 (2012) 21374–21381, https://doi.org/10.1021/jp3061976.

[65] J. Zhao, H. Yu, Z. Liu, M. Ji, L. Zhang, G. Sun, Supercritical deposition route of preparing Pt/graphene composites and their catalytic performance toward methanol electrooxidation, J. Phys. Chem. C 118 (2014) 1182–1190, https://doi.org/10.1021/jp402620p.

[66] S. Wolff, M. Crone, T. Muller, M. Enders, S. Bräse, M. Türk, Preparation of supported Pt nanoparticles by supercritical fluid reactive deposition: Influence of precursor, substrate and pressure on product properties, J. Supercrit. Fluids 95 (2014) 588–596, https://doi.org/10.1016/j.supflu.2014.08.034.

[67] S.E. Bozbağ, T. Gümüşoğlu, S. Yılmaztürk, C.J. Ayala, M. Aindow, H. Deligöz, C. Erkey, Electrochemical performance of fuel cell catalysts prepared by supercritical deposition: effect of different precursor conversion routes, J. Supercrit. Fluids 97 (2015) 154–164, https://doi.org/10.1016/j.supflu.2014.08.014.

[68] K. Matsuyama, M. Motomura, T. Kato, T. Okuyama, H. Muto, Catalytically active Pt nanoparticles immobilized inside the pores of metal organic framework using supercritical CO_2 solutions, Microporous Mesoporous Mater. 225 (2016) 26–32, https://doi.org/10.1016/j.micromeso.2015.12.005.

[69] E. Daş, S. Alkan Gürsel, L. Işikel Şanli, A. Bayrakçeken Yurtcan, Comparison of two different catalyst preparation methods for graphene nanoplatelets supported platinum catalysts, Int. J. Hydrogen Energy 41 (2016) 9755–9761, https://doi.org/10.1016/j.ijhydene.2016.01.111.

[70] J.W. Deal, P. Le, C.B. Corey, K. More, C.W. West, Water-gas shift reaction on alumina-supported Pt-CeOx catalysts prepared by supercritical fluid deposition, J. Supercrit. Fluids 119 (2017) 113–121, https://doi.org/10.1016/j.supflu.2016.08.016.

[71] C. Jiménez, J. García, R. Camarillo, F. Martínez, J. Rincón, Electrochemical CO_2 reduction to fuels using Pt/CNT catalysts synthesized in supercritical medium, Energy Fuels 31 (2017) 3038–3046, https://doi.org/10.1021/acs.energyfuels.6b03017.

[72] E. Daş, S. Alkan Gürsel, L. Işıkel Şanlı, A. Bayrakçeken Yurtcan, Thermodynamically controlled Pt deposition over graphene nanoplatelets: effect of Pt loading on PEM fuel cell performance, Int. J. Hydrogen Energy 42 (2017) 19246–19256, https://doi.org/10.1016/j.ijhydene.2017.06.108.

[73] Ş.B. Barım, A. Bayrakçeken, S.E. Bozbağ, L. Zhang, R. Kızılel, M. Aindow, C. Erkey, Control of average particle size of carbon aerogel supported platinum nanoparticles by supercritical deposition, Microporous Mesoporous Mater. 245 (2017) 94–103, https://doi.org/10.1016/j.micromeso.2017.01.037.

[74] F.E. Sarac Oztuna, S.B. Barim, S.E. Bozbag, H. Yu, M. Aindow, U. Unal, C. Erkey, Graphene aerogel supported Pt electrocatalysts for oxygen reduction reaction by supercritical deposition, Electrochim. Acta 250 (2017) 174–184, https://doi.org/10.1016/j.electacta.2017.08.067.

[75] A. Bayrakçeken Yurtcan, E. Daş, Chemically synthesized reduced graphene oxide-carbon black based hybrid catalysts for PEM fuel cells, Int. J. Hydrogen Energy 43 (2018) 18691–18701, https://doi.org/10.1016/j.ijhydene.2018.06.186.

[76] Z. Sun, Z. Liu, B. Han, Y. Wang, J. Du, Z. Xie, G. Han, Fabrication of ruthenium-carbon nanotube nanocomposites in supercritical water, Adv. Mater. 17 (2005) 928–932, https://doi.org/10.1002/adma.200400839.

[77] Y. Zhang, D. Kang, M. Aindow, C. Erkey, Preparation and characterization of ruthenium/carbon aerogel nanocomposites via a supercritical fluid route, J. Phys. Chem. B 109 (2005) 2617–2624, https://doi.org/10.1021/jp0467595.

[78] Y. Zhang, C. Erkey, Preparation of supported metallic nanoparticles using supercritical fluids: a review, J. Supercrit. Fluids 38 (2006) 252–267, https://doi.org/10.1016/j.supflu.2006.03.021.
[79] C.H. Yen, H.W. Lin, C.-S. Tan, Hydrogenation of bisphenol A—using a mesoporous silica based nano ruthenium catalyst Ru/MCM-41 and water as the solvent, Catal. Today 174 (2011) 121–126, https://doi.org/10.1016/j.cattod.2011.01.050.
[80] Y. Zhao, J. Zhang, J. Song, J. Li, J. Liu, T. Wu, P. Zhang, B. Han, Ru nanoparticles immobilized on metal-organic framework nanorods by supercritical CO_2-methanol solution: highly efficient catalyst, Green Chem. 13 (2011) 2078–2082, https://doi.org/10.1039/C1GC15340D.
[81] Y. Zhang, H. Jiang, Y. Wang, M. Zhang, Synthesis of highly dispersed ruthenium nanoparticles supported on activated carbon via supercritical fluid deposition, Ind. Eng. Chem. Res. 53 (2014) 6380–6387, https://doi.org/10.1021/ie404155u.
[82] Z. Chen, J. Lu, Y. Ai, Y. Ji, T. Adschiri, L. Wan, Ruthenium/graphene-like layered carbon composite as an efficient hydrogen evolution reaction electrocatalyst, ACS Appl. Mater. Interfaces 8 (2016) 35132–35137, https://doi.org/10.1021/acsami.6b09331.
[83] B.I. Lee, D. Bae, J.K. Kang, H. Kim, S.H. Byeon, Synthesis of SBA-15 supported Rh nanoparticles with high loading density and its catalytic hydrogenation of phenol in supercritical carbon dioxide, Bull. Korean Chem. Soc. 30 (2009) 1701–1702.
[84] I. Ushiki, N. Takahashi, T. Shimizu, Y. Sato, M. Ota, R.L. Smith, H. Inomata, Adsorption equilibria of rhodium acetylacetonate with MCM-41, MSU-H, and HMS silica substrates in supercritical carbon dioxide for preparing catalytic mesoporous materials, J. Supercrit. Fluids 120 (2017) 240–248, https://doi.org/10.1016/j.supflu.2016.05.032.
[85] Q.H. Do, C. Zeng, C. Zhang, B. Wang, J. Zheng, Supercritical fluid deposition of vanadium oxide on multi-walled carbon nanotube Buckypaper for supercapacitor electrode application, Nanotechnology 22 (2011) 365402, https://doi.org/10.1088/0957-4484/22/36/365402.
[86] V. Aggarwal, L.F. Reichenbach, M. Enders, T. Muller, S. Wolff, M. Crone, M. Türk, S. Bräse, Influence of perfluorinated end groups on the SFRD of [Pt(cod)Me(CnF2*n*+1)] onto porous Al_2O_3 in CO_2 under reductive conditions, Chem. A Eur. J. 19 (2013) 12794–12799, https://doi.org/10.1002/chem.201301191.
[87] C. Erkey, Supercritical Fluids and Organometallic Compounds, first ed., Elsevier BV, Amsterdam, 2011.
[88] J.J. Watkins, T.J. McCarthy, Polymer/metal nanocomposite synthesis in supercritical CO_2, Chem. Mater. 7 (1995) 1991–1994, https://doi.org/10.1021/cm00059a001.
[89] B. Wong, S. Yoda, S.M. Howdle, The preparation of gold nanoparticle composites using supercritical carbon dioxide, J. Supercrit. Fluids 42 (2007) 282–287, https://doi.org/10.1016/j.supflu.2007.03.005.
[90] K.S. Morley, P.C. Marr, P.B. Webb, A.R. Berry, F.J. Allison, G. Moldovan, P.D. Brown, S.M. Howdle, Clean preparation of nanoparticulate metals in porous supports: a supercritical route, J. Mater. Chem. 12 (2002) 1898–1905, https://doi.org/10.1039/b111111f.
[91] R.K. Boggess, L.T. Taylor, D.M. Stoakley, A.K.S. Clair, Highly reflective polyimide films created by supercritical fluid infusion of a silver additive, J. Appl. Polym. Sci. 64 (1997) 1309–1317, https://doi.org/10.1002/(SICI)1097-4628(19970516)64:7<1309::AID-APP10>3.0.CO;2-S.
[92] S.D. Gittard, D. Hojo, G.K. Hyde, G. Scarel, R.J. Narayan, G.N. Parsons, Antifungal textiles formed using silver deposition in supercritical carbon dioxide, J. Mater. Eng. Perform. 19 (2010) 368–373, https://doi.org/10.1007/s11665-009-9514-7.
[93] V.N. Bagratashvili, A.O. Rybaltovsky, N.V. Minaev, P.S. Timashev, V.V. Firsov, V.I. Yusupov, Laser-induced atomic assembling of periodic layered nanostructures of silver

nanoparticles in fluoro-polymer film matrix, Laser Phys. Lett. 7 (2010) 401–404, https://doi.org/10.1002/lapl.200910159.

[94] T. Hasell, L. Lagonigro, A.C. Peacock, S. Yoda, P.D. Brown, P.J.A. Sazio, S.M. Howdle, Silver nanoparticle impregnated polycarbonate substrates for surface enhanced Raman spectroscopy, Adv. Funct. Mater. 18 (2008) 1265–1271, https://doi.org/10.1002/adfm.200701429.

[95] T. Nakanishi, K. Hata, S. Katoh, T. Moriyoshi, Infusion of silver nanoparticles into transparent plastics using supercritical carbon dioxide, Kobunshi Ronbunshu 62 (2005) 183–189, https://doi.org/10.1295/koron.62.183.

[96] O. Sawai, Y. Oshima, Mechanism of silver nano-particles formation on α-alumina using supercritical water, J. Mater. Sci. 43 (2008) 2293–2299, https://doi.org/10.1007/s10853-007-2031-x.

[97] E. Said-Galiyev, L. Nikitin, R. Vinokur, M. Gallyamov, M. Kurykin, O. Petrova, B. Lokshin, I. Volkov, A. Khokhlov, New chelate complexes of copper and iron: synthesis and impregnation into a polymer matrix from solution in supercritical carbon dioxide, Ind. Eng. Chem. Res. 39 (2000) 4891–4896, https://doi.org/10.1021/ie000251g.

[98] E.-B. Kim, J.-T. Kim, S.-Y. Kim, C.-S. Ju, Synthesis of copper-poly tetrafluoro-ethylene composites by supercritical impregnation process, Korean J. Chem. Eng. 28 (2011) 440–444, https://doi.org/10.1007/s11814-010-0374-5.

[99] N. Martinez, K. Hisada, I. Tabata, K. Hirogaki, S. Yonezawa, T. Hori, The effectiveness of thermal treatment for development of conductive metalized aramid fiber using supercritical fluid carbon dioxide—fiber-metal adhesive strength improvement, J. Supercrit. Fluids 56 (2011) 322–329, https://doi.org/10.1016/j.supflu.2010.10.011.

[100] R. Jiang, Y. Zhang, S. Swier, X. Wei, C. Erkey, H.R. Kunz, J.M. Fenton, Preparation via supercritical fluid route of Pd-impregnated Nafion membranes which exhibit reduced methanol crossover for DMFC, Electrochem. Solid-State Lett. 8 (2005) A611, https://doi.org/10.1149/1.2050527.

[101] S. Yoda, A. Hasegawa, H. Suda, Y. Uchimaru, K. Haraya, T. Tsuji, K. Otake, Preparation of a platinum and palladium/polyimide nanocomposite film as a precursor of metal-doped carbon molecular sieve membrane via supercritical impregnation, Chem. Mater. 16 (2004) 2363–2368, https://doi.org/10.1021/cm0349250.

[102] J. Fu, M. Wang, S. Wang, X. Wang, H. Wang, L. Hu, Q. Xu, Supercritical carbon dioxide-assisted preparation of palladium nanoparticles on cyclotriphosphazene-containing polymer nanospheres, Appl. Surf. Sci. 257 (2011) 7129–7133, https://doi.org/10.1016/j.apsusc.2011.03.069.

[103] S. Tengsuwan, M. Ohshima, Environmentally benign electroless nickel plating using supercritical carbon-dioxide on hydrophilically modified acrylonitrile-butadiene-styrene, Appl. Surf. Sci. 311 (2014) 189–200, https://doi.org/10.1016/j.apsusc.2014.05.040.

[104] O.P. Parenago, P.S. Timashev, E.A. Karakhanov, A.L. Maximov, A.E. Lazhko, A.V. Zolotukhina, V.N. Bagratashvili, Obtaining of highly-active catalysts of unsaturated compounds hydrogenation by using supercritical carbon dioxide, J. Supercrit. Fluids 140 (2018) 387–393, https://doi.org/10.1016/j.supflu.2018.07.010.

[105] M. Ge, S.E. Bozbag, C.J. Ayala, M. Aindow, C. Erkey, Synthesis of Ru/PDMS nanocomposites via supercritical deposition, Mater. Chem. Phys. 180 (2016) 1–4, https://doi.org/10.1016/j.matchemphys.2016.05.058.

[106] M.R. Mauricio, F.C.G. Manso, M.H. Kunita, D.S. Velasco, A.C. Bento, E.C. Muniz, G.M. de Carvalho, A.F. Rubira, Synthesis and characterization of ZnO/PET composite using supercritical carbon dioxide impregnation technology, Compos. Part A Appl. Sci. Manuf. 42 (2011) 757–761, https://doi.org/10.1016/j.compositesa.2011.03.002.

[107] Y. Zhang, C. Erkey, Preparation of platinum-nafion-carbon black nanocomposites via a supercritical fluid route as electrocatalysts for proton exchange membrane fuel cells, Ind. Eng. Chem. Res. 44 (2005) 5312–5317, https://doi.org/10.1021/ie049177i.
[108] A. Gruger, A. Régis, T. Schmatko, P. Colomban, Nanostructure of Nafion membranes at different states of hydration: an IR and Raman study, Vib. Spectrosc. 26 (2001) 215–225, https://doi.org/10.1016/S0924-2031(01)00116-3.
[109] D. Kim, J. Sauk, J. Byun, K.S. Lee, H. Kim, Palladium composite membranes using supercritical CO_2 impregnation method for direct methanol fuel cells, Solid State Ionics 178 (2007) 865–870, https://doi.org/10.1016/j.ssi.2007.02.034.
[110] H. Gunes, Y. Özbakir, S.B. Barim, H. Yousefzadeh, S.E. Bozbag, C. Erkey, A remarkable class of nanocomposites: aerogel supported bimetallic nanoparticles, Front. Mater. 7 (2020) 18, https://doi.org/10.3389/fmats.2020.00018.
[111] J.E. Mondloch, E. Bayram, R.G. Finke, A review of the kinetics and mechanisms of formation of supported-nanoparticle heterogeneous catalysts, J. Mol. Catal. A Chem. 355 (2012) 1–38, https://doi.org/10.1016/J.MOLCATA.2011.11.011.
[112] R. Ferrando, J. Jellinek, R.L. Johnston, Nanoalloys: from theory to applications of alloy clusters and nanoparticles, Chem. Rev. 108 (2008) 845–910, https://doi.org/10.1021/cr040090g.
[113] O.A. Oviedo, E.P.M. Leiva, in: M.M. Mariscal, O.A. Oviedo, E.P.M. Leiva (Eds.), Thermodynamic Modeling of Metallic Nanoclusters, Springer, New York, NY, 2013, pp. 305–350, https://doi.org/10.1007/978-1-4614-3643-0_10.
[114] B.N. Wanjala, J. Luo, R. Loukrakpam, B. Fang, D. Mott, P.N. Njoki, M. Engelhard, H.R. Naslund, J.K. Wu, L. Wang, O. Malis, C.-J. Zhong, Nanoscale alloying, phase-segregation, and core-shell evolution of gold-platinum nanoparticles and their electrocatalytic effect on oxygen reduction reaction, Chem. Mater. 22 (2010) 4282–4294, https://doi.org/10.1021/cm101109e.
[115] M.S. Nashner, A.I. Frenkel, D.L. Adler, J.R. Shapley, R.G. Nuzzo, Structural characterization of carbon-supported platinum-ruthenium nanoparticles from the molecular cluster precursor PtRu5C(CO)16, J. Am. Chem. Soc. (1997), https://doi.org/10.1021/JA971039F.
[116] M. Sankar, N. Dimitratos, P.J. Miedziak, P.P. Wells, C.J. Kiely, G.J. Hutchings, Designing bimetallic catalysts for a green and sustainable future, Chem. Soc. Rev. 41 (2012) 8099–8139, https://doi.org/10.1039/c2cs35296f.
[117] J.R. Kitchin, J.K. Nørskov, M.A. Barteau, J.G. Chen, Role of strain and ligand effects in the modification of the electronic and chemical properties of bimetallic surfaces, Phys. Rev. Lett. 93 (2004) 156801, https://doi.org/10.1103/PhysRevLett.93.156801.
[118] J.G. Bosco, J.P. Humbert, M.P. Chen, Design of Heterogeneous Catalysts: New Approaches based on Synthesis, Characterization and Modeling, Wiley-VCH, 2009, https://doi.org/10.1002/9783527625321.
[119] A. Winkler, D. Ferri, R. Hauert, Influence of aging effects on the conversion efficiency of automotive exhaust gas catalysts, Catal. Today 155 (2010) 140–146, https://doi.org/10.1016/J.CATTOD.2008.11.021.
[120] M.H. Wiebenga, C.H. Kim, S.J. Schmieg, S.H. Oh, D.B. Brown, D.H. Kim, J.-H. Lee, C.H.F. Peden, Deactivation mechanisms of Pt/Pd-based diesel oxidation catalysts, Catal. Today 184 (2012) 197–204, https://doi.org/10.1016/J.CATTOD.2011.11.014.
[121] R. Ghosh Chaudhuri, S. Paria, Core/shell nanoparticles: classes, properties, synthesis mechanisms, characterization, and applications, Chem. Rev. 112 (2012) 2373–2433, https://doi.org/10.1021/cr100449n.
[122] J. Greeley, J.K. Nørskov, A general scheme for the estimation of oxygen binding energies on binary transition metal surface alloys, Surf. Sci. 592 (2005) 104–111, https://doi.org/10.1016/J.SUSC.2005.07.018.

[123] S. Koh, P. Strasser, Electrocatalysis on bimetallic surfaces: modifying catalytic reactivity for oxygen reduction by voltammetric surface dealloying, J. Am. Chem. Soc. 129 (2007) 12624–12625, https://doi.org/10.1021/ja0742784.

[124] V.R. Stamenkovic, B. Fowler, B.S. Mun, G. Wang, P.N. Ross, C.A. Lucas, N.M. Markovic, Improved oxygen reduction activity on Pt3Ni(111) via increased surface site availability, Science (80-) 315 (2007) 493–497, https://doi.org/10.1126/science.1135941.

[125] P. Mani, R. Srivastava, P. Strasser, Dealloyed Pt-Cu core-shell nanoparticle electrocatalysts for use in PEM fuel cell cathodes, J. Phys. Chem. C 112 (2008) 2770–2778, https://doi.org/10.1021/jp0776412.

[126] C.F.A. Negre, C.G. Sánchez, Optical properties of metal nanoclusters from an atomistic point of view, in: M.M. Mariscal, O.A. Oviedo, E.P.M. Leiva (Eds.), Met. Clust. Nanoalloys From Model. to Appl, Springer, New York, NY, 2013, pp. 105–157, https://doi.org/10.1007/978-1-4614-3643-0_4.

[127] C.H. Yen, K. Shimizu, Y.Y. Lin, F. Bailey, I.F. Cheng, C.M. Wai, Chemical fluid deposition of Pt-based bimetallic nanoparticles on multiwalled carbon nanotubes for direct methanol fuel cell application, Energy Fuels 21 (2007) 2268–2271, https://doi.org/10.1021/ef0606409.

[128] A.J. Byrd, K.K. Pant, R.B. Gupta, Hydrogen production from glucose using Ru/Al_2O_3 catalyst in supercritical water, Ind. Eng. Chem. Res. 46 (2007) 3574–3579, https://doi.org/10.1021/ie070241g.

[129] M. Watanabe, T. Akimoto, E. Kondoh, Synthesis of platinum-ruthenium alloy nanoparticles on carbon using supercritical fluid deposition, ECS J. Solid State Sci. Technol. 2 (2012) M9–M12, https://doi.org/10.1149/2.020301jss.

[130] A. Bayrakçeken, B. Cangül, L.C. Zhang, M. Aindow, C. Erkey, PtPd/BP2000 electrocatalysts prepared by sequential supercritical carbon dioxide deposition, Int. J. Hydrogen Energy 35 (2010) 11669–11680, https://doi.org/10.1016/j.ijhydene.2010.08.059.

[131] S.R. Puniredd, S. Weiyi, M.P. Srinivasan, Pd-Pt and Fe-Ni nanoparticles formed by covalent molecular assembly in supercritical carbon dioxide, J. Colloid Interface Sci. 320 (2008) 333–340, https://doi.org/10.1016/j.jcis.2007.11.034.

[132] C. Aymonier, A. Denis, Y. Roig, M. Iturbe, E. Sellier, S. Marre, F. Cansell, J.L. Bobet, Supported metal NPs on magnesium using SCFs for hydrogen storage: interface and interphase characterization, J. Supercrit. Fluids 53 (2010) 102–107, https://doi.org/10.1016/j.supflu.2010.02.012.

[133] M. Alibouri, S.M. Ghoreishi, H.R. Aghabozorg, Effect of supercritical deposition synthesis on dibenzothiophene hydrodesulfurization over $NiMo/Al_2O_3$ nanocatalyst, AIChE J. 55 (2009) 2665–2674, https://doi.org/10.1002/aic.11867.

[134] P. Kolla, A. Smirnova, Methanol oxidation and oxygen reduction activity of PtIrCo-alloy nanocatalysts supercritically deposited within 3D carbon aerogel matrix, Electrochim. Acta 182 (2015) 20–30, https://doi.org/10.1016/j.electacta.2015.09.033.

[135] Y. Zhou, C.H. Yen, Y.H. Hu, C. Wang, X. Cheng, C.M. Wai, J. Yang, Y. Lin, Making ultra-fine and highly-dispersive multimetallic nanoparticles in three-dimensional graphene with supercritical fluid as excellent electrocatalyst for oxygen reduction reaction, J. Mater. Chem. A 4 (2016) 18628–18638, https://doi.org/10.1039/C6TA08508C.

[136] W. Yu, Y.-P. Hsu, C.-S. Tan, Synthesis of rhodium-platinum bimetallic catalysts supported on SBA-15 by chemical fluid deposition for the hydrogenation of terephthalic acid in water, Appl. Catal. B Environ. 196 (2016) 185–192, https://doi.org/10.1016/j.apcatb.2016.05.034.

[137] Y. Zhou, X. Cheng, C.H. Yen, C.M. Wai, C. Wang, J. Yang, Y. Lin, Synthesis of an excellent electrocatalyst for oxygen reduction reaction with supercritical fluid: graphene cellular

monolith with ultrafine and highly dispersive multimetallic nanoparticles, J. Power Sources 347 (2017) 69–78, https://doi.org/10.1016/j.jpowsour.2017.02.044.

[138] S.B. Barim, S.E. Bozbag, H. Yu, R. Kızılel, M. Aindow, C. Erkey, Mesoporous carbon aerogel supported PtCu bimetallic nanoparticles via supercritical deposition and their dealloying and electrocatalytic behaviour, Catal. Today 310 (2018) 166–175, https://doi.org/10.1016/j.cattod.2017.09.023.

[139] D. Hu, H. Xu, Z. Yi, Z. Chen, C. Ye, Z. Wu, H.F. Garces, K. Yan, Green CO_2-assisted synthesis of mono- and bimetallic Pd/Pt nanoparticles on porous carbon fabricated from sorghum for highly selective hydrogenation of furfural, ACS Sustain. Chem. Eng. 7 (2019) 15339–15345, https://doi.org/10.1021/acssuschemeng.9b02665.

[140] E. Daş, S. Alkan Gürsel, A. Bayrakçeken Yurtcan, Pt-alloy decorated graphene as an efficient electrocatalyst for PEM fuel cell reactions, J. Supercrit. Fluids 165 (2020) 104962, https://doi.org/10.1016/j.supflu.2020.104962.

[141] S.B. Barim, S.E. Bozbag, B. Deljoo, M. Aindow, C. Erkey, Highly active carbon supported PtCu electrocatalysts for PEMFCs by in situ supercritical deposition coupled with electrochemical dealloying, Fuel Cells 20 (2020) 285–299, https://doi.org/10.1002/fuce.201900087.

[142] K. Matsuyama, N. Tomiyasu, K. Inoue, R. Yokomizo, T. Okuyama, H. Muto, Catalytically active PdRu and CuRu bimetallic nanoparticle formation in the mesoporous SiO_2 by supercritical CO_2-assisted immobilization, J. Supercrit. Fluids 160 (2020) 104818, https://doi.org/10.1016/j.supflu.2020.104818.

[143] Y. Lin, X. Cui, C.H. Yen, C.M. Wai, PtRu/carbon nanotube nanocomposite synthesized in supercritical fluid: a novel electrocatalyst for direct methanol fuel cells, Langmuir. 21 (2005) 11474–11479, https://doi.org/10.1021/la051272o.

[144] Q. Peng, D. Hojo, K.J. Park, G.N. Parsons, Low temperature metal oxide film deposition and reaction kinetics in supercritical carbon dioxide, Thin Solid Films 516 (2008) 4997–5003, https://doi.org/10.1016/j.tsf.2007.10.057.

[145] Y. Zhao, K. Jung, T. Momose, Y. Shimogaki, Smooth and conformal TiO_2 thin-film formation using supercritical fluid deposition, ECS J. Solid State Sci. Technol. 2 (2013) N191–N195, https://doi.org/10.1149/2.003311jss.

[146] X. Shan, D.P. Schmidt, J.J. Watkins, Study of natural convection in supercritical CO_2 cold wall reactors: simulations and experiments, J. Supercrit. Fluids 40 (2007) 84–92, https://doi.org/10.1016/j.supflu.2006.03.023.

[147] A. Cabañas, D.P. Long, J.J. Watkins, Deposition of gold films and nanostructures from supercritical carbon dioxide, Chem. Mater. 16 (2004) 2028–2033, https://doi.org/10.1021/cm034739u.

[148] E.T. Hunde, J.J. Watkins, Reactive deposition of cobalt and nickel films from their metallocenes in supercritical carbon dioxide solution, Chem. Mater. 16 (2004) 498–503, https://doi.org/10.1021/cm034433n.

[149] A. O'Neil, J.J. Watkins, Reactive deposition of conformal ruthenium films from supercritical carbon dioxide, Chem. Mater. 18 (2006) 5652–5658, https://doi.org/10.1021/cm060142d.

[150] A. O'Neil, J.J. Watkins, Reactive deposition of conformal metal oxide films from supercritical carbon dioxide, Chem. Mater. 19 (2007) 5460–5466, https://doi.org/10.1021/cm070288s.

[151] Y. Zhao, K. Jung, Y. Shimoyama, Y. Shimogaki, T. Momose, Conformal bismuth titanate formation using supercritical fluid deposition, ECS J. Solid State Sci. Technol. 6 (2017) P483–P488, https://doi.org/10.1149/2.0011708jss.

[152] G. Tekeli, Photoelectrocatalytic Water Splitting on Tungsten (VI) Oxide Films Prepared by Supercritical Deposition in a Cold-Wall Reactor and by a Hydrothermal Technique, Koç University, 2018.

[153] Y. Zong, J.J. Watkins, Deposition of copper by the H2-assisted reduction of Cu(tmod)2 in supercritical carbon dioxide: kinetics and reaction mechanism, Chem. Mater. 17 (2005) 560–565, https://doi.org/10.1021/cm048665d.

[154] N. Nazem, L.T. Taylor, A.F. Rubira, Metallized poly(etherether ketone) films achieved by supercritical fluid impregnation of a silver precursor followed by thermal curing, J. Supercrit. Fluids 23 (2002) 43–57, https://doi.org/10.1016/S0896-8446(01)00130-9.

[155] C. Park, J.-W. Kim, G. Sauti, J. Ho Kang, C.S. Lovell, L.J. Gibbons, S.E. Lowther, P.T. Lillehei, J.S. Harrison, N. Nazem, L.T. Taylor, Metallized nanotube polymer composites via supercritical fluid impregnation, J. Polym. Sci. B Polym. Phys. 50 (2012) 394–402, https://doi.org/10.1002/polb.23015.

[156] A. Cabañas, D.P. Long, J.J. Watkins, Deposition of gold films and nanostructures from supercritical carbon dioxide, Chem. Mater. 16 (2004) 2028–2033, https://doi.org/10.1021/cm034739u.

[157] D. Mesguich, C. Aymonier, J.-M. Bassat, F. Mauvy, E. You, J.J. Watkins, Low-temperature deposition of undoped ceria thin films in $scCO_2$ as improved interlayers for IT-SOFC, Chem. Mater. 23 (2011) 5323–5330, https://doi.org/10.1021/cm2012846.

[158] E.T. Hunde, J.J. Watkins, Reactive deposition of cobalt and nickel films from their metallocenes in supercritical carbon dioxide solution, Chem. Mater. 16 (2004) 498–503, https://doi.org/10.1021/cm034433n.

[159] A. Cabañas, X. Shan, J.J. Watkins, Alcohol-assisted deposition of copper films from supercritical carbon dioxide, Chem. Mater. 15 (2003) 2910–2916, https://doi.org/10.1021/cm021814c.

[160] H. Ohde, S. Kramer, S. Moore, C.M. Wai, Low-temperature deposition of conformal copper films in supercritical CO_2 by catalytic hydrogen reduction of copper hexafluoroacetylacetonate, Chem. Mater. 16 (2004) 4028–4031, https://doi.org/10.1021/cm049542w.

[161] J.M. Blackburn, D.P. Long, A. Cabañas, J.J. Watkins, Deposition of conformal copper and nickel films from supercritical carbon dioxide, Science (80-) 294 (2001) 141–145, https://doi.org/10.1126/science.1064148.

[162] E. Kondoh, H. Kato, Characteristics of copper deposition in a supercritical CO_2 fluid, Microelectron. Eng. 64 (2002) 495–499, https://doi.org/10.1016/S0167-9317(02)00826-2.

[163] E. Kondoh, K. Shigama, Deposition of Cu thin films from supercritical carbon dioxide using hexafluoroacetylacetnatecopper, Thin Solid Films 491 (2005) 228–234, https://doi.org/10.1016/j.tsf.2005.06.005.

[164] E. Kondoh, J. Fukuda, Deposition kinetics and narrow-gap-filling in Cu thin film growth from supercritical carbon dioxide fluids, J. Supercrit. Fluids 44 (2008) 466–474, https://doi.org/10.1016/j.supflu.2007.12.004.

[165] A. Cabañas, J.M. Blackburn, J.J. Watkins, Deposition of Cu films from supercritical fluids using Cu(I) β-diketonate precursors, Microelectron. Eng. 64 (2002) 53–61, https://doi.org/10.1016/S0167-9317(02)00783-9.

[166] B. Zhao, M.-T. Zhao, Y.-F. Zhang, J.-H. Yang, Deposition of Cu seed layer film by supercritical fluid deposition for advanced interconnects, Chinese Phys. B 22 (2013) 64217, https://doi.org/10.1088/1674-1056/22/6/064217.

[167] N.E. Fernandes, S.M. Fisher, J.C. Poshusta, D.G. Vlachos, M. Tsapatsis, J.J. Watkins, Reactive deposition of metal thin films within porous supports from supercritical fluids, Chem. Mater. 13 (2001) 2023–2031, https://doi.org/10.1021/cm000837t.

[168] A. O'Nei, J.J. Watkins, Reactive deposition of conformal ruthenium films from supercritical carbon dioxide, Chem. Mater. 18 (2006) 5652–5658, https://doi.org/10.1021/cm060142d.

[169] C.F. Karanikas, J.J. Watkins, Kinetics of the ruthenium thin film deposition from supercritical carbon dioxide by the hydrogen reduction of Ru(tmhd)2cod, Microelectron. Eng. 87 (2010) 566–572, https://doi.org/10.1016/j.mee.2009.08.011.
[170] E. Ogel, M. Casapu, D.E. Doronkin, R. Popescu, H. Störmer, C. Mechler, G. Marzun, S. Barcikowski, M. Türk, J.-D. Grunwaldt, Impact of preparation method and hydrothermal aging on particle size distribution of Pt/γ-Al_2O_3 and its performance in CO and NO oxidation, J. Phys. Chem. C 123 (2019) 5433–5446, https://doi.org/10.1021/acs.jpcc.8b11065.
[171] A.M. Gänzler, M. Casapu, A. Boubnov, O. Müller, S. Conrad, H. Lichtenberg, R. Frahm, J.D. Grunwaldt, Operando spatially and time-resolved X-ray absorption spectroscopy and infrared thermography during oscillatory CO oxidation, J. Catal. 328 (2015) 216–224, https://doi.org/10.1016/j.jcat.2015.01.002.
[172] A. Boubnov, S. Dahl, E. Johnson, A.P. Molina, S.B. Simonsen, F. Morales Cano, S. Helveg, L.J. Lemus-Yegres, J.-D. Grunwaldt, Structure-activity relationships of Pt/Al_2O_3 catalysts for CO and NO oxidation at diesel exhaust conditions, Appl. Catal. B Enrivon. 126 (2012) 315–325, https://doi.org/10.1016/j.apcatb.2012.07.029.

Chapter 8

Synthesis of metal oxide nanoparticles

Continuous hydrothermal synthesis (CHTS) using sub and supercritical water has been recognized as a "green" process for the production of metal oxide nanoparticles. This process exploits the strong variations in the thermo-physical properties of water around its critical point and is based on the mixing of a cold aqueous metal salt solution with hot compressed water. Such sub/supercritical aqueous systems are also called "hydrothermal solutions." First, this chapter will give an introduction into the basic principles of the process. Based on this, the current state is described and selected experimental and modeling results are presented and discussed. At the end the main conclusions and further perspectives are summarized.

8.1 Introduction

Nanotechnology is regarded worldwide as one of the future technologies of the 21st century and in the past decade has led to a large number of novel applications in the fields of so-called "nanostructured materials." Due, among other things, to the increasing discussion about safety and the effects on humans and the environment, the manufacture of nanoscale products is of particular importance. As a result, there is an increasing need for detailed knowledge of the processes and mechanisms that take place during the manufacturing process.

In principle, the so-called "top-down" or "bottom-up" approach can be considered for the production of nanostructured materials; the latter is particularly suitable for the production or synthesis of nanoscale particles with defined properties. The use of supercritical fluids (SCFs) enables the efficient production of nanoparticles with defined particle size and size distribution as well as the targeted modification of surface properties so that these particles can also be functionalized [1–4].

Synthesis of Nanostructured Materials in Near and/or Supercritical Fluids
https://doi.org/10.1016/B978-0-444-64089-5.00008-1

8.2 Basics of continuous hydrothermal synthesis (CHTS) in sub/supercritical water

As a "green solvent," water has advantages from an ecological, economic and safety point of view. Supercritical H_2O (scH_2O) is used as a solvent or working medium in many chemical and technical applications. In steam power plants liquid H_2O is used at temperatures up to 600 K and superheated steam can reach temperatures of 993 K [5, 6] e.g., with the "Benson-Steam-Generator." Above its critical point ($p_C = 22.1$ MPa, $T_C = 647.1$ K, and $\rho_C = 322$ kg m^{-3}), H_2O can be continuously converted by compression without phase change from the gas state to states with liquid-like density. This so-called supercritical phase has remarkable properties, which differ greatly from those under ambient conditions. Usually the molecular and ionic properties of most substances change only marginally with pressure and temperature changes in the state range considered here, whereas the hydrogen bonds typical for H_2O react very sensitively to them. For example, the increase in the electrical conductivity of H_2O at high pressures and temperatures indicates a transition from molecular to ionic states [5]. Furthermore, as shown in Table 8.1 and Fig. 8.1, the properties of H_2O, such as density, dielectric constant, ion product and viscosity can be changed in a wide range by varying pressure and temperature [7, 8]. This enables the adjustment of process conditions under which scH_2O is completely miscible with nonpolar organic substances, while at subcritical temperatures polar and ionic substances can be dissolved in high concentrations.

The dielectric constant, ε, influences the solvent behavior and the ionic dissociation of salts in H_2O. Therefore, reliable experimental data for ε which cover a large range of pressure and temperature are crucial for a large number of applications. The variation of the dielectric constant with pressure and temperature is depicted in Fig. 8.2. It is obvious that ε decreases with increasing temperature and increases with increasing pressure and that the high value of

TABLE 8.1 Selected data of pure H_2O as a function of pressure and temperature [7, 8].

p (MPa)	T (K)	ρ (kg m^{-3})	ε	K_W
0.1	298	997	78	10^{-14}
30	523	826	28	10^{-11}
30	673	359	6	10^{-15}
50	673	578	12	10^{-12}
0.1	673	0.32	1	10^{-57}

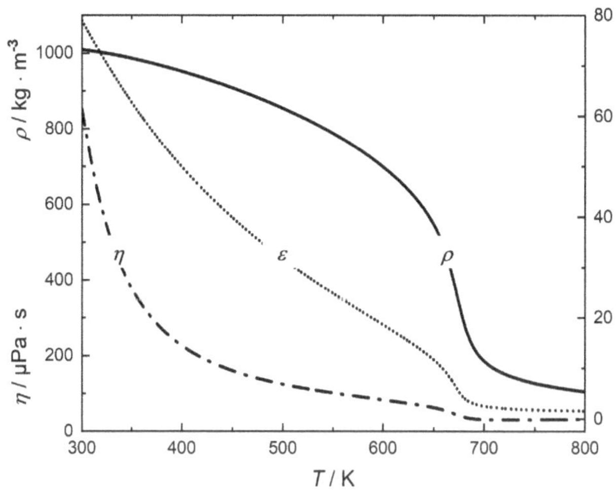

FIG. 8.1 Influence of temperature on density, viscosity and dielectric constant of H_2O at 30 MPa. *(Data from NIST Chemistry WebBook, NIST Standard Reference Database Number 69, https://webbook.nist.gov/chemistry/, 10.18434/T4D303; U. Grigull, Dielektrizitätskonstante und Ionenprodukt von Wasser und Wasserdampf, BKW 35 (1983) 6.)*

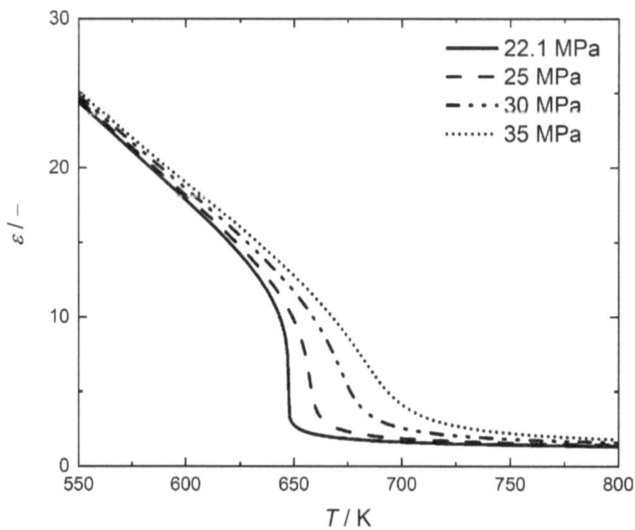

FIG. 8.2 Influence of temperature and pressure on the dielectric constant of H_2O at sub and supercritical conditions [8].

ε (i.e., = 80, see Fig. 8.1) occurs only in a small region at ambient temperatures. Such high values at normal conditions result from preferred dipole orientations in the water structure [5] while the substantial decrease is caused by the breakdown of extensive hydrogen bonding under these high temperatures and low densities.

From Fig. 8.2 follows that in the sub and supercritical region and temperatures above 550 K the dielectric constant decreases from 25 at 550 K down to values between 15 and 3 at 650 K and close to 2 at 800 K. These values are typical to those of dipolar liquids, such as acetone or nonpolar liquids such as hexane at ambient conditions and sufficiently high to dissolve and ionize electrolytes, but also enables miscibility with nonpolar solutes.

As discussed later in Section 8.2.2 in detail the significant decrease of the dielectric constant of H_2O in the near-critical region leads to a significant increase in the reaction rate and to a decreased metal salt and metal oxide solubility at supercritical conditions. This strong enhancement of the reaction kinetics can be explained by the fact that a decrease of the dielectric constant causes an increase in reaction rate.

Another key property in many applications is the dissociation equilibrium $H_2O \leftrightharpoons H^+ + OH^-$, which can be characterized by the ion product K_W:

$$K_W = a(H^+) \cdot a(OH^-)$$

where a refers to the activities of the species. Fig. 8.3 shows the ion product at high pressures and temperatures. In general, the ion product increases slightly with temperature from 10^{-14} at ambient conditions up to 10^{-11} at 550 K. It is obvious that above temperatures of around 600 K, the ion product decreases dramatically with increasing temperature. At 25 MPa the ion product decreases from 10^{-11} at 550 K to around 10^{-23} at 800 K. It is worthy to notice that this effect is much more pronounced at higher pressures than at low pressures. At 650 K the ion product decreases from 10^{-13} at 35 MPa to 10^{-21} at 22.1 MPa

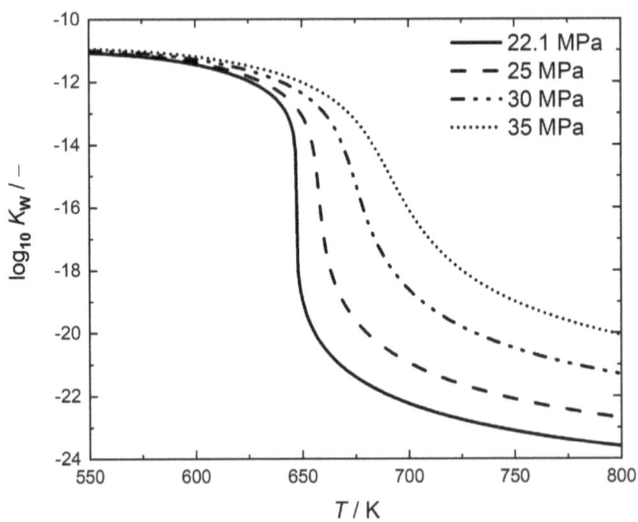

FIG. 8.3 Influence of temperature and pressure on the ion product of H_2O at sub and supercritical conditions [8].

while at 800 K the ion product varies between 10^{-20} and ca. 10^{-24}. Thus, at high pressures the ion product in sub and supercritical H_2O can be some orders of magnitude higher than at ambient conditions.

8.2.1 Phase behavior of binary aqueous systems

Fig. 8.4 shows the vapor-pressure curve for pure H_2O and the critical loci for various binary systems with H_2O. Note that the critical lines terminate the two-phase regions. On the low-temperature side a two-phase region exists, while on the right-hand side of these lines a homogeneous phase is formed at any composition. From the practical point of view, such lines are of great importance as borderline, above which a chemical reaction can be performed in homogeneous phases with high concentrations of the reactants.

For H_2, a small and hard quadrupolar molecule, the critical curve starts from the critical point of pure water with a positive slope and only little changes in the critical temperature. In opposite thereto a weak temperature minimum occurs for Xe, while with increasing stronger interactions, the minima become more pronounced, as observed for e.g., C_6H_6 or CO_2, that have strong quadrupole moments. As a result, the homogenous region becomes large. An exception is the binary system H_2O/NaCl, caused by the fact that both the melting point and the critical point of the solute is higher than that of water. Fig. 8.4 shows that the critical temperature of this mixture initially increases continuously [5] and that the critical line probably extends without interruption to the estimated critical point of NaCl near 2700 K as discussed in more detail below.

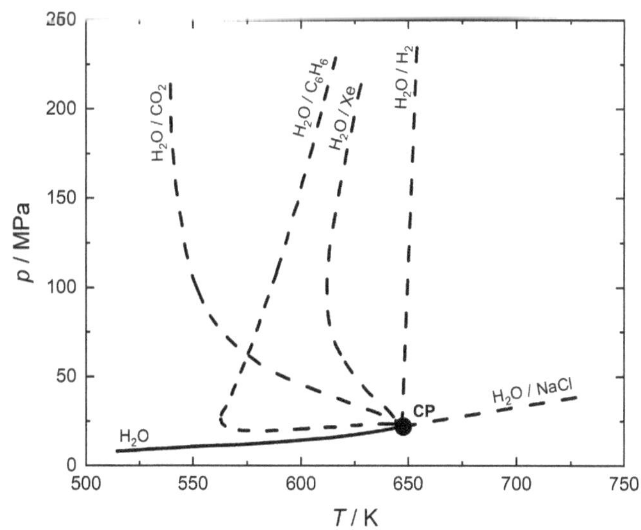

FIG. 8.4 Phase behavior of binary aqueous systems [5]. The *solid line* near the bottom left represents the vapor-pressure curve of pure H_2O that terminates in the critical point (CP).

In order to select the optimum process parameters, the phase equilibrium of the respective binary (or ternary) material system must be known in addition to the knowledge of the desired product properties. Information on the phase behavior and solubility of the metal salts in cold, sub and supercritical H_2O is therefore essential [3, 5, 9]. Since even for simple salts such as NaCl or Na_2CO_3 the melting temperature of 1074 or 1127 K respectively, is above the critical temperature of pure H_2O, H_2O is the most volatile component in such mixtures, in contrast to nonelectrolyte systems. This leads to very complex phase diagrams (cf. Fig. 8.5), which can often contain several solid phases in addition to solid-liquid equilibria (s-l) [3]. It has been discussed in detail by Voisin et al. that the different salts have been subdivided by Marshall with regard to their solubility behavior in H_2O into type 1 and type 2 [10]. Following this approach, type 1 salts generally have a higher solubility in scH_2O even at

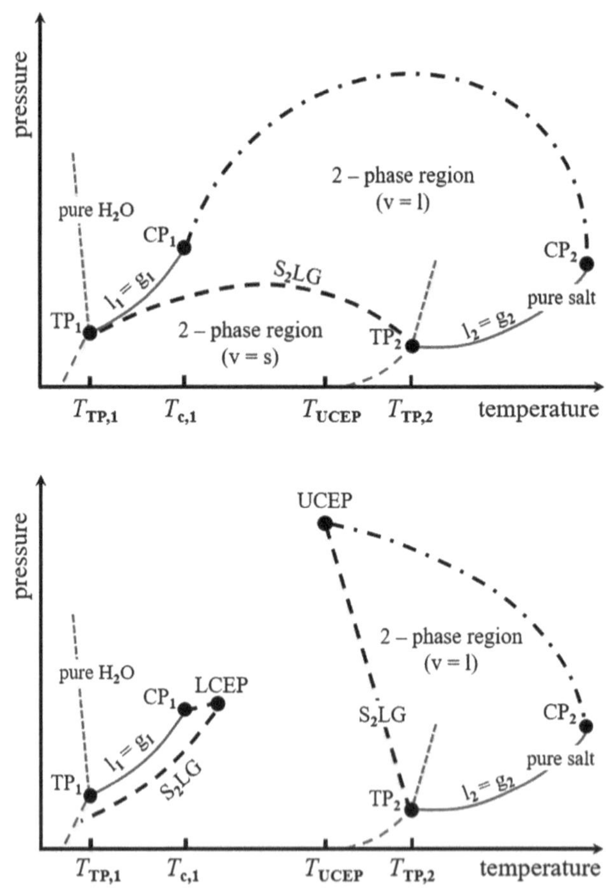

FIG. 8.5 *p-T* projection for a type 1 salt (upper scheme) and a type 2 salt (bottom scheme).

sub or supercritical conditions than type 2 salts. For example, Voisin et al. show that at 25 MPa and 700 K the solubility of $NaNO_3$ (type 1) in H_2O is more than two orders of magnitude higher than the solubility of Na_2SO_4 (type 2).

In general, however, it can be stated that for the selection of the process conditions it is decisive whether the vapor pressure curve of the saturated solution, i.e., the three-phase line (S_2LG) of the binary system, always runs below the critical line or intersects it. As depicted in Fig. 8.5, for a type 1 salt, the critical locus is continuous and takes place between the critical points of the two components CP_1 and CP_2. In opposite thereto for the type 2 salt the binary critical curve is interrupted by the two S_2LG lines, leading to the lower (LCEP) and upper (UCEP) critical end point (see Fig. 8.5). Typically, type 1 salts have a melting temperature below 1100 K (e.g., $NaNO_3$ = 582 K) while type 2 salts are characterized by melting temperatures above 1100 K (e.g., K_2SO_4 = 1432 K). Some typical representatives of these salts are listed in Table 8.2.

It is shown in Fig. 8.6, that the solubility of Na_2CO_3 in H_2O at approx. 25 MPa and 600 K is about 8 wt%; while at 725 K only 0.003 wt% Na_2CO_3 can be dissolved in H_2O [11]. This behavior can be explained by the temperature dependence of the dielectric constant of H_2O mentioned above. This determines the solvent behavior of H_2O as well as the ion dissociation of salts and is 80 at 300 K; such a high value results from preferred dipole orientations in the water structure [5]. In the vicinity of the critical point, the value for the dielectric constant is about 6 [8] which corresponds to that of an organic solvent. The reason for this is the significantly reduced density of H_2O and the resulting lower number of H_2 bridge bonds. Thus, the ability to dissolve and dissociate electrolytes decreases strongly [5]. As the temperature increases, the dielectric constant decreases to a value of about 2 at 800 K that is almost identical to that of the nonpolar solvent *n*-hexane (= 2) [5].

TABLE 8.2 Classification of different salts with regard to their solubility in H_2O [5, 10].

Type 1	Type 2
$BaBr_2$, $BaCl_2$, $Ba(NO_3)_2$	BaF_2
$CaBr_2$, $CaCl_2$, CaI_2, CsBr, CsCl, CsF	CaF_2, $CaSO_4$
KBr, KCl, K_2CO_3, KF, K_3PO_4	K_2SO_4
	$MgSO_4$
LiBr, LiCl	Li_2CO_3, LiF, Li_3PO_4, Li_2SO_4
NaBr, NaCl, $NaNO_3$, Na_2SO_4	Na_2CO_3, NaF, Na_3PO_4, Na_2SO_4
RbBr, RbCl, Rb_2CO_3, RbF, Rb_2SO_4	

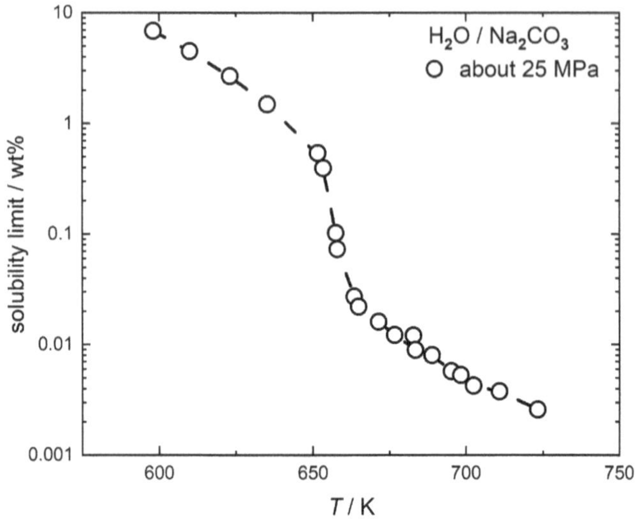

FIG. 8.6 Influence of temperature on the solubility of Na_2CO_3 in H_2O at about 25 MPa [11].

More details about the extensive experimental results of different working groups can be found in various publications [5, 9–14]. From the author's point of view, the most current and very comprehensive overviews of this topic can be found in the publications of Xu et al., Voisin et al. and Liu et al., published between 2015 and 2017 [9, 10, 12].

All the solubility data regarding inorganic salts in sub and supercritical H_2O are summarized in Table 8.3. At this point it is worth to notice that this overview of solubility data available in the literature is based on the review articles [10, 12] and an own additional literature research that covers the period from the middle of 2017 to end of 2019. The data listed in Table 8.3 illustrate that 24 studies on the phase behavior of binary systems but only 5 studies on ternary systems are published so far in literature. Thus, it is obvious that especially experimental investigations on the solubility behavior of salt mixtures under sub and supercritical conditions need to be regarded more intensively. Therefore, it is necessary to establish a complete solubility database for inorganic compounds-water systems.

In general, the reported solubility data for the various salts show that there is a linear relationship between solubility and H_2O density and an approximate linear increase with increasing water density.

From Table 8.3 follows that in 6 papers experimental results for the phase equilibrium of the $H_2O/NaCl$ system (Type 1 salt) and in 8 papers experimental data for the H_2O/Na_2SO_4 system (Type 2 salt) under sub and supercritical conditions are reported; they are depicted in Fig. 8.7.

The tendency of NaCl solubility in H_2O measured by various researchers shows a similar trend. Moreover, the solubility data varied from 10^{-8} to

TABLE 8.3 Experimental solubility data of inorganic compounds in sub and supercritical water.

No.	Compound	Number of papers	*p* (MPa)	*T* (K)	Refs.
1	KCl	2	9–24	643–679	[10, 12]
2	NaCl	6	1–24	593–823	[10, 12]
3	LiCl	1	19–24	660–692	[10, 12]
4	$MgCl_2$	1	19–23	666–668	[10, 12]
5	$CaCl_2$	1	19–23	669–687	[10, 12]
6	Na_2SO_4	8	19–30	583–683	[10, 12]
7	K_2SO_4	1	25	646–655	[10, 12]
8	$MgSO_4$	1	25	646–655	[10, 12]
9	$Al_2(SO_4)_3$	1	15–30	619–675	[10, 12]
10	$NaNO_3$	1	17–23	663–679	[10, 12]
11	KNO_3	1	20–24	673–682	[10, 12]
12	$LiNO_3$	1	18–23	663–680	[10, 12]
13	NaH_2PO_4	1	20–24	671–694	[10, 12]
14	Na_2HPO_4	1	21–23	668–680	[10, 12]
15	K_2HPO_4	1	25–31	673–723	[10, 12]
16	KOH	2	22–30	653–798	[10, 12]
17	NaOH	1	28	673	[10, 12]
18	$Mg(OH)_2$	3	1–25	673–823	[10, 12]
19	$CaCO_3$	1	24	698–723	[10, 12]
20	Na_2CO_3	3	24–30	593–773	[10, 12, 15]
21	H_3BO_3	1	24	713	[10, 12]
22	K_3PO_4	1	30	593–773	[15]
23	CuO	2	8–42	330–824	[16, 17]
24	PbO	1	26–34	523–773	[17]
25	$NaCl/Na_2SO_4$	2	20–25	593–763	[18, 19]
26	Na_2SO_4/Na_2CO_3	1	25	650–680	[20]
27	Na_3PO_4/K_2SO_4	1	30	623–681	[20]
28	Na_3PO_4/Na_2SO_4	1	30	623–681	[20]
29	KPO_4/Na_2SO_4	1	30	623–681	[11]

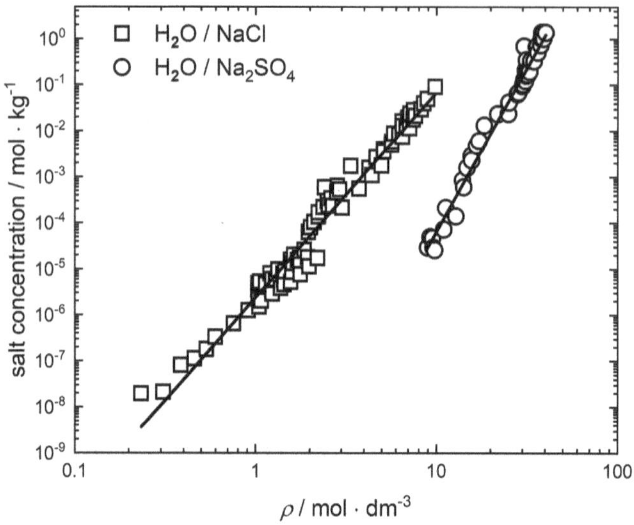

FIG. 8.7 Solubility of NaCl and Na_2SO_4 in sub and supercritical H_2O. *(Data from Q. Liu, X. Ding, B. Du, T. Fang, Multi-phase equilibrium and solubilities of aromatic compounds and inorganic compounds in sub- and supercritical water: a review, Crit. Rev. Anal. Chem. 47 (2017) 513–523.)*

10^{-1} mol kg^{-1} in the density range from 0.2 to around 10 mol dm^{-3}, which indicated that the difference of solubility of NaCl in H_2O is significant. The solubility of Na_2SO_4 in H_2O shows a similar linear trend, however an up to four orders of magnitude lower solubility is observed in the identical sub and supercritical density range (e.g., 10^{-5} vs 10^{-1} mol kg^{-1} at a H_2O density of 10 mol dm^{-3}).

Since most of the solubility data of the other inorganic compounds were only measured by one researcher or at a single pressure and temperature condition (cf. Table 8.3), these data are not plotted. However, a summarizing graphical overview of the majority of the solubility data for inorganic salts vs H_2O density available in literature is given in Ref. [10].

Experimental data published by Hearn et al. [16] and Sue et al. [17] showed that the solubility of CuO in sub and supercritical H_2O varies significantly around the critical point. Fig. 8.8 shows that under isobaric conditions the solubility increases with increasing temperature up to a maximum at around 600 K, followed by a steep decrease in solubility when the temperature increases into the supercritical region. This decrease is caused by the low density and dielectric constant at high pressures and temperatures higher than T_C (cf. Fig. 8.1).

Among others, reliable solubility data are one key property for the design and understanding of the particle formation processes. However, especially under supercritical conditions, solubility data are not always

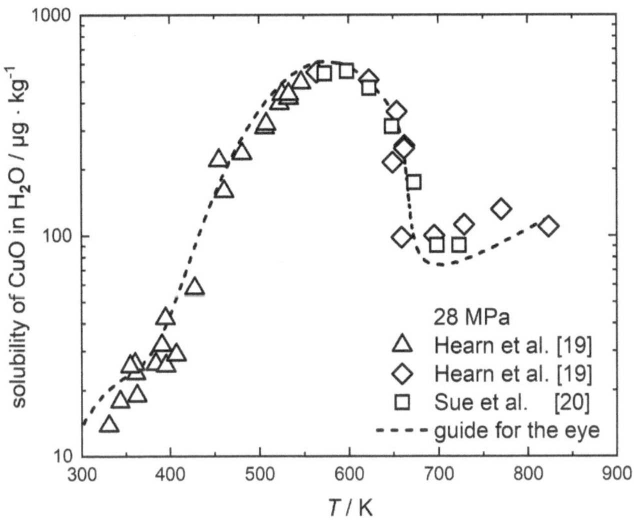

FIG. 8.8 Solubility of CuO in sub and supercritical H_2O. *(Data from B. Hearn, M. Hunt, A. Hayward, Solubility of cupric oxide in pure subcritical and supercritical water, J. Chem. Eng. Data 14 (1969) 442–447; K. Sue, Y. Hakuta, R.L. Smith, T. Adschiri, K. Arai, Solubility of lead(II) oxide and copper(II) oxide in subcritical and supercritical water, J. Chem. Eng. Data 44 (1999) 1422–1426.)*

easy to measure and the experimental set-ups become expensive, complex and unhandy. As pointed out in various reviews and papers (e.g. [21, 22]) models based on equations of state, together with different mixing rules, are most widely used to correlate and predict the solubility behavior in supercritical fluids. Therefore, the accurate knowledge of the required solute data is essential. However, the common group-contribution methods are mostly empirical and often lead to inconsistent and unreliable results. Thus, due to the lack of information on these data, various models are often used for the correlation of experimental solubility data. A more detailed compilation and critical evaluation of different models such as the numerous density based models as well as the revised Helgeson-Kirkham-Flowers (rHKF), and Sue-Adschiri-Arai (SAA) model for the description of salt solubility and the equilibrium constant can be found in various papers (e.g. [9, 10, 21–23]).

8.2.2 Kinetics of phase transition and particle formation[a]

For the description of the kinetics of particle formation, the nucleation rate J, i.e., the number of nuclei formed per unit time and volume, is normally used.

a. Revised and extended version of Chapter 3.4 in Ref. [3]; Copyright © (2014) Elsevier.

According to the classical nucleation theory J is composed of three different terms (K, Ω, ln S) and given by Türk [3]:

$$J = K \cdot \exp\left(\frac{16 \cdot \pi}{3} \cdot \frac{\Omega}{(\ln S)^2}\right)$$

with

$$K = \Theta \cdot \alpha_C \cdot v_{2,s} \cdot (p \cdot y_2 \cdot N_A)^2 \cdot \sqrt{\frac{2 \cdot \sigma}{k \cdot T}} \quad (8.1)$$

with

$$\Omega = \frac{(\sigma^3 \cdot v_{2,s}^2)}{(k \cdot T)^3}$$

In Eq. (8.1) K is the preexponential factor that is directly proportional to the square of the solute's mole fraction, y_2, Θ is the nonisothermal factor (= 1 in case of dilute mixtures), α_C is the condensation coefficient (= 0.1), $v_{2,S}$ is the molecular volume of the solid phase, ρ is the density of the mixture, N_A is the Avogadro number, σ is the interfacial tension of the solute, and k is the Boltzmann's constant. The term Ω is only a function of the pure solute properties while the term $[1/(\ln S)^2]$ depends strongly on the degree of supersaturation, S, and therewith on the phase behavior of the system of interest. Thus, as discussed below in brief, solubility and therewith the degree of supersaturation influence the particle formation process, i.e., the nucleation and condensation rate to a different extent.

From Eq. (8.1) follows that the nucleation rate depends more strongly on S than does the rate of condensation, which is linear in S. In addition, it is obvious that in case of high supersaturation the preexponential factor K is the relevant value for the nucleation rate since the exponential factor becomes unity. Own calculations show that the preexponential factor K is in the range of 10^{+25}–10^{+23} cm^{-3} s^{-1} for typical CHTS conditions (25 MPa $\leq p \leq$ 30 MPa and 573 K $\leq T \leq$ 673 K) while Nadimpalli et al. published a value of 10^{+28} cm^{-3} s^{-1} [24]. Thus, from Eq. (8.1) follows that a lower solubility should result in lower nucleation rates and therewith in smaller particles due to delayed precipitation and /or a lower number of particle collisions that inhibit particle growth. These considerations can be descriptively interpreted by taking into account the reciprocal of the nucleation rate, which is the average time between the formation of nuclei in a unit volume. For example, in 1 cm^3, one nucleus is formed every microsecond at $J = 1 \times 10^{+6}$ s^{-1} cm^{-3}, while every microsecond approximately 10^{+6} nuclei are formed at $J = 1 \times 10^{+12}$ s^{-1} cm^{-3}. A more detailed discussion about the influence of supersaturation on nucleation rate can be found in literature [3, 25].

It will be discussed in detail below, that a higher process temperature favors the formation of uniform nanoparticles. This holds particularly in the supercritical region due to the sudden decrease of the metal oxide solubility in scH_2O. In addition to this, the particle size is strongly influenced by the reaction rate k or strictly speaking by the dielectric factor $(1/\varepsilon - 1/\varepsilon_0)$ [26, 27]:

$$\ln k = \ln k_0 - \left(\frac{E}{R \cdot T}\right) + \left(\frac{\omega}{R \cdot T}\right) \cdot \left(\frac{1}{\varepsilon} - \frac{1}{\varepsilon_0}\right) \tag{8.2}$$

In Eq. (8.2) k_0 is the reaction rate at dielectric constant ε_0, E is the activation energy, R is the gas constant, T is the absolute temperature, and ω is a constant determined by the relevant reaction system. From Eq. (8.2) follows that in the near critical region the reaction rate is extremely sensitive to the variation of temperature since the dielectric constant at a constant pressure (i.e., $p = 25$ MPa) decreases from around 22 for the subcritical temperature $T_3 = 573$ K to about 5.5 at $T_3 = 673$ K (cf. Fig. 8.1). Thus, the dielectric factor has a strong influence on the solubility of metal oxides (cf. Fig. 8.8) and the reaction rate as shown in Fig. 8.10.

Fig. 8.9 shows the calculated dielectric factor as a function of temperature and pressures ranging from 25 to 35 MPa.

For these calculations 30 MPa and 473 K were chosen as the reference state for ε_0. It is obvious that for temperatures below the critical temperature the dielectric factor is small enough to be neglected. In opposite thereto, a strong increase is observed as the critical temperature is exceeded. For example, the dielectric factor increases from 0.1 at 25 MPa and 650 K to 0.6 at 775 K. However, at higher pressure the dielectric factor is reduced to 0,45 at 35 MPa and 775 K.

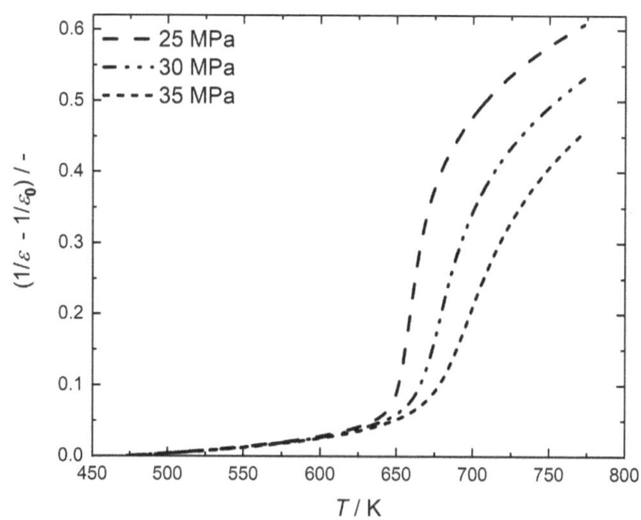

FIG. 8.9 Influence of temperature and pressure on the dielectric factor for H_2O.

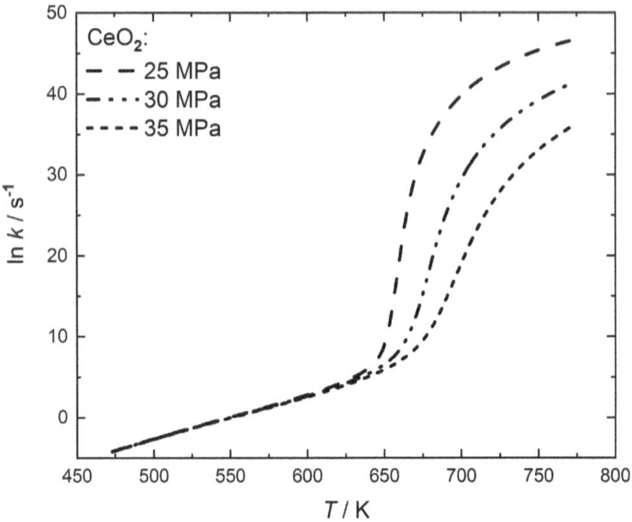

FIG. 8.10 Influence of temperature and pressure on the reaction rate k. *(The curves were calculated with Eq. (8.2) using the data published by N. Aoki, A. Sato, H. Sasaki, A.A. Litwinowicz, G. Seong, T. Aida, D. Hojo, S. Takami, T. Adschiri, Kinetics study to identify reaction-controlled conditions for supercritical hydrothermal nanoparticle synthesis with flow-type reactors, J. Supercrit. Fluids 110 (2016) 161–166.)*

For the synthesis of CeO_2 in a continuous T-mixer Aoki et al. used the experimentally in the temperature range of 473–613 K obtained rate constants to determine the parameters of Eq. (8.2) to $k_0 = 2.8 \times 10^{+7}\ s^{-1}$, $E = 84$ kJ mol^{-1}, and $\omega = 4.5 \times 10^{+2}$ kJ mol^{-1} [28]. Fig. 8.10 shows for three different pressures the influence of the temperature on the first-order reaction rate constant k. The curves were calculated using the data given by Aoki et al. and illustrate the strong influence of the temperature, especially at $T > T_C$ on the reaction rate k. Obviously, the plot shows a weak increase of the reaction rate constant k below the critical temperature, but increases significantly for temperatures higher than the critical one.

Furthermore, it is important to notice that for supercritical temperatures the reaction rate constant k increases significantly with decreasing pressure. This is caused by the increase of the dielectric factor with decreasing pressure which is shown in Fig. 8.9 and discussed in detail in various publications (e.g. [26, 27]). From Eq. (8.2) follows that a high dielectric constant and thus low value for the dielectric factor leads to a low reaction rate, while low values for ε lead to high dielectric factors and therewith high reaction rates. The former one leads to larger particles while the latter one intensifies the formation of nanoparticles.

It is worthy to notice that in addition to the process parameters temperature and concentration, the velocity constant of a reaction k can also be influenced by the pressure as shown below.

$$\left(\frac{\partial \ln k}{\partial p}\right)_T = -\frac{\Delta v}{R \cdot T} \tag{8.3}$$

From Eq. (8.3) follows that the reaction rate is proportional to the molar activation volume Δv, which depends on two quantities: the difference between the molar volumes (or sizes) of the reactants and the transition state and the strength of the interactions between the reactants and the solvent molecules. Usually in incompressible liquids typical values of Δv are in the range from -50 to $+30\ cm^3\ mol^{-1}$, while for scH_2O values up to $1000\ cm^3\ mol^{-1}$ can be achieved. Due to these high values for the molar activation volume, which are more than 30 times higher than those of incompressible liquids, even a small change in pressure leads to a significant change in the reaction rate in scH_2O [4, 29].

As mentioned in Section 8.2, an important key feature for a large number of applications is the dissociation equilibrium ($H_2O \leftrightarrows H^+ + OH^-$) that is characterized by the ion product $K_W = a(H^+)\cdot a(OH^-)$. At ambient conditions, $K_W = 10^{-14}$, but initially increases with increasing temperature or density and reaches values in the range of 10^{-11} at 25 MPa and temperatures around 525 K (cf. Fig. 8.3). Above the critical temperature, however, the ion product decreases dramatically with increasing temperature and reaches values around 10^{-20} at 700 K [8]. Thus, depending on pressure and temperature in scH_2O, either heterolytic or homolytic reactions can be favored.

However, in contrast to density, dielectric constant and salt solubility, the viscosity of scH_2O is comparatively low and only slightly dependent on temperature above 500 K. The high molecular mobility resulting from the low viscosity accelerates chemical reactions controlled by mass transfer or diffusion and thus eliminates inhibitions of mass transport. Due to these properties, scH_2O is of great interest for many technically important reactions such as the oxidative decomposition of organic substances by means of the supercritical water oxidation (SCWO) process as solvent and/or reaction medium (see overview in Refs. [30–34]).

8.2.3 Analyzing methods

Usually, the particles synthesized by CHTS are characterized offline with the following measurement and analysis methods: Raman Spectroscopy, Differential Centrifugal Sedimentation (DCS), Electronic Light Scattering (ELS), X-Ray Diffraction (XRD), Transmission Electron Microscopy (TEM) and X-Ray Photoelectron Spectroscopy (XPS) [4]. Nevertheless, the use of fast online and/or in situ measurement techniques such as Electrochemical

Impedance Spectroscopy (EIS), Small Angle X-ray Scattering (SAXS), Extended X-ray Absorption Fine Structure (EXAFS) and/or X-ray Absorption Near-Edge Structure (XANES), Neutron radiography (NT) under hydrothermal conditions is absolutely necessary for the in-depth understanding of the relationship between the numerous process parameters, the reaction and particle formation kinetics and the achieved product properties. However, so far only isolated approaches and results have been published in the literature [35–37].

8.3 Experimental procedure

The basic principle of the continuous hydrothermal synthesis is based on the fact that the properties of H_2O such as density, viscosity and dielectric constant change greatly in the vicinity of the critical point and can therefore produce metal oxide nanoparticles without the addition of organic solvents such as strong bases or surfactants. Furthermore, the exceptional and adjustable thermo-physical properties of H_2O allow the targeted control of the particle formation process with regard to size, morphology and surface chemistry of the synthesized particles as well as the production of in situ functionalized metal oxide nanoparticles (e.g. [3, 26–28, 38–40]). As schematically illustrated in Fig. 8.11 the CHTS process developed by Adschiri et al. is based on the mixing of a cold aqueous salt solution with sub or supercritical H_2O [26, 39].

As an example, it is shown below, that hydrothermal reactions include hydrolysis of the metal salt and dehydration of the hydrous oxide. Thus, the reaction scheme and the formation of solid CeO_2 nanoparticles can be described as follows:

$$\begin{aligned} &\text{Hydrolysis}: \ \mathrm{Ce(NO_3)}_x + x\mathrm{H_2O} \rightleftarrows \mathrm{Ce(OH)}_x + x\mathrm{HNO_3} \\ &\text{Dehydration}: \ \mathrm{Ce(OH)}_x \rightleftarrows \mathrm{CeO}_{0.5x} + 0.5x \cdot \mathrm{H_2O} \\ &\text{Equilibrium}: \ \mathrm{CeO_2(aq)} \rightleftarrows \mathrm{CeO_2(S)} \end{aligned}$$

It should be noted that the equilibrium in dehydrogenation is shifted in the direction of the products by an increase in temperature, i.e., the formation of the metal oxides is favored by high temperatures [38].

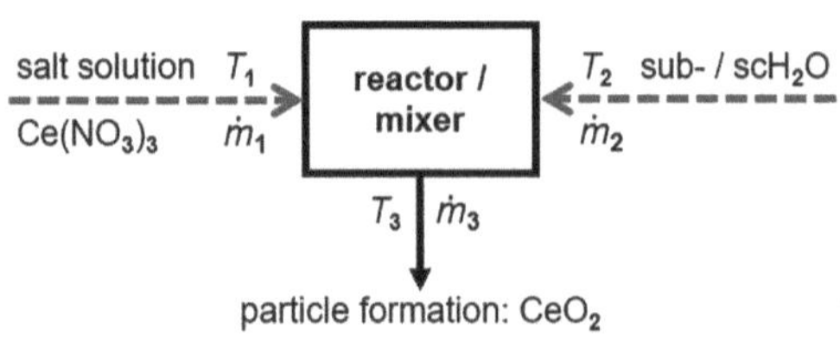

FIG. 8.11 Basic of continuous hydrothermal synthesis. *(Modified from M. Türk, Particle Formation With Supercritical Fluids: Challenges and Limitations, first ed., Elsevier, Amsterdam, 2014; T. Adschiri, Y. Hakuta, K. Arai, Hydrothermal synthesis of metal oxide fine particles at supercritical conditions, Ind. Eng. Chem. Res. 39 (2000) 4901–4907.)*

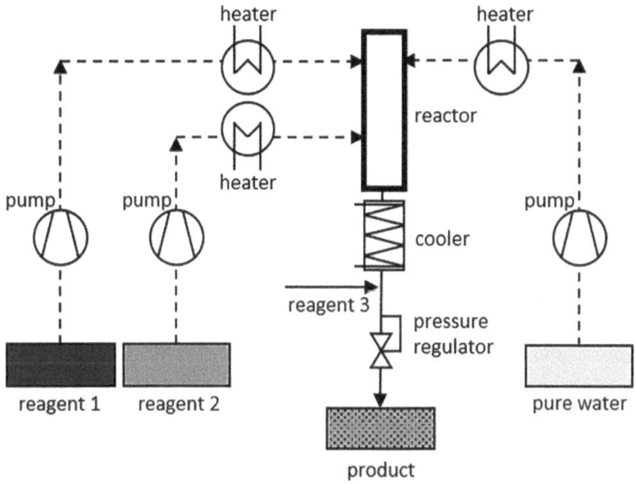

FIG. 8.12 CHTS apparatus for producing submicron metal oxide particles in sub and supercritical H_2O. *(Modified from M. Türk, Particle Formation With Supercritical Fluids: Challenges and Limitations, first ed., Elsevier, Amsterdam, 2014.)*

The schematic representation of a typical pilot plant for particle synthesis using CHTS shows Fig. 8.12. Typically, such a pilot plant allows investigations under the following process conditions: The pressure in the system is $\leq$ 40 MPa at temperatures $\leq$723 K. The total mass flow (cold metal salt stream plus hot water stream) varies between 3 and 12 kg h^{-1}; the mixing ratio (= cold metal salt stream/hot water stream) is usually in the range between 0.05 and 0.5. At the reactor outlet the reaction mixture is rapidly cooled down to ambient temperature.

8.3.1 Experimental results

Adschiri et al. showed for the first time in 1992 that the production of AlO(OH) with particle sizes in the range of 5–20 nm is possible by means of CHTS [39]. Fig. 8.13 shows the number of articles published since 1990, the year that the Web of Science database indicated a first hit from a search on "supercritical water & hydrothermal" and "supercritical water & hydrothermal synthesis." In addition, a selection of metallic nano-materials synthesized by the CHTS process is summarized in the very extensive review article written by Darr et al. [41]. Obviously, a large number of publications have been published, so that only a brief and therefore certainly incomplete overview of the most important results of the numerous experimental studies from the author's point of view can be given here.

The results of a large number of systematic investigations on the synthesis of metal oxide nanoparticles carried out in the Adschiri working group show that

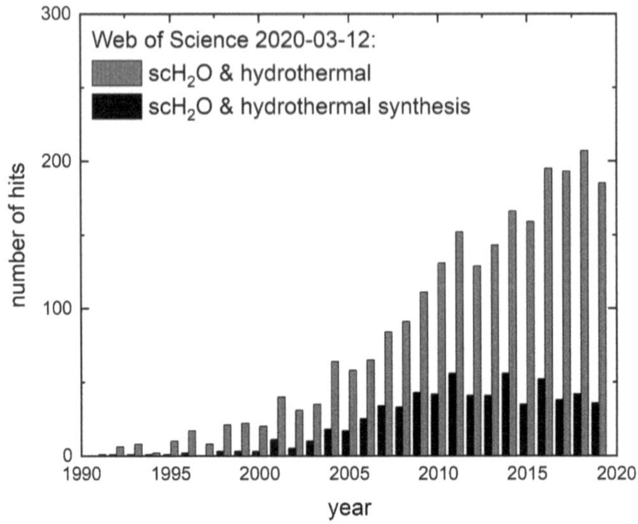

FIG. 8.13 Number of papers on reactions in scH_2O.

the most varied mono- and bimetal oxide particles in the size range from 5 nm to 1000 nm can be produced by CHTS [26–28, 38, 39]. The application of these particles is not limited to catalysis, sensor technology, semiconductors etc. but is also used as cathode material for lithium ion batteries such as $LiMn_2O_4$, $LiCoO_2$ and $LiFePO_4$ [40]. Furthermore, results of investigations on reaction kinetics under hydrothermal conditions show that the reaction rate depends strongly on the reaction temperature [28, 38]. For example, at 30 MPa the reaction rate constant for the formation of AlO(OH) and CeO_2 at supercritical temperatures is up to 5 orders of magnitude higher than under subcritical conditions [38].

The results of various investigations performed by Daschner de Tercero et al. [42–44] show that CHTS enables the synthesis of $BaTiO_3$, CeO_2, Fe_xO_y particles with a mean particle size between 10 and 40 nm. In these experiments, particle concentrations in the range from 0.08 to 5.7 g dm^{-3} were achieved in the product suspension.

Sugioka et al. investigated the influence of mixer geometry, flow control and mixing ratio on the synthesis of CeO_2 particles [45]. They used a T-mixer and varied the flow control as well as the ratio of cold, saline aqueous solution to hot water flow in the range of 0.125 to 0.5. The results of these investigations show that a high total mass flow leads to nanoparticles ($<$ 50 nm) and a mass flow ratio of 0.125, compared to 0.5, leads to a significantly wider particle size distribution.

Studies carried out by Slostowski et al. at 24.5 MPa and 573 and 673 K show that the synthesis of lanthanum (La), praseodymium (Pr) and/or zirconium (Zr) substituted CeO_2 nanoparticles with a defined composition is possible [46].

Furthermore, this study proposes an overview of the influences of residence time, temperature and concentration of precursor over the proportion of the substitution element in the CeO_2 nanostructures.

Sue et al. showed that at 30 MPa and 673 K Pr doped $CaTiO_3$ nanoparticles with different Ca/Ti ratios and a mean particle size of 22 nm can be synthesized. Furthermore, the results of the experimental investigations show that the increase of the residence time and the KOH concentration in scH_2O leads to an increase of the Ca/Ti ratio and to larger particles [47].

The results of the work of Hayashi and Hakuta show that the reaction temperature and the reaction rate can influence the chemical composition of the synthesized particles. Thus, at pressures around 30 MPa and temperatures between 550 and 650 K γ-AlO(OH) particles were formed, while at temperatures above 650 K γ-Al_2O_3 particles were preferably formed [48].

While the "conventional" CHTS process generally uses a single-stage mixer, Hellstern et al. proposed the use of a two-stage mixer as a promising modification [49, 50]. This process variant allows, for example, the addition of protective colloids after the first mixer (see Fig. 8.11, reagent 2) to prevent particle agglomeration. Furthermore, by adding a further reactant or salt solution in the second mixer, the synthesis of a second particle species, of particles doped with a second substance, of core-shell particles with a defined layer thickness or of composite materials with a specific particle morphology can be specifically controlled [50].

CHTS experiments were performed by Türk and Schüssler at 30 MPa with $Ce(NO_3)_3$ and different salt concentrations (0.005–0.15 mol dm^{-3}), mass flow rates (35, 45 and 55 g min^{-1}) and temperatures (573 and 673 K) [51]. In order to investigate the influence of the anion on particle size additional experiments were conducted at $T_3 = 523$ K with $Ce_2(SO_4)_3$. The obtained results are summarized in Fig. 8.14, which shows the relationship between the mean particle size and the residence time, t_R. Obviously, the mean particle size was found to decrease with decreasing residence time from around 40 nm and $t_R = 1.5$ s to 14 nm and $t_R = 0.5$ s due to an improvement in mixing intensity and less time available for particle growth. In addition, particle size increases with decreasing reaction temperature, i.e., at $T_3 = 573$ K large but uniform particles with $d_{50} = 41$ nm were produced while at a temperature of $T_3 = 673$ K nanoparticles with a mean particle size of 19 nm were synthesized. Furthermore, the results show that at constant process conditions i.e., constant residence time the mean particle size increases with increasing salt and therewith reactant concentrations from around $d_{50} = 12$ nm and $c_{salt} = 0.005$ mol dm^{-1} up to $d_{50} = 20$ nm at the two higher concentrations that is in accordance with the classical nucleation theory [3]. However, it is worthy to notice that the synthesis of metal oxide nanoparticles must be considered as a complex system, consisting of reaction, nucleation, and particle growth stages. In addition, it must be taken into account, that under these conditions the reaction equilibrium of metal nitrates shifts toward metal hydroxides or metal oxides, which

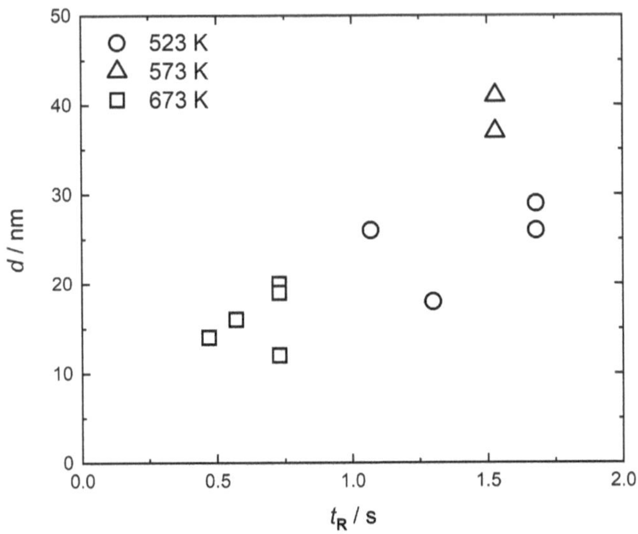

FIG. 8.14 CeO_2 particle size as a function of residence time. The data were obtained at 30 MPa by varying the flow rate, temperature, salt concentration and type [51].

subsequently precipitate from solution according to the degree of supersaturation [38].

For further information regarding the materials investigated, the process and reaction conditions as well as the size and morphology of the synthesized particles, please refer to the extensive review written by Darr et al. [41]. On the basis of this review article and own literature research, the overview shown in Fig. 8.15 has been compiled. This figure shows that the majority of the investigations were focused on the synthesis of monometallic particles. Furthermore, from Fig. 8.15 follows that the synthesis of bi- and multimetallic nanoparticles by CHTS is an area with a lot of potential and much work is needed in this area.

Based on a multitude of experimental investigations it can be stated that the pressure and temperature in the mixer or reactor as well as the mixer geometry and the ratio of cold metal salt stream to total mass stream were identified as the process parameters essential for the size of the particles, their size distribution and morphology. Furthermore, it was shown that the increase of the total mass flow, and thus the reduction of the residence time, leads to a mono-modal particle size distribution. Such an improved knowledge is needed for an optimized reactor design and for successful process scale-up.

More details of the different concepts that are currently employed in the synthesis of metallic/metal oxide nanoparticles using sub and/or supercritical H_2O can be found in various review articles (e.g., see without any claim to completeness Refs. [1–4, 41, 52–54]).

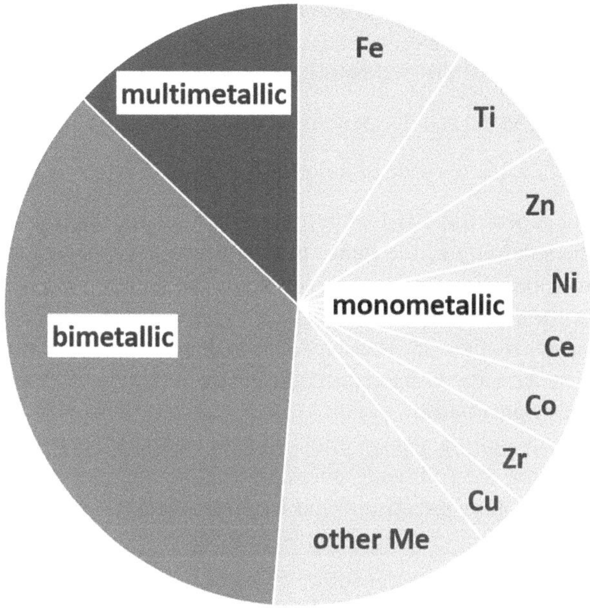

FIG. 8.15 Overview of number of publications dealing with the synthesis of different metal oxide particles via CHTS.

8.3.2 Modeling results

While experimental investigations have been carried out for more than two decades, comparatively few results of theoretical investigations have been published so far. The experimental results discussed and presented above show that the properties of metal oxide particles produced by CHTS are influenced by a variety of process parameters. Therefore, the targeted combination of experiment and modeling is necessary in order to reduce the complex experimental investigations; i.e., the predictions from the model must be checked with reliable experimental data and the model has to be improved if necessary. In the long term, this approach would have the advantage that it would also be possible to transfer the results or findings obtained to other similar material systems and questions.

To obtain a complete description of the CHTS process, different fields have to be discussed and considered: dissolution of the metal salt in cold H_2O, followed by mixing with the pure hot H_2O, and resulting in the synthesis of the metal oxides. Furthermore, it is also important to note that a complete model should take into consideration the complex geometry inside the mixing and reactor unit since these circumstances affect the fluid dynamics and therewith temperature, density, velocity and mass flow rate in the system.

The complete theoretical description of the mechanism for the formation of metal oxides via CHTS requires e.g., in case of the system H_2O/NaCl the profound knowledge of the two equilibrium constants for:

$$\text{Metal salt dissociation}\ NaCl_{aq} \rightleftarrows Na^+ + Cl^-$$

$$\text{Metal oxide dissolution}\ NaCl_{aq} \rightleftarrows NaCl_s$$

In sub and supercritical H_2O. Until now, various models are available for estimating the solubility of the ionic species of interest. Thereby the dissociation reaction equilibrium constant K, the mass, and the charge balance have to be calculated for a given reaction system. The experimental data can be correlated with various models (cf. Section 8.2.1) to describe the solubility behavior, i.e., increase or decrease with temperature and the dramatic decrease in solubility around the critical point.

Aqueous and solid-aqueous equilibria of electrolyte systems can be expressed by using the following equation for the equilibrium constant K of any general reaction at pressure p and temperature T. In case of the system H_2O/NaCl the constants can be simplified written as:

$$\text{(a) Dissociation}\ K_1 = \frac{a(Na^+) \times a(Cl^-)}{a(NaCl_{aq})} \tag{8.4}$$

$$\text{(b) Dissolution}\ K_2 = \frac{a(NaCl_{aq})}{a(NaCl_s)} \tag{8.5}$$

where a refers to the activities of the species.

Calculated results of the NaCl dissociation, K_1 (full line), and dissolution constants, K_2 (dot-dashed line), at 23 MPa and temperatures between 300 and 1000 K are depicted in Fig. 8.16.

Thereby, the SAA model was used for the estimation of the equilibrium constants at sub and supercritical conditions [55]. In case of the dissociation constant it is obvious that there is only a weak temperature influence at subcritical temperatures but that a strong decrease in the supercritical region is observed. This can be explained be the strong decrease of the dielectric constant and the ion product of H_2O (cf. Figs. 8.2 and 8.3).

In opposite thereto, a continuous decrease of the dissolution constant with increasing temperature is shown in Fig. 8.16 and a step near the critical temperature is observed. However, it must be considered that the influence of the temperature on the equilibrium constant is much more pronounced for the dissociation than for the dissolution.

In addition, the subsequent precipitation of the metal oxides from a solution according to the degree of supersaturation as well as the energetics and kinetics of nucleus formation, particle formation, and growth must be considered. In general, under acidic conditions where hydrothermal synthesis takes place, the solubility of metal oxides is relatively high at subcritical conditions.

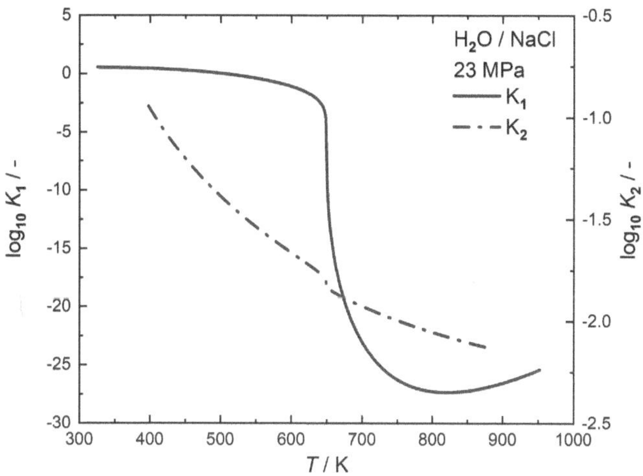

FIG. 8.16 Influence of temperature on the dissociation (left *y*-axis, *full line*) and dissolution (right *y*-axis, *dot-dashed line*) constants [51].

Thus, under such conditions, a low nucleation rate can be expected due to the low supersaturation. In opposite thereto, the reaction rate is much higher (see Fig. 8.10) while the solubility of CuO is lower under super than subcritical conditions (see Fig. 8.8). According to the classical nucleation theory, the low solubility causes high supersaturations and therewith high nucleation rates. The extremely fast reaction rate in combination with a very high nucleation rate at supercritical conditions leads to smaller particles compared to particles synthesized under subcritical conditions. Therefore, from the author's point of view, the most important publications and their essential results on the CHTS process are presented and discussed in Section 9.3.

Interim summary: H_2O can form a homogeneous phase with inorganic and organic substances, and H_2O itself works as solvent and reaction media. These unique properties of sub and supercritical H_2O enable the addition of substantial amounts of different organic molecules to form homogeneous reaction mixtures. This opens the possibility to synthesize nanocrystals by in situ functionalization with organic capping molecules, which offers the possibility for controlling the particle formation process with regard to the size and morphology as well as to control and modify the surface chemistry of the nanoparticles. In addition, the CHTS process enables the synthesis of organic-inorganic hybrid particles by in situ functionalization with organic molecules. Furthermore, because CHTS is quite fast this process can, compared with classical particle formation processes, treat much greater volumes of reactants and products. For a large number of different substances, CHTS enables the rapid and continuous synthesis of metal oxide nanoparticles.

The reasons why small and uniform nanoparticles are formed under supercritical conditions are

(a) High supersaturation caused by the low solubility and the extremely fast reaction rate.
(b) Absence of Ostwald ripening, which could not occur in scH_2O as a result of the extremely low solubility of the metal oxide.
(c) Increasing the flow velocity causes an improvement in mixing intensity and less time available for particle growth.

8.4 Summary and outlook

In contrast to the formation and functionalization of organic particles with $scCO_2$ by means of rapid expansion of supercritical solution (RESS), supercritical anti solvent (SAS) or particle generation from gas saturated solution (PGSS) processes, which can already be carried out continuously and in industrial dimensions, the processes for the continuous production and functionalization of metal oxide particles in scH_2O still require considerable research and development. It can be stated that the properties of the nanoparticles produced using CHTS can be influenced by the process parameters pressure, temperature, chemical nature of the salt and its concentration, and thus the solution equilibrium as well as the type of process control, in particular the mixer and reactor geometry. Since scH_2O is completely miscible with a large number of organic compounds and H_2O molecules can act as catalysts for organic reactions, organic surface-modified metal oxide nanoparticles in the range from 2 to 10 nm could be synthesized by CHTS. These organic-inorganic hybrid nanoparticles could be well dispersed in organic solvents and form a lattice structure. Due to their unique properties, these hybrid materials have potential applications in various technologies related to water and energy supply and efficiency, semiconductor materials, etc. [40, 42].

However, at present, there is a lack of reliable methods for process design and control to guarantee consistent product quality, particle size and morphology. Therefore, the process and the relevant plant components such as the geometry of the mixer and the reactor and the flow control must be optimized by further systematic investigations of the relationships between the above-mentioned process parameters and the achieved product properties (size, shape and structure as well as surface properties). For this purpose, for example, the increased use and further development of micro-reactor systems are absolutely necessary. Such systems should then be able to image the occurring multiphase flows exactly and to describe the individual physico-chemical processes during nanoparticle synthesis very precisely. Furthermore, the integration of faster and more reliable in situ and / or online measurement techniques such as EIS, SAXS, EXAFS, XANES and/or NT for process analysis and control is necessary under the given demanding reaction conditions of CHTS. This integration of sensor technology into the micro-reactor system represents a major challenge. However, it should then be possible, at least for selected model reactions, to map the individual partial steps during multistage nanoparticle synthesis with spatial and temporal resolution and thus to identify and understand the essential reaction mechanisms.

Furthermore, the results of such work would also make an important contribution to a deeper understanding of the basic chemical mechanisms, phase transition phenomena as well as nucleation and growth processes in reactive flows under supercritical reaction conditions and thus provide answers to the numerous open questions. Based on these findings, an energetic optimization of the entire process would then be possible.

References

[1] S. Marre, Y. Roig, C. Aymonier, Supercritical microfluidics: opportunities in flow-through chemistry and materials science, J. Supercrit. Fluids 66 (2012) 251–264.

[2] C. Aymonier, A. Loppinet-Serani, H. Reverchon, Y. Garrabos, F. Cansell, Review of supercritical fluids in inorganic materials science, J. Supercrit. Fluids 38 (2006) 242–251.

[3] M. Türk, Particle Formation With Supercritical Fluids: Challenges and Limitations, first ed., Elsevier, Amsterdam, 2014.

[4] A. Loppinet-Serani, C. Aymonier, F. Cansell, Supercritical water for environmental technologies, J. Chem. Technol. Biotechnol. 85 (2010) 583–589.

[5] H. Weingärtner, E.U. Franck, Supercritical water as a solvent, Angew. Chem. Int. Ed. 44 (2005) 2672–2692.

[6] G. Brunner, Supercritical process technology related to energy and future directions—an introduction, J. Supercrit. Fluids 96 (2015) 11–20.

[7] NIST Chemistry WebBook, NIST Standard Reference Database Number 69, 2018, https://webbook.nist.gov/chemistry/ https://doi.org/10.18434/T4D303.

[8] U. Grigull, Dielektrizitätskonstante und Ionenprodukt von Wasser und Wasserdampf, BKW 35 (1983) 6.

[9] D. Xu, C. Huang, S. Wang, G. Lin, Y. Guo, Salt deposition problems in supercritical water oxidation, Chem. Eng. J. 279 (2015) 1010–1022.

[10] T. Voisin, A. Erriguible, D. Ballenghien, D. Mateos, D. Kunegel, F. Cansell, C. Aymonier, Solubility of inorganic salts in sub- and supercritical hydrothermal environment: application to SCWO processes, J. Supercrit. Fluids 120 (2017) 18–31.

[11] M.S. Khan, S.N. Rogak, Solubility of Na_2SO_4, Na_2CO_3 and their mixture in supercritical water, J. Supercrit. Fluids 30 (2004) 359–373.

[12] Q. Liu, X. Ding, B. Du, T. Fang, Multi-phase equilibrium and solubilities of aromatic compounds and inorganic compounds in sub- and supercritical water: a review, Crit. Rev. Anal. Chem. 47 (2017) 513–523.

[13] M. Schubert, J.W. Regler, F. Vogel, Continuous salt precipitation and separation from supercritical water. Part 1: type 1 salts, J. Supercrit. Fluids 55 (2010) 99–112.

[14] M. Schubert, J.W. Regler, F. Vogel, Continuous salt precipitation and separation from supercritical water. Part 2. Type 2 salts and mixtures of two salts, J. Supercrit. Fluids 52 (2010) 113–124.

[15] G. Lemoine, H.A. Turc, A. Leybros, J.C. Ruiz, Y.S. de Gelicourt, H. Muhr, Na_2CO_3 and K_3PO_4 solubility measurements at 30 MPa in near-critical and supercritical water using conductimetry and high pressure calorimetry, J. Supercrit. Fluids 130 (2017) 91–96.

[16] B. Hearn, M. Hunt, A. Hayward, Solubility of cupric oxide in pure subcritical and supercritical water, J. Chem. Eng. Data 14 (1969) 442–447.

[17] K. Sue, Y. Hakuta, R.L. Smith, T. Adschiri, K. Arai, Solubility of lead(II) oxide and copper(II) oxide in subcritical and supercritical water, J. Chem. Eng. Data 44 (1999) 1422–1426.

[18] T. Voisin, A. Erriguible, C. Aymonier, Influence of multiphasic systems on salt(s) solubility in supercritical water: the case of NaCl and NaCl-Na_2SO_4, J. Supercrit. Fluids 152 (2019). UNSP 104567.

[19] M.M. DiPippo, K. Sako, J.W. Tester, Ternary phase equilibria for the sodium chloride-sodium sulfate-water system at 200 and 250 bar up to 400 degrees C, Fluid Phase Equilib. 157 (1999) 229–255.

[20] M. Schubert, J. Aubert, J.B. Mueller, F. Vogel, Continuous salt precipitation and separation from supercritical water. Part 3: interesting effects in processing type 2 salt mixtures, J. Supercrit. Fluids 61 (2012) 44–54.

[21] M. Skerget, Z. Knez, M. Knez-Hrncic, Solubility of solids in sub- and supercritical fluids: a review, J. Chem. Eng. Data 56 (2011) 694–719.

[22] F. Masoodiyeh, M.R. Mozdianfard, J. Karimi-Sabet, Solubility estimation of inorganic salts in supercritical water, J. Chem. Thermodyn. 78 (2014) 260–268.

[23] J. Chrastil, Solubility of solids and liquids in supercritical gases, J. Phys. Chem. 86 (1982) 3016–3021.

[24] N.K.V. Nadimpalli, R. Bandyopadhyaya, V. Runkana, A coupled CFD-PBM and thermodynamic analysis of continuous supercritical hydrothermal synthesis of nanoparticles, J. Supercrit. Fluids 136 (2018) 169–179.

[25] M. Türk, Influence of thermodynamic behaviour and solute properties on homogeneous nucleation in supercritical solutions, J. Supercrit. Fluids 18 (2000) 169–184.

[26] T. Adschiri, Y. Hakuta, K. Arai, Hydrothermal synthesis of metal oxide fine particles at supercritical conditions, Ind. Eng. Chem. Res. 39 (2000) 4901–4907.

[27] T. Adschiri, Y. Hakuta, K. Sue, K. Arai, Hydrothermal synthesis of metal oxide nanoparticles at supercritical conditions, J. Nanopart. Res. 3 (2001) 227–235.

[28] N. Aoki, A. Sato, H. Sasaki, A.A. Litwinowicz, G. Seong, T. Aida, D. Hojo, S. Takami, T. Adschiri, Kinetics study to identify reaction-controlled conditions for supercritical hydrothermal nanoparticle synthesis with flow-type reactors, J. Supercrit. Fluids 110 (2016) 161–166.

[29] C.A. Eckert, B.L. Knutson, P.G. Debenedetti, Supercritical fluids as solvents for chemical and materials processing, Nature 383 (1996) 313–318.

[30] A. Kruse, E. Dinjus, Hot compressed water as reaction medium and reactant: properties and synthesis reactions, J. Supercrit. Fluids 39 (2007) 362–380.

[31] A. Kruse, E. Dinjus, Hot compressed water as reaction medium and reactant: 2. Degradation reactions, J. Supercrit. Fluids 41 (2007) 361–379.

[32] G. Brunner, Near critical and supercritical water. Part I. Hydrolytic and hydrothermal processes, J. Supercrit. Fluids 47 (2009) 373–381.

[33] G. Brunner, Near and supercritical water. Part II: oxidative processes, J. Supercrit. Fluids 47 (2009) 382–390.

[34] A. Kruse, N. Dahmen, Water—a magic solvent for biomass conversion, J. Supercrit. Fluids 96 (2015) 36–45.

[35] M. Bremholm, H. Jensen, S.B. Iversen, B.B. Iversen, Reactor design for in situ X-ray scattering studies of nanoparticle formation in supercritical water syntheses, J. Supercrit. Fluids 44 (2008) 385–390.

[36] M. Bremholm, M. Felicissimo, B.B. Iversen, Time-resolved in situ synchrotron X-ray study and large-scale production of magnetite nanoparticles in supercritical water, Angew. Chem. Int. Ed. 48 (2009) 4788–4791.

[37] G. Philippot, K.M.O. Jensen, M. Christensen, C. Elissalde, M. Maglione, B.B. Iversen, C. Aymonier, Coupling in situ synchrotron radiation with ex situ spectroscopy characterizations to study the formation of $Ba_{1-x}Sr_xTiO_3$ nanoparticles in supercritical fluids, J. Supercrit. Fluids 87 (2014) 111–117.

[38] T. Adschiri, Supercritical hydrothermal synthesis of organic-inorganic hybrid nanoparticles, Chem. Lett. 36 (2007) 1188–1193.

[39] T. Adschiri, K. Kanazawa, K. Arai, Rapid and continuous hydrothermal crystallization of metal-oxide particles in supercritical water, J. Am. Ceram. Soc. 75 (1992) 1019–1022.

[40] T. Adschiri, Y.W. Lee, M. Goto, S. Takami, Green materials synthesis with supercritical water, Green Chem. 13 (2011) 1380–1390.

[41] J.A. Darr, J.Y. Zhang, N.M. Makwana, X.L. Weng, Continuous hydrothermal synthesis of inorganic nanoparticles: applications and future directions, Chem. Rev. 117 (2017) 11125–11238.

[42] M. Daschner de Tercero, Near-critical and supercritical hydrothermal flow synthesis of metal oxide nanoparticles and hybrid metal oxide nanoparticles presenting clickable anchors (Ph.D. thesis), Karlsruhe Institute of Technology, Germany, 2014, https://doi.org/10.5445/KSP/1000043063.

[43] M. Daschner de Tercero, I.G. Martinez, M. Herrmann, M. Bruns, C. Kübel, S. Jennewein, U. Fehrenbacher, L. Barner, M. Türk, Synthesis of in situ functionalized iron oxide nanoparticles presenting alkyne groups via a continuous process using near-critical and supercritical water, J. Supercrit. Fluids 82 (2013) 83–95.

[44] M. Daschner de Tercero, C. Roder, U. Fehrenbacher, U. Teipel, M. Türk, Continuous supercritical hydrothermal synthesis of iron oxide nanoparticle dispersions and their characterization, J. Nanopart. Res. 16 (2014) 2350.

[45] K. Sugioka, K. Ozawa, M. Kubo, T. Tsukada, S. Takami, T. Adschiri, K. Sugimoto, N. Takenaka, Y. Saito, Relationship between size distribution of synthesized nanoparticles and flow and thermal fields in a flow-type reactor for supercritical hydrothermal synthesis, J. Supercrit. Fluids 109 (2016) 43–50.

[46] C. Slostowski, S. Marre, J.M. Bassat, C. Aymonier, Synthesis of cerium oxide-based nanostructures in near- and supercritical fluids, J. Supercrit. Fluids 84 (2013) 89–97.

[47] K. Sue, S.I. Kawasaki, T. Sato, D. Nishio-Hamane, Y. Hakuta, T. Furuya, Continuous hydrothermal synthesis of Pr-doped $CaTiO_3$ nanoparticles from a TiO_2 Sol, Ind. Eng. Chem. Res. 55 (2016) 7628–7634.

[48] H. Hayashi, Y. Hakuta, Hydrothermal synthesis of metal oxide nanoparticles in supercritical water, Materials 3 (2010) 3794–3817.

[49] H.L. Hellstern, J. Becker, P. Hald, M. Bremholm, A. Mamakhel, B.B. Iversen, Development of a dual-stage continuous flow reactor for hydrothermal synthesis of hybrid nanoparticles, Ind. Eng. Chem. Res. 54 (2015) 8500–8508.

[50] H.L. Hellstern, A. Mamakhel, M. Bremholm, B.B. Iversen, Core–shell nanoparticles by silica coating of metal oxides in a dual-stage hydrothermal flow reactor, Chem. Commun. 52 (2016) 3434–3437.

[51] M. Türk, C. Schüssler, 2019, unpublished data.
[52] A.S. Teja, P.-Y. Koh, Synthesis, properties, and applications of magnetic iron oxide nanoparticles, Prog. Cryst. Growth Charact. Mater. 55 (2009) 22–45.
[53] O. Pascu, S. Marre, C. Aymonier, Creation of interfaces in composite/hybrid nanostructured materials using supercritical fluids, Nanotechnol. Rev. 4 (2015) 487–515.
[54] M.K. Mitchell Lane, J.B. Zimmerman, Controlling metal oxide nanoparticle size and shape with supercritical fluid synthesis, Green Chem. 21 (2019) 3769–3781.
[55] K. Sue, T. Adschiri, K. Arai, Predictive model for equilibrium constants of aqueous inorganic species at subcritical and supercritical conditions, Ind. Eng. Chem. Res. 41 (2002) 3298–3306.

Chapter 9

Modeling of particle formation in supercritical fluids (SCF)

In approximately more than 95% of all publications on SCF based particle formation processes the influence of respective characteristic process conditions (pressure, temperature and precursor concentration) on the resulting product properties (particle size, size distribution and morphology) of the synthesized particles has been investigated experimentally. Nevertheless, due to the large number of (often unknown) interactions between the different operating conditions, such type of analysis and interpretation of the obtained results is often difficult. Therefore, findings received from the simulation and theoretical analysis coupled with the comparison of the experimental results can be very helpful for the interpretation of achieved product properties as well as process and product optimization.

9.1 Introduction

All models and approaches described in the following subchapters are useful to identify and to explain the influence of the respective characteristic process parameters on product properties. Independent from the particle formation process of interest (i.e., supercritical fluid reactive deposition, SFRD, or continuous hydrothermal synthesis, CHTS), the SCF acts as solvent, reaction and separation media. Generally, important process parameters are: phase behavior at pre and postexpansion (reaction) conditions, flow and mixing conditions, amount and chemical nature of the precursor. However, until today, the various models are able to predict correctly the variation of particle size and morphology with the changing process parameters, but fail very often in predicting the absolute value of the particle size. The reason therefore is basically the insufficient knowledge about interfacial tension and transport properties under sub or supercritical conditions, aggregation phenomena such as secondary nucleation or particle coalescence or even the complex solvent-precursor interactions. Furthermore, modeling results demonstrate the importance of the knowledge of the solvent-precursor and of the solvent-product interactions, i.e., solubility behavior and the critical locus, for the design of such particle formation processes [1].

Synthesis of Nanostructured Materials in Near and/or Supercritical Fluids
https://doi.org/10.1016/B978-0-444-64089-5.00021-4

9.2 SFRD process[a]

For the SFRD process, some abbreviations in common use are given below (see e.g. [2–5]); nevertheless, the basic principle of the deposition and particle formation process remains the same (for more details see Chapter 7).

- CFD: chemical fluid deposition
- SCD: supercritical deposition
- SCFD: supercritical fluid deposition
- SFCD: supercritical fluid chemical deposition
- SFRD: supercritical fluid reactive deposition

The *experimental* procedure of the SFRD process is performed usually in the batch mode and this technique involves the dissolution of a metallic precursor in a SCF (usually CO_2) and the exposure of a substrate to the solution. After sufficient time for adsorption of the precursor on the substrate, the precursor is transformed to its metal form. The reduction methods used are either chemical reduction in the SCF with a reducing agent, such as H_2, or thermal reduction in the SCF, or after depressurization thermal decomposition in an inert atmosphere, or chemical conversion with hydrogen or air. Since both, CO_2 and the reaction products are in the gaseous state, phase separation can be easily realized and a solvent free **and** ready to use product is obtained without any further post processing.

The complete *theoretical* description of the SFRD process requires considering the phase behavior, i.e., dissolution of the precursor in supercritical (sc) CO_2, sorption/adsorption of the precursor from $scCO_2$ onto the substrate, conversion of the adsorbed precursor to its metal form, and kinetics and energetics of nucleus formation, particle formation and growth.

One of the key properties in such particle formation processes is the phase behavior of the precursor in CO_2 (i.e., solubility in and melting point depression under CO_2) since the adsorption behavior is strongly influenced by the precursor concentration in the fluid phase. It is worthy to notice that the influence of temperature and pressure on the equilibrium adsorbed amount, i.e., uptake, is depicted in Fig. 9.1 and can be summarized as follows:

(a) At constant pressure a temperature increase leads to the lowering of the adsorbed amount.
(b) At constant temperature, the adsorbed amount decreases with an increase in pressure.
(c) Far above the critical point the influence of pressure and/or temperature on the adsorbed amount is weak.

---

a. Revised and extended version of Chapter 8.4 in Ref. [1]; Copyright © (2014) Elsevier.

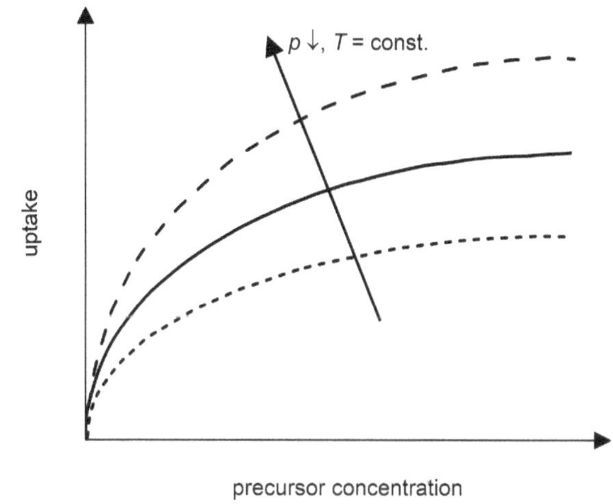

FIG. 9.1 General tendencies in uptake as a function of precursor concentration in $scCO_2$.

Thus for the successful deposition of metallic nanoparticles on or inside different substrates the knowledge of the thermodynamics and kinetics of the adsorption behavior is essential.

Aymonier et al. clarified recently that only few papers were dedicated to the simulation of the synthesis of inorganic materials in supercritical CO_2 [2]. In opposite thereto and as pointed out in a couple of review articles (see e.g. [3–6]), a large number of experiments are performed in order to obtain detailed information about the influence of SFRD process conditions such as type and amount of precursor in CO_2 and substrate properties on the obtained product characteristics. However, as mentioned and discussed below, at least a few number of papers investigated the adsorption behavior, whereas theoretical studies that address the full description of the mechanisms and kinetics of particle formation in the SFRD process are scarce.

From a more general point of view, two approaches are commonly employed for the synthesis of supported mono- and bimetallic nanoparticles by SFRD: a thermodynamically-controlled approach and a kinetically-controlled approach.

The first approach is strongly influenced by the thermodynamics of precursor adsorption onto the substrate that is mainly related to the solubility, i.e., concentration of the precursor in CO_2. As shown by Erkey's group [7–9] and discussed in detail below, the adsorption behavior of precursors which have high solubilities can be represented by Langmuir type isotherms. On the other hand, the behavior of precursors which have low solubilities in CO_2, can be described by the adsorption isotherms from Henry's type. Thus, it is possible to control the uptake, i.e., the precursor loading on the substrate, and therewith the degree of coverage according to the adsorption isotherms. So, the particle size can be influenced by the uptake since higher uptakes lead to larger particles. For example, Cangul et al. have shown that the adsorption behavior of $Pd(acac)_2$ and of $Pt(cod)me_2$ on BP2000 in $scCO_2$ differs significantly [8]. The adsorption isotherm for the first-mentioned precursor is linear whereas the isotherm for the second one is highly nonlinear and can be described by a modified Langmuir model. Similar findings were reported from Bozbag et al. for $Pt(cod)me_2$ on Resorcinol-formaldehyde aerogel (RFA) in CO_2 [9]. Thus, these investigations clarify that the adsorption behavior is strongly influenced by the complex nature of the precursor-substrate interaction and that a deeper understanding of the kinetics and thermodynamics of the adsorption and/or sorption process is essential for the further development of the SFRD technology.

In the second approach, the kinetics of the precursor transformation depends strongly from the chemical nature of the selected co-solvent. Thereby a co-solvent (e.g., methanol (MeOH), ethanol (EtOH), isopropanol (iPrOH)) is added to a $scCO_2$/precursor mixture in order to increase the precursors solubility and the density of the supercritical mixture. Several investigations show, that using iPrOH as co-solvent reduces the reduction kinetics of $Cu(hfac)_2{\cdot}H_2O$ over 2 orders of magnitude (2×10^{-4} to 3.2×10^{-2} min^{-1}) compared with using MeOH or EtOH as co-solvent [10–12]. Thereby it is worthy to notice that both, increasing temperature and residence time lead to larger particles while the smallest particles were obtained for the lowest reaction rate constant (2×10^{-4} min^{-1}). Therefore, the main modeling results obtained for the formation and growth of Cu nanoparticles in such CO_2/alcohol mixtures will be summarized below.

Based on batch and continuous experiments, Pessey et al. investigated the thermal decomposition of a Cu precursor which was dissolved in a supercritical CO_2/EtOH mixture [10]. In the simulation it was assumed that the precursor decomposition reaction is instantaneous and that liquid-like coalescence of the particles is the dominant growth mechanism while aggregation was not considered. The results show that experimental and calculated values are of the same order of magnitude. Furthermore, modeling results show that the initial precursor concentration has a strong influence on particle size and morphology. Similar to the experimental findings described in the Chapter 7 these modeling results show that a low initial concentration lead to spherical nanostructured particles while in case of higher concentrations "polyhedral" particles are obtained.

Erriguible et al. used a mono-disperse model in order to predict the growth of inorganic nanostructured particles in supercritical fluids [11]. The model considers the two mechanisms coalescence and subsequent aggregation; thereby the population balance equation (PBE) is solved by assuming a mono-disperse distribution. Furthermore, it is assumed that the chemical transformation of a metal precursor is instantaneous. The validation with experimental data obtained from the synthesis of metal Cu particles in a supercritical CO_2/EtOH mixture shows, that the simple model is able to predict the evolution of particle size as a function of the process operating parameters such as pressure, temperature, precursor concentration and residence time.

Marre et al. investigated the influence of different alcohols (e.g., MeOH or EtOH) on the reduction kinetics of $Cu(hfac)_2{\cdot}H_2O$ in $scCO_2$/alcohol/H_2 mixtures [12]. The results are used to model the formation of silica supported Cu nanoparticles at low temperatures (373 K to 398 K) and short residence times (< 120 min). Besides the (known) decomposition kinetics of the precursor, the model is based on a bimodal particle growth process: homogeneous nucleation in the supercritical media, i.e., continuous generation of nuclei and a fast heterogeneous growth due to deposition, coagulation and coalescence of the nuclei on the surface of the silica particles. The comparison with experimental results shows that the bottleneck of such processes is the continuous generation of nuclei in the reaction media and that such a formation and growth process is kinetically controlled.

Similar to the approach described above, Sierra-Pallares et al. [13] used computational fluid dynamics (CFD) and PBE calculations to estimate the size of TiO_2 nanoparticles synthesized from organometallic precursors using $scCO_2$ as reaction media in combination with EtOH as reactant. The authors suggested a pseudo-first order kinetic that has been proposed for mass and population balances. Thereby, the influence of Reynolds (*Re*) and Damköhler (*Da*) number on the particle diameter was investigated and the obtained modeling results were compared with experimental data. The *Da* number represents the ratio of reaction to mixing rate. A parametric study of the kinetic constant shows that for faster reactions ($Da \gg 10^{-3}$) the particle diameter increases when the Reynolds number is increased.

Müller developed a simple model for the description of the formation of metallic nanoparticles synthesized via SFRD [14]. Thereby, it is assumed that particle growth is a pure coagulation process of two-dimensional islands according to the Smoluchowski theory. The model allows to calculate the evolution over time of the particle number concentration, the mean particle diameter and the particle size distribution. This approach requires the knowledge about the degree of coverage, θ_{exp}, which can be determined from the loading of the substrate, surface area of the pure substrate, the projection area and molar mass of the metal particle, the reduction time and the two coagulation parameters. As depicted in Fig. 9.2, the modeling results are in good agreement with the experimental data. Furthermore, it is obvious that a higher degree of coverage, i.e., higher loading, results in larger particles and in case of some substrates in a broader size distribution.

Thus, in **summary**, independent from the use of a co-solvent or not, such SCF based particle formation processes enable the deposition of small (<5 nm) mono- and bimetallic particles. Thereby, modeling results show clearly that the average particle size and size distribution of such supported nanoparticles can be affected by the type and amount of the co-solvent (if used)

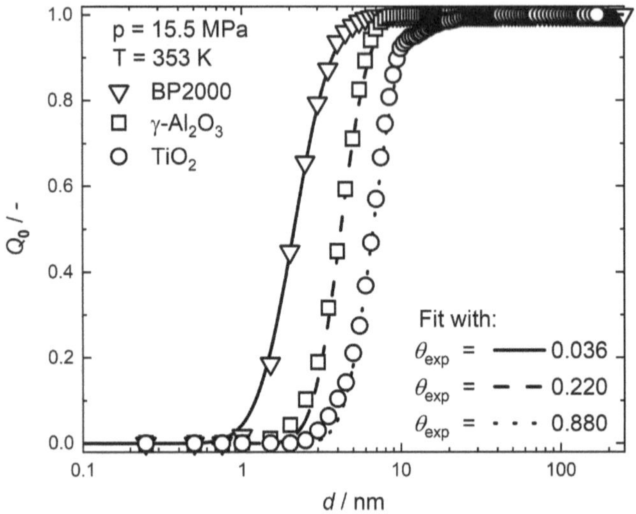

FIG. 9.2 Comparison between experimental obtained *(symbols)* and calculated Pt particle size distribution *(lines)* [14].

as well as a precursor in the system, precursor reduction method and condition, and surface properties (surface area and chemical nature) of the substrate. In agreement with the previous findings described in literature [3] a lower reactant concentration in the initial mixture results in smaller particles and a narrower size distribution.

9.3 CHTS process[b]

The continuous hydrothermal synthesis (CHTS) of metal oxide nanoparticles in sub and supercritical H_2O (scH_2O) has been investigated experimentally for more than 25 years from various research groups and is based on the mixing of a cold aqueous metal salt solution with hot compressed water [15, 16]. As described in Section 8.2 this process exploits the strong variations in the properties of water around the critical point ($p_C = 22.1$ MPa, $T_C = 647.1$ K, $\rho_C = 322$ kg m^{-3}) since scH_2O has very unusual and tunable properties. The diffusion rates are high and the viscosities low, mass transfer properties are similar to those of gases, density and solvating capabilities close to liquids together with the complete miscibility with a large number of substances, make scH_2O to an excellent process medium. Simplified, the synthesis of the metal oxides occurs in two consecutive reaction steps [17]:

- Hydrolysis of the metal ions and formation of metal hydroxides.
- Dehydration of the metal hydroxides to metal oxides and water.

b. Revised and extended version of Chapter 8.5 in Ref. [1]; Copyright © (2014) Elsevier.

Followed by the precipitation of the metal oxides due to dehydration of the hydroxide.

In order to obtain a complete description of the CHTS process, different fields have to be discussed and considered: dissolution of the metal salt in cold H_2O, followed by mixing with the sub or supercritical H_2O, two step synthesis of the metal oxides, and kinetics and energetics of nucleus formation, particle formation and growth. In addition, a complete model should take into consideration the complex geometry and fluid dynamics inside the mixing and reactor unit.

With regard to the CHTS process, the size and morphology of the synthesized nanoparticles can be controlled by various process parameters such as salt concentration, mixing behavior and residence time since these parameters control the nucleation and growth mechanism. Therefore, a detailed modeling of the CHTS process is important for an improved understanding of the mechanisms of particle formation and growth, which enables to optimize process conditions with regard to particle size and distribution. Thus, the reliable determination of the particle size distribution requires CFD models coupled with Population Balance (PB) models. Furthermore, this knowledge is needed for an optimized reactor design and for successful process scale-up.

In the following, the most important publications from the author's point of view are mentioned and the essential results achieved are briefly presented. First comprehensive results on the synthesis of CeO_2 and AlO(OH) were published in 2001 by Adschiri et al. [17]. Thereby the authors used the Sue-Adschiri-Arai (SAA) model which is based on the revised Helgeson-Kirkham-Flowers (rHKF) model for the calculation of equilibrium constants and solubilities for a large number of inorganic species in a wide range of pH, pressure (up to 50 MPa), and temperature (300 to 1250 K) [18, 19]. Due to the elimination of the partial molar volume in the rHKF model, the SAA model leads to simplified equations for the equilibrium constant. Thus, the concentration of dissolved species can be estimated by solving the system of nonlinear equations of the relevant equilibrium constants, mass and charge balances relation, as well as formulations for the ion activity coefficients. This enables the estimation of the supersaturation, i.e., the driving force for precipitation. Moreover, in agreement with experimental results, the modeling results show that larger particles are formed at subcritical reaction temperatures due to Ostwald ripening. In opposite thereto the formation of smaller particles at supercritical temperatures is caused by the increased reaction rate due to the lower salt solubility and the resulting higher supersaturation and thus increased nucleation rate. Furthermore, a faster heating of the aqueous salt solution leads to a higher supersaturation rate and thus to smaller particles.

The effect of different starting materials such as $Cu(NO_3)_3$, $Fe(NO_3)_3$, $Ni(NO_3)_2$, $ZrO(NO_3)_2$ and the concentration on particle size and the crystallinity of the synthesized metal oxide nanoparticles was analyzed on the basis of estimated metal oxide solubility and supersaturation by Sue et al. [20]. In this investigation the authors used the rHKF model for the calculation of the dissolution constants of metal oxides and the association reactions of metal cations and anions. The driving force for precipitation, the supersaturation S, is defined as:

$$S = \frac{(C_0 - C)}{C}$$

In this equation, C_0 and C denote metal nitrate concentration of the starting solutions and estimated metal oxide solubility at process conditions, respectively. Thereby the nitrite concentration varies from 0.001–0.1 mol kg^{-1} and the metal oxide solubility ranges from 10^{-18} (ZrO_2) to 10^{-4} (CuO) leading to supersaturations in the range from around 10 to 10^{+14}. The determined correlation between supersaturation and particle size is shown in Fig. 9.3. Obviously, a low solubility leads to high supersaturation that causes high nucleation rates and thus small particles. This behavior is well known from nucleation theory [1]. Furthermore, based on the obtained results the authors suggest that the crystallinity of the particles increases with decreasing supersaturation due to slower precipitation. However, the results show that for the precise control and tuning of the particle size, other underlying factors such as the geometry of the mixer unit have to be considered in a complete model. Since homogeneous mixing is crucial for uniform particle nucleation and growth and therewith the formation of small and uniform particles in the CHTS process

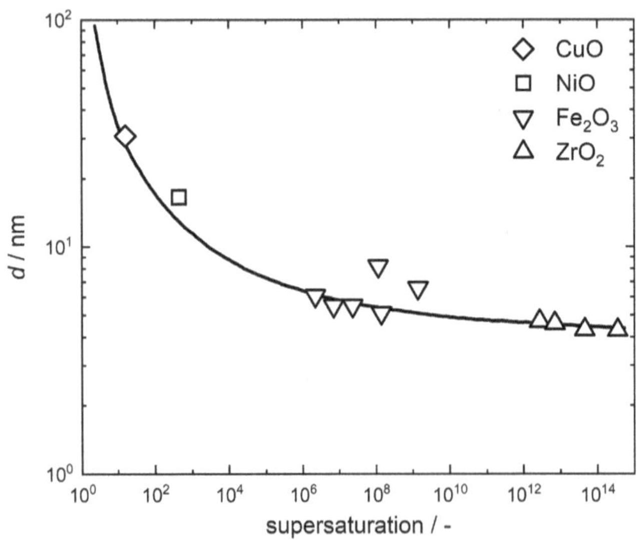

FIG. 9.3 Influence of supersaturation on particle size at $p = 25$–37.5 MPa and 673 K. *(Data from K. Sue, M. Suzuki, K. Arai, T. Ohashi, H. Ura, K. Matsui, Y. Hakuta, H. Hayashi, M. Watanabe, T. Hiaki, Size-controlled synthesis of metal oxide nanoparticles with a flow-through supercritical water method, Green Chem. 8(7) (2006) 634–638.)*

a large number of experimental studies has been performed. Among others different mixing units, in situ measurements under ambient or supercritical conditions and several studies on the mixing step including CFD simulations were carried out and are described briefly below.

Sierra-Pallares et al. [21] investigated the influence of different flow conditions in a micro-mixer by means of CFD and the so-called Reynolds Averaged Navier Stokes (RANS) approximation using the fundamental equation of Wagner and Pruss [22]. In agreement with the experimental results already described in Section 8.3.1, the simulation results show that high turbulences in the mixer and thus short mixing times or an effective mixing of the cold and hot water flow lead to smaller particles. Furthermore, the obtained modeling results suggest that a strong link exists between the particle size distribution (PSD) of the particles synthesized by CHTS and the residence time distribution (RTD).

Chen et al. presented results from simulation studies of the reaction and precipitation in a CHTS process considering reaction kinetics, thermodynamics, nucleation and PB modeling technique [23]. Thereby, in the PB models, particle growth mechanism and particle aggregation have also been taken into account. Based on this, the influence of temperature, aggregation and growth rate on the PSD under supercritical conditions was investigated. Besides this, a case study of simulating the production of nano-size TiO_2 particles in a CHTS system was performed. The modeling results suggest that TiO_2 particles in the range of 25–30 nm can be produced under steady-state conditions in the reactor. Furthermore, the influence of the system temperature on PSD seems to be not significant under supercritical conditions. Due to the high nucleation rates and short residence time, particle growth can be neglected. Additional simulation results showed that the enlarged mean particle size is the result of agglomeration.

Leybros et al. presented a CFD model coupled with a population balance approach for the formation of ZnO nanoparticles in scH_2O [24]. Models for salt solubility, reaction kinetics, nucleation and growth are included. The comparison with experimental results shows a good agreement in general trends but does not yet match exactly with the measured mean particle sizes.

Sugioka et al. carried out theoretical investigations as well as experimental ones as mentioned in Section 8.3.1 using the example of the hydrothermal synthesis of CeO_2 from $Ce(NO_3)_3$ [25, 26]. The three-dimensional, turbulent and transient flow and temperature profiles in a T-mixer and reactor were described with the commercial software ANSYS FLUENT 12.1. The required thermophysical properties of pure H_2O (density, heat capacity, viscosity and thermal conductivity) were taken from the database of the National Institute of Standards and Technology (NIST) and implemented in ANSYS FLUENT in the form of temperature-dependent polynomials (each at $p =$ constant). In these investigations, the flow conditions in the T-mixer as well as the ratio of cold to hot water flow was varied and the flow and temperature profile in the mixer was characterized by neutron tomography. The results of the theoretical investigations and their comparison with the experimental results show that in particular the particle size distribution is influenced by the velocity and temperature profile in the reactor and thus by the given flow conditions in the T-mixer.

Ma et al. used CFD in combination with a PB model for the investigation of hydrothermal fluid flow and heat transfer, as well as the mixing behavior in a counter-current and a confined reactor used for the synthesis of metal nanoparticles and to predict the particle size distribution [27]. One result of this investigation shows that the predicted temperature distributions for both types of reactors were found to be in good agreement with the measured data. Furthermore, the obtained results show that the mixing behavior in the counter-current reactor is significantly slower than in the confined jet mixer which results in a more uniform particle size distribution.

In a second study Ma et al. investigated the scale-up from laboratory to pilot plant scale of a confined jet mixer [28]. Thereby CFD simulation was used for the calculation of the fluid flow field and temperature profiles. In addition, the mixing behavior between the supercritical H_2O jet and precursor streams with different scale-up ratios was examined. Again, a good agreement between the calculated temperature profiles and experimental data was observed. Furthermore, the obtained results demonstrate that both, the laboratory and the pilot plant, show a similar reactor performance and therewith similar particle sizes under the investigated conditions. This result can be explained by the very high *Re* numbers ($\gg$10,000) and mean residence times which are in the same order of magnitude.

Similar experimental results have been observed by Aoki et al. for the synthesis of CeO_2 nanoparticles [29]. Thereby, the particle size was independent from the *Re* number at values higher than 2.5×10^4 (at 753 K).

Zielke et al. used the results from the CFD simulation for the optimization of the mixer geometry and based on this the design of a CHTS reactor with a two-stage mixer [30]. This made it possible to synthesize pure YSZ ($Y_{0.16}Zr_{0.84}O_2$) and pure NiO nanoparticles as well as particles consisting of YSZ and NiO phases.

Sierra-Pallares et al. carried out RTD experiments at low *Re* numbers ($360 < Re < 2130$) using a modified CHTS apparatus which enables the injection of a chromophoric tracer [31, 32]. In addition, CFD calculations were conducted using a mixer geometry which represent the experimental conditions. The flow and mixing of fluids under sub and supercritical conditions were governed by the equations for mass, momentum and energy balance in combination with a special approach for modeling turbulence. The results of the experimental tracer studies which were performed at low *Re* numbers and hot water temperatures in the range from 373.15 to 673.15 K show, that the RTD profiles are strongly influenced by the mixing behavior and turbulent diffusion in the reactor. The comparison between the experimental and calculated RTD profiles show that the proposed CFD model is able to predict the basic trends with sufficient accuracy.

Nadimpalli et al. developed a coupled CFD-PB model and a Gibbs-energy minimization based thermodynamic model for the simulation of the supercritical hydrothermal synthesis of CeO_2 nanoparticles [33]. In this investigation, the thermodynamic model was used to determine the solubility of CeO_2 as a function of temperature. The modeling results were compared with published experimental data for (a) the radial temperature of the reaction mixture near the T-junction and (b) CeO_2 PSD, obtained for three different flow configurations at various reactant flow rates. The comparison between experimental and modeling data shows, that the PSD of the product is broader, when back-mixing is dominant, whereas it is narrower when mixing is limited over a small sector. Furthermore, the PSD is wider at low flow rates of the reactants and becomes narrower when the flow rate increases. Based on the flow configuration and operating conditions the overall synthesis process can be either reaction controlled or mixing controlled. Furthermore, relationships between the mean particle size and the process conditions were derived which is useful for optimizing the process conditions for the synthesis of nanoparticles with desired properties.

Recently, the hydrothermal synthesis of Bi_2O_3 nanoparticles was investigated by Mozdianfard et al. [34]. The authors used Response Surface Methodology (RSM) and PBE to investigate the influence of the process variables (i.e., temperature, reactant concentration and pH) on the mean particle size and PSD of Bi_2O_3 particles. For the optimal process conditions, the comparison between experimental (19.63 nm $\pm$ 2.32) and calculated PSD (20.28 nm $\pm$ 1.82) shows good agreement and confirm the validity of the applied model.

In a first approach Türk and Schüssler calculated the temperature and velocity field in a tubular flow reactor numerically under typical experimental conditions using the software Analysis System (ANSYS CFD) that is based on the Finite-Volume Method (FVM) for the discretization of equations of fluid dynamics [35]. Five models for different mixing configurations (Counter, Bending, Straight, Y-30° and Cyclone model) were created and afterwards the thermo-physical properties of H_2O were implemented for simulating the mixing process of a cold and a hot H_2O stream. The pressure and temperature dependences of the thermo-physical properties of H_2O, i.e., density, viscosity, thermal conductivity, and specific heat, were obtained from the NIST database for a temperature range of 300 to 700 K [22]. These properties were piecewise curve-fitted over several temperature ranges for each constant pressure of 25, 30, and 35 MPa.

The aim of this investigation was to figure out the effects of various CHTS process parameters such as the system's pressure, hot water stream's temperature, mixing ratio and total mass flow rate on the temperature and velocity profiles for different reactor configurations.

The obtained numerical results for the temperature and velocity field (see Fig. 9.4) were compared and validated with experimental data from CHTS experiments which were performed at 25, 30 and 35 MPa, temperatures in the range from 575 to 673 K and different mixing ratios (see Table 9.1). The comparison between experimental and calculated mixing temperature shows deviations less than 4% for T_3.

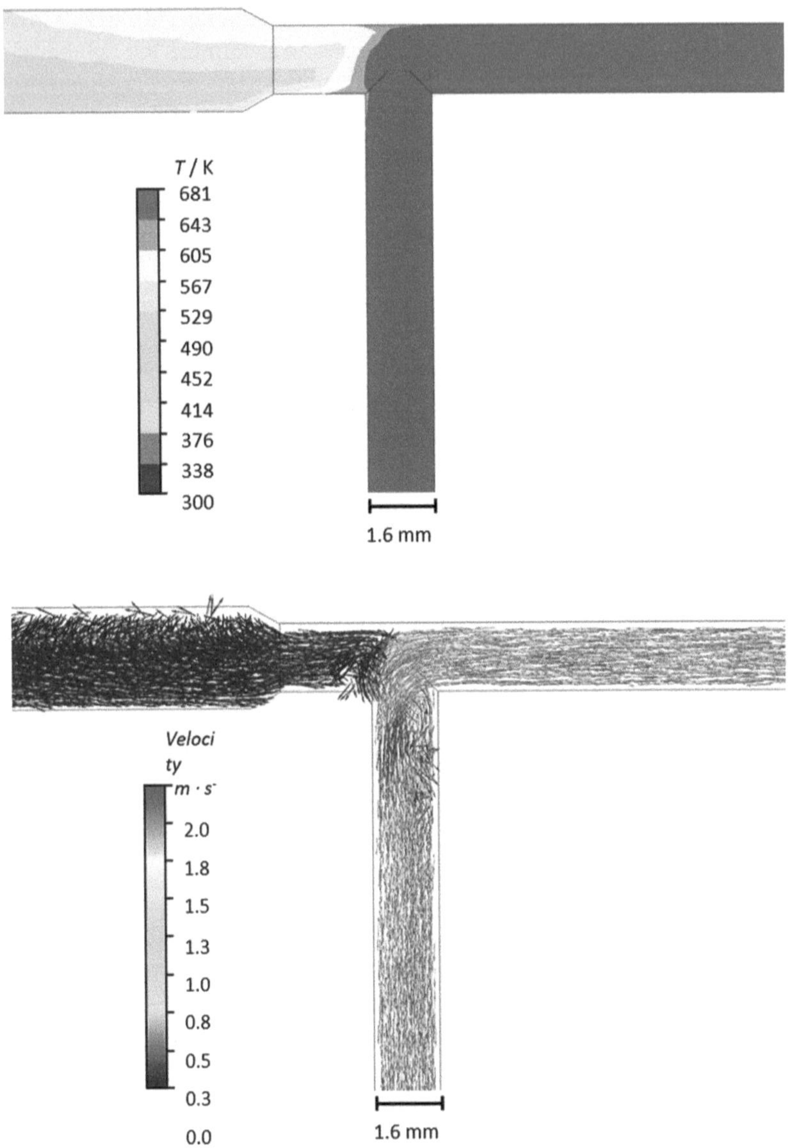

FIG. 9.4 Calculated temperature (above) and velocity (below) distribution in the mixing unit where $p = 35$ MPa, $T_1 = 300$ K, $T_2 = 673$ K, total mass flow rate $\dot{m}_3 = 35$ g min^{-1} and mixing ratio $a = 0.05$ [36].

In a next step the model was extended to validate experimental PSD with modeling results. As described by Sierra-Pallares et al. the RTD curves can be obtained by simulating an injection of an inert tracer into the system [21]. Therefore, CFD was coupled with a Particle Balance Model (PBM) for

TABLE 9.1 Comparison between experimental (exp) and modeling (sim) results of mixing a cold and a hot H_2O steam (Counter model [36]).

p (MPa)	T_2 (K)	a	$T_{3,exp}$ (K)	$T_{3,sim}$ (K)	$(\Delta T_3/T_{3,exp})$ (%)
25	573	0.05	544	552	1.5
25	573	0.20	518	525	1.3
25	573	0.35	495	501	1.1
25	673	0.05	661	664	0.4
25	673	0.35	647	649	0.3
35	583	0.05	552	563	1.9
30	583	0.05	555	565	1.7
35	583	0.35	502	508	1.1
35	673	0.05	647	664	2.6
30	673	0.35	607	617	1.5
35	673	0.35	588	611	3.8

estimating the PSD of particles synthesized under different CHTS process conditions. The behavior of the tracer is modeled by fixing the fluid flow field and using an additional equation to model the concentration of tracer.

It is important to note that the main deficit of such a particular distribution is the lack of information relating to the absolute values of the particle size. Nevertheless, this method is important for the prediction of the influence of any process parameter on the size of the synthesized particles and their distribution. Therefore, RTD curves are calculated by changing the following process parameters (see Table 9.1):

- total mass flow rate
- temperature of the hot H_2O stream
- mixing ratio
- system pressure

The calculated RTD curves for different setups are shown in Fig. 9.5. Note, that RTD curves represent the most probable distribution of time that solid or fluid materials stay inside a continuous flow system and is a crucial index for the understanding the flow profile.

Thereby, in accordance with the experimental results (cf. Fig. 8.14 in Section 8.3.1), the following trends were observed: Increasing the temperature T_2, the total mass flow rate and/or the mixing ratio between the aqueous metal salt solution and the hot scH_2O stream leads to a shorter residence times which

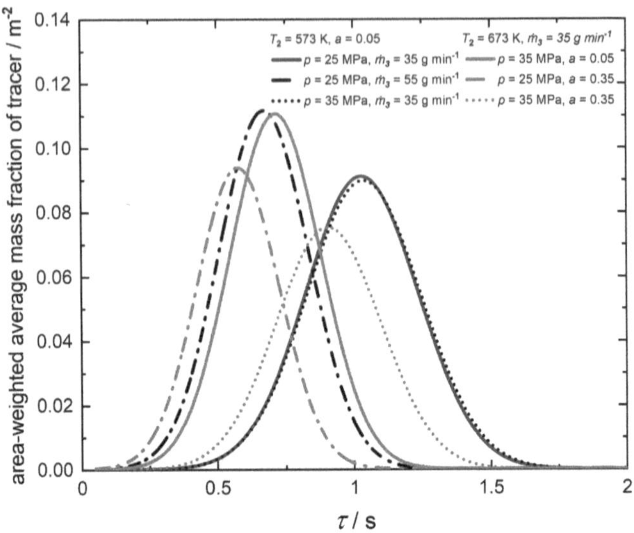

FIG. 9.5 Influence of various process parameters on RTD curves [36].

lead to smaller particles. In opposite thereto, increasing reactant concentration or lower mass flow rate result in a longer residence time in the reactor which leads to larger particles since agglomeration is favored.

Lümmen and Kvamme [37] and Nahtigal et al. [38] investigated the kinetics of particle formation, growth and nucleation in scH_2O by molecular dynamics simulations (MDS) at different system temperatures and densities. Since the understanding of the kinetics of the salt association and agglomeration processes is important for an improved understanding of the CHTS process, an aqueous NaCl solution was used as an example for a nonreacting model system. Thereby the H_2O-H_2O interaction was modeled by the rigid extended single point charge model. For the interaction of the sodium and chloride ions with the H_2O molecules, the authors used a model given in literature. Modeling results show that the initial particle growth by monomer and dimer addition onto ion pairs occurs immediately (within a few 100 ps) after the jump from ambient to supercritical conditions. Thereafter, the particles grow by particle-particle collisions. The nucleation rates are calculated from the time development of the cluster size distribution and the model leads to large values for the nucleation rate in the range up to $10^{+36}\ m^{-3}\ s^{-1}$. Thus, it is not clear if the observed particle formation is induced either by homogeneous nucleation or spinodal decomposition.

More details of the different concepts that are currently employed in modeling the synthesis of metallic/metal oxide nanoparticles (NPs) using SCFs, the different boundary conditions and the obtained numerical results can be found in the original literature and in various review articles [e.g., see without any claim to completeness Refs. (e.g. [25, 26, 28, 29, 32–34, 39–42])].

The approach in modeling the CHTS process can be summarized as follows: A detailed description of the CHTS process is an extended approach proposed by Helfgen et al. [43] modeling particle formation and growth without chemical reaction and by Chen et al. [23] who considers chemical reaction under sub and supercritical conditions. It consists of the following tasks described below. Every single step corresponds to a deeper understanding of the underlying physical and chemical phenomena of the relationship between the process conditions and the mechanism of particle formation during CHTS.

Task 1: Determination of the flow field (i.e., temperature, density, velocity) based on the flow rate of the cold and the hot H_2O stream and its temperatures, the total flow rate, the mixer and reactor configuration. Thereby the model should include information about the geometric dimensions of the mixing and reactor device. As the H_2O/salt system is dilute, calculations can be performed for the pure H_2O. Thus, the fundamental conservation equations to calculate the flow field are the general mass, momentum and energy balances plus reliable thermo-physical data for pure H_2O. Thereby it is important to validate the model by comparing the measured and calculated mixing temperatures for all process configurations.

Task 2: The simulation of the flow field results in temperature, density, and velocity profiles. These profiles can be used to calculate RTD curves and to investigate the influence of total flow rate, temperature, mixing ratio and pressure on the prospective PSDs of the nanoparticles synthesized by the CHTS process. Such RTD curves can be received by simulating an injection of an inert tracer into the system. Thereby the behavior of the tracer is modeled by fixing the fluid flow field and using an additional equation to model the concentration of the tracer, with a diffusion coefficient equal to the turbulent diffusivity. Nevertheless, it must be considered that RTD curves offer only information about the changes in the particle size distributions when the process conditions are varied.

Task 3: A reaction model for the hydrothermal synthesis is needed for calculation of the reaction rate constant under various flow rates, process pressures and temperatures. Thereby first order kinetics can be assumed because the concentration of the salt is significantly lower than the concentration of H_2O which can be considered as constant. Usually the reaction rate is expressed by the so called "Born type equation" (see Eq. (8.2) in Section 8.2.2) which is mainly influenced by the frequency factor and activation energy of the considered reaction and the dielectric constant of H_2O at the current process conditions.

Task 4: Modeling the thermodynamic equilibrium constant at the current process pressure and temperature to determine solubility profiles and degree of supersaturation in the system. This can be achieved by using the Sue-Adschiri-Arai (SAA) model proposed by Sue et al. [18, 19] that could predict successfully both reaction equilibrium and also metal oxide solubility in sub and supercritical H_2O. The model consists of the revised Helgeson-Kirkham-Flowers (rHKF) model with a modified nonelectrostatic term.

Task 5: For the calculation of the nucleation rate (see Eq. (8.1) in Section 8.2.2), an equation based on the classical nucleation theory arguments but extended to fluid-phase nonideality is used to model the nucleation process in aqueous solutions under sub and supercritical conditions. Thereby the supersaturation is obtained from the SAA model and used to calculate the maximum value of Gibbs energy for spherical particles, the critical nuclei size and number concentration of the critical nuclei.

Task 6: Finally, particle formation and growth are calculated using the general dynamic equation (GDE). Particle formation and growth is considered by solving the GDE for simultaneous nucleation, condensation, and coagulation. Therefore, different ways to solve the GDE are available in literature: One common approach is the method of moments, proposed by Friedlander [44] and Pratsinis [45]. More details about the various concepts that are currently used can be found in various articles (e.g., see without any claim to completeness Refs. [10–13, 43–45]) and in the section below.

9.4 Supplementary explanations on the application of the general dynamic equation[c]

Once nucleation occurs, particles grow by coagulation, which is the growth by the collision of particles, and by condensation, which is the deposition of molecules on the particles surface as pictured in Fig. 9.6.

Such a process of particle formation and growth can be described by using the GDE for simultaneous nucleation, condensation, and coagulation [43–45]:

$$\frac{\partial n_v(vt)}{\partial t} = \underbrace{J(v_{\mathrm{crit}}) \cdot \delta(v - v_{\mathrm{crit}})}_{\text{particle formation}} - \underbrace{\frac{\partial(G \cdot \mathrm{n}_v(vt))}{\partial v}}_{\text{condensation}}$$

$$+ \underbrace{\frac{1}{2}\int_0^v \beta(v - v^+v^+) \cdot n_v(v - v^+t) \cdot n_v(v^+t)dv^+}_{\text{coagulation}} \underbrace{- n_v(vt)\int_0^v \beta(vv^+) \cdot n_v \cdot (v^+t)dv^+}_{\text{coagulation}} \tag{9.1}$$

The term on the left-hand side in Eq. (9.1) describes the change of the particle concentration n_v within the limit of particle volume v to $v + dv$. The first term on the right-hand side of Eq. (9.1) considers the homogeneous nucleation of particles with the volume v_{crit} at a given nucleation rate J and Delta function δ. The second term consider the variation of the particle volume by condensation

c. Revised and extended version of Chapter 3.5 in Ref. [1]; Copyright © (2014) Elsevier.

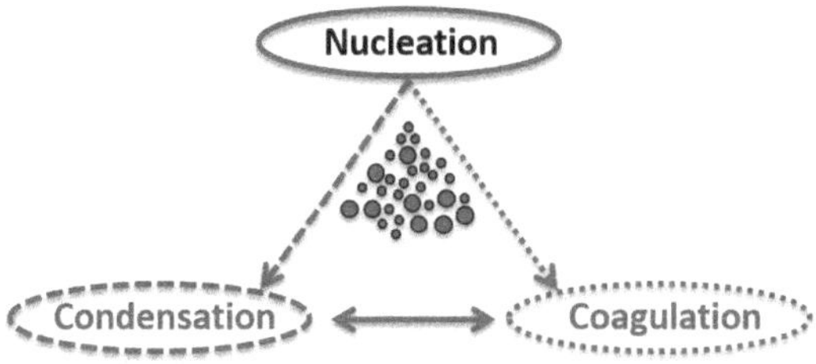

FIG. 9.6 Schematic representation of the mechanism of precipitation and particle growth.

through the condensation rate G. Particle growth due to coagulation is considered by the last two terms with the coagulation coefficient β. Thereby the third term describes the increase of a particle by coagulation of a particle of volume v with a particle of volume $v - v^+$. This term has to be multiplied by ½ since every particle is considered twice. The last term accounts for the decrease of particles in the range from v^+ to $(v^+ + dv^+)$ due to coagulation with any other particle of volume v^+. A common approach to solve the GDE is the method of moments [45]. These equations can be simplified by assuming a certain particle distribution, e.g., a lognormal size distribution and one obtains three differential equations for the zeroth (M_0), first (M_1) and second (M_2) moment. Thereby it is worthy to notice that M_0 is influenced only by nucleation and coagulation while the condensation process does not affect the number concentration. The first moment M_1 describes the volume change of the solid in the gas phase due to nucleation and condensation while coagulation does not play any role as it does not affect the total solid volume concentration. In opposite thereto, the second moment M_2 considers all three mechanisms: nucleation, condensation and coagulation.

From these three equations the characteristic quantities of a particle collective i.e., number concentration of the particles N, the mean particle diameter $d_{50,0}$ and the standard deviation σ can be obtained from the three moments [43]:

$$N = M_0$$

$$d_{50,0} = \left(\frac{6}{\pi} \frac{M_1^2}{M_0^{3/2} \cdot M_2^{1/2}} \right)^{1/3} \tag{9.2}$$

$$\ln^2 \sigma = \frac{1}{9} \ln \left(\frac{M_0 \cdot M_2}{M_1^2} \right)$$

A typical modeling result is depicted in Fig. 9.7 and show the temporal change of the concentration N and mean particle diameter $d_{50,0}$. It is obvious that at the initial stage the particles grow very fast while the number concentration decreases simultaneously.

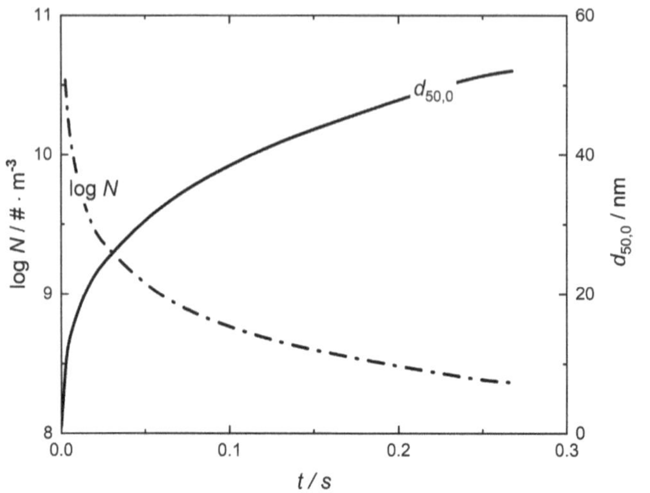

FIG. 9.7 Influence of residence time on the calculated number concentration (left *y*-axis) and mean particle size (right *y*-axis).

Note, as the measured particle sizes are normally given as number volume averaged diameters d, the geometric diameters $d_{50,0}$ can be converted to d according to:

$$\ln d = \ln d_{50,0} + 1.5\ \ln^2\sigma \tag{9.3}$$

The standard deviation of all calculations tends toward $\sigma = 1.3$ which is a result of both the one-dimensional assumption and coagulation as the only mechanism of particle growth.

Interim summary: Compared to the different modeling approaches which deal with the description of the micronization of organic particles by CO_2-based particle formation processes, theoretical studies that address the mechanism and kinetics of particle synthesis with CHTS are at an early stage. Coupling of process conditions, i.e., mixing geometry, pressure, temperature, salt concentration and pH with solubility data and/or chemical reaction models enables to identify the best operating conditions and predict particle size distributions based on the chemical nature of the salt and solubility of metal oxides in sub and supercritical H_2O. The results of these investigation indicate that besides pressure and temperature the pH value is an additional parameter that can affect size and shape of the final metal oxide nanoparticles.

In addition, the modeling results show that a decrease of the salt concentration in the initial mixture and an increase of the total mass flow rate lead to smaller particles and a narrower size distribution. Thus, since these findings agree with experimental results (see Section 8.3.1) the models proposed by Adschiri et al. [17], Ma et al. [27, 28], Sierra-Pallares et al. [31, 32], and Türk and Schüssler [35] are a first step toward a quantitative understanding of the CHTS process. Nevertheless, from a more general point of view, it must be noted that the theoretical work on the CHTS process published so far in literature often only considers partial aspects and not the complete particle formation process.

References

[1] M. Türk, Particle formation with supercritical fluids: challenges and limitations, in: E. Kiran (Ed.), Supercritical Fluid Science and Technology Series, first ed., vol. 6, Elsevier, Amsterdam, 2014.

[2] C. Aymonier, G. Philippot, A. Erriguible, Playing with chemistry in supercritical solvents and the associated technologies for advanced materials by design, J. Supercrit. Fluids 134 (2018) 184–196.

[3] M. Türk, C. Erkey, Synthesis of supported nanoparticles in supercritical fluids by supercritical fluid reactive deposition: current state, further perspectives and needs, J. Supercrit. Fluids 134 (2018) 176–183.

[4] Y. Zhang, C. Erkey, Preparation of supported metallic nanoparticles using supercritical fluids: a review, J. Supercrit. Fluids 38 (2006) 252–267.

[5] C. Erkey, Preparation of metallic supported nanoparticles and films using supercritical fluid deposition, J. Supercrit. Fluids 47 (2009) 517–522.

[6] S.E. Bozbag, D. Sanli, C. Erkey, Synthesis of nanostructured materials using supercritical CO_2: part II. Chemical transformations, J. Mater. Sci. 47 (2012) 3469–3492.

[7] Y. Zhang, B. Cangul, Y. Garrabos, C. Erkey, Thermodynamics and kinetics of adsorption of bis(2,2,6,6-tetramethyl-3,5-heptanedionato) (1,5-cyclooctadiene) ruthenium (II), J. Supercrit. Fluids 44 (2008) 71–77.

[8] B. Cangul, L.C. Zhang, M. Aindow, C. Erkey, Preparation of carbon black supported Pd, Pt and Pd–Pt nanoparticles using supercritical CO_2 deposition, J. Supercrit. Fluids 50 (2009) 82–90.

[9] S.E. Bozbag, N.S. Yasar, L.C. Zhang, M. Aindow, C. Erkey, Adsorption of Pt(cod)me_2 onto organic aerogels from supercritical solutions for the synthesis of supported platinum nanoparticles, J. Supercrit. Fluids 56 (2011) 105–113.

[10] V. Pessey, R. Garriga, F. Weill, B. Chevalier, J. Etourneau, F. Cansell, Control of particle growth by chemical transformation in supercritical CO_2/ethanol mixtures, J. Mater. Chem. 12 (2002) 958–965.

[11] A. Erriguible, F. Mariasi, F. Cansell, C. Aymonier, Monodisperse model to predict the growth of inorganic nanostructured particles in supercritical fluids through a coalescence and aggregation mechanism, J. Supercrit. Fluids 48 (2009) 79–84.

[12] S. Marre, A. Erriguible, A. Perdomo, F. Cansell, F. Marias, C. Aymonier, Kinetically controlled formation of supported nanoparticles in low temperature supercritical media for the development of advanced nanostructured materials, J. Phys. Chem. C 113 (2009) 5096–5104.

[13] J. Sierra-Pallares, E. Alonso, I. Montequi, M.J. Cocero, Particle diameter prediction in supercritical nanoparticle synthesis using three-dimensional CFD simulations. Validation for anatase titanium dioxide production, Chem. Eng. Sci. 64 (2009) 3051–3059.

[14] S.A. Müller, Untersuchungen zur Herstellung geträgerter mono- und bimetallischer Nanopartikel mithilfe von überkritischem CO_2, Karlsruhe Institute of Technology (KIT), Germany, 2015, https://doi.org/10.5445/KSP/1000051502.

[15] Y. Hakuta, H. Hayashi, K. Arai, Fine particle formation using supercritical fluids, Curr. Opinion Solid State Mater. Sci. 7 (2003) 341–351.

[16] T. Adschiri, Highlight review: supercritical hydrothermal synthesis of organic-inorganic hybrid nanoparticles, Chem. Lett. 36 (10) (2007) 1188–1193.

[17] T. Adschiri, Y. Hakuta, K. Sue, K. Arai, Hydrothermal synthesis of metal oxide nanoparticles at supercritical conditions, J. Nanopart. Res. 3 (2001) 227–234.

[18] K. Sue, T. Adschiri, K. Arai, Estimation of metal oxide solubility in sub- and supercritical water, in: Proc. Book Joint 6th ISHR & 4th ICSTR, Kochi, Japan, 2001.

[19] K. Sue, T. Adschiri, K. Arai, Predictive model for equilibrium constants of aqueous inorganic species at subcritical and supercritical conditions, Ind. Eng. Chem. Res. 41 (2002) 3298–3306.

[20] K. Sue, M. Suzuki, K. Arai, T. Ohashi, H. Ura, K. Matsui, Y. Hakuta, H. Hayashi, M. Watanabe, T. Hiaki, Size-controlled synthesis of metal oxide nanoparticles with a flow-through supercritical water method, Green Chem. 8 (7) (2006) 634–638.

[21] J. Sierra-Pallares, D.L. Marchisio, E. Alonso, M.T. Parra-Santos, F. Castro, M.J. Cocero, Quantification of mixing efficiency in turbulent supercritical water hydrothermal reactors, Chem. Eng. Sci. 66 (2011) 1576–1589.

[22] W. Wagner, A. Pruss, The IAPWS formulation 1995 for the thermodynamic properties of ordinary water substance for general and scientific use, J. Phys. Chem. Ref. Data 31 (2002) 387–535.

[23] M. Chen, C.Y. Ma, T. Mahmud, J.A. Darr, X.Z. Wang, Modelling and simulation of continuous hydrothermal flow synthesis process for nano-materials manufacture, J. Supercrit. Fluids 59 (2011) 131–139.

[24] A. Leybros, R. Piolet, M. Ariane, H. Muhr, F. Bernhard, F. Demoisson, CFD simulation of ZnO nanoparticle precipitation in a supercritical water synthesis reactor, J. Supercrit. Fluids 70 (2012) 17–26.

[25] K. Sugioka, K. Ozawa, T. Tsukada, S. Takami, T. Adschiri, K. Sugimoto, N. Takenaka, Y. Saito, Neutron radiography and numerical simulation of mixing behavior in a reactor for supercritical hydrothermal synthesis, AICHE J. 60 (2014) 1168–1175.

[26] K. Sugioka, K. Ozawa, M. Kubo, T. Tsukada, S. Takami, T. Adschiri, K. Sugimoto, N. Takenaka, Y. Saito, Relationship between size distribution of synthesized nanoparticles and flow and thermal fields in a flow-type reactor for supercritical hydrothermal synthesis, J. Supercrit. Fluids 109 (2016) 43–50.

[27] C.Y. Ma, M. Chen, X.Z. Wang, Modelling and simulation of counter-current and confined jet reactors for hydrothermal synthesis of nano-materials, Chem. Eng. Sci. 109 (2014) 26–37.

[28] C.Y. Ma, J.J. Liu, Y. Zhang, X.Z. Wang, Simulation for scale-up of a confined jet mixer for continuous hydrothermal flow synthesis of nanomaterials, J. Supercrit. Fluids 98 (2015) 211–221.

[29] N. Aoki, A. Sato, H. Sasaki, A.A. Litwinowicz, G. Seong, T. Aida, D. Hojo, S. Takami, T. Adschiri, Kinetics study to identify reaction-controlled conditions for supercritical hydrothermal nanoparticle synthesis with flow-type reactors, J. Supercrit. Fluids 110 (2016) 161–166.

[30] P. Zielke, Y. Xu, S.B. Simonsen, P. Norby, R. Kiebach, Simulation, design and proof-of-concept of a two-stage continuous hydrothermal flow synthesis reactor for synthesis of functionalized nano-sized inorganic composite materials, J. Supercrit. Fluids 117 (2016) 1–12.

[31] J. Sierra-Pallares, T. Huddle, E. Alonso, F.A. Mato, J.G. Serna, M. Cocero, E. Lester, Prediction of residence time distributions in supercritical hydrothermal reactors working at low Reynolds numbers, Chem. Eng. J. 299 (2016) 373–385.

[32] J. Sierra-Pallares, T. Huddle, J.G. Serna, E. Alonso, F. Mato, I. Shvets, O. Luebben, M.J. Cocero, E. Lester, Understanding bottom-up continuous hydrothermal synthesis of nanoparticles using empirical measurement and computational simulation, Nano Res. 9 (11) (2016) 3377–3387.

[33] N.K.V. Nadimpalli, R. Bandyopadhyaya, V. Runkana, A coupled CFD-PBM and thermodynamic analysis of continuous supercritical hydrothermal synthesis of nanoparticles, J. Supercrit. Fluids 136 (2018) 169–179.

[34] M.R. Mozdianfard, F. Masoodiyeh, J. Karimi-Sabet, Supercritical water hydrothermal synthesis of Bi_2O_3 nanoparticles: process optimization using response surface methodology based on population balance equation, J. Supercrit. Fluids 136 (2018) 144–156.

[35] M. Türk, Ch. Schüssler, (2019) unpublished results.

[36] S.K. Ponusamy, Theoretical Study of the Mixing Behavior During Continuous Hydrothermal Synthesis (CHTS) of Metal Oxide Nanoparticles (Master thesis), in: Karlsruhe Institute of Technology (KIT), 2018.

[37] N. Lümmen, B. Kvamme, Kinetics of NaCl nucleation in supercritical water investigated by molecular dynamics simulations, Phys. Chem. Chem. Phys. 9 (2007) 3251–3260.

[38] I.G. Nahtigal, A.Y. Zasetsky, I.M. Svishchev, Nucleation of NaCl nanoparticles in supercritical water: molecular dynamics simulations, J. Phys. Chem. B 112 (2008) 7537–7543.

[39] I. Shvab, R.J. Sadus, Atomistic water models: aqueous thermodynamic properties from ambient to supercritical conditions, Fluid Phase Equilib. 407 (2016) 7–30.

[40] N.T.K. Thanh, N. Maclean, S. Mahiddine, Mechanisms of nucleation and growth of nanoparticles in solution, Chem. Rev. 114 (2014) 7610–7630.

[41] J.A. Darr, J. Zhang, N.M. Makwana, X. Weng, Continuous hydrothermal synthesis of inorganic nanoparticles: applications and future directions, Chem. Rev. 117 (2017) 11125–11238.

[42] F.J.G. Ortiz, A. Kruse, The use of process simulation in supercritical fluids applications, React. Chem. Eng. 5 (2020) 424–451.

[43] B. Helfgen, M. Türk, K. Schaber, Hydrodynamic and aerosol modelling of the rapid expansion of supercritical solutions (RESS-process), J. Supercrit. Fluids 26 (2003) 225–242.

[44] S.K. Friedlander, Smoke, Dust and Haze, Fundamentals of Aerosol Behavior, Wiley, New York, London, Sydney, Toronto, 1977.

[45] S.E. Pratsinis, Simultaneous nucleation, condensation, and coagulation in aerosol reactors, J. Colloid Interface Sci. 124 (2) (1988) 416–427.

Chapter 10

Future research needs and developments

At present and in future, discovering environment-friendly and renewable energy sources, energy generation, and energy savings is one of the major challenges. Furthermore, manufacturing as well as service industries must increasingly attempt to avoid the production, use, and release of harmful substances into our environment. Therefore, one consequence is that new processes must be environmentally friendly when employed at a large scale.

For improvement of the technological and economic feasibility of particle formation processes in which the supercritical fluid acts either as solvent, and/or reaction and/or separation media and their optimization, it is crucial to achieve a deeper understanding of the influence of the various operating conditions and process parameters on the obtained product characteristics. The successful attainment of this challenging task requires an intensive interdisciplinary cooperation between researchers from different but complementary disciplines as well as fundamental and experimental researchers.

The following chapters summarize what is still missing for further advances in SCF-based particle formation processes.

10.1 Synthesis of mono and bimetallic nanoparticles[a]

With regard to the synthesis of supported nanostructured materials by SFRD experimental and theoretical investigations of the kinetics and thermodynamics of the adsorption or sorption processes are very beneficial for the development of this technology. Furthermore, information about the pure and impregnated substrate characteristics ranging from e.g., pore volume, pore diameter, porosity, specific surface area, pore size distribution to surface properties should be known.

a. Revised and extended version from Chapters 3 and 4 in M. Türk, C. Erkey, Synthesis of supported nanoparticles in supercritical fluids by supercritical fluid reactive deposition: current state, further perspectives and needs, J. Supercrit. Fluids 134 (2018) 176–183, Copyright © (2017) Elsevier.

Synthesis of Nanostructured Materials in Near and/or Supercritical Fluids
https://doi.org/10.1016/B978-0-444-64089-5.00019-6

It is evident that in a stirred high-pressure vessel the time to reach equilibrium depends strongly, i.e., inversely proportional, to the binary diffusion coefficient D_{12} that is typically at least one order of magnitude higher than in organic liquids. The most important dependency is, that D_{12} changes stronger with pressure than with temperature. Note, it is important to realize that experimental D_{12} data are only available for a limited number of precursors.

Exploring the influence of reduction conditions on the properties of the synthesized mono- and bimetallic NPs and the investigation of the effects of different precursors on the morphology of the particles and their product properties (e.g., catalytic) will gain further importance.

Besides this, there is a need for a deeper understanding and insight into formation and growth mechanisms as well as coarsening of NPs on solid substrates. Due to the fact that both, mono- and bimetallic NPs are extremely active, preventing their aggregation, growth and coarsening is a substantial problem that must be addressed. In addition, a further development of modeling nucleation, cluster formation and growth at a molecular level is essential.

Furthermore, highly desirable are fast online and in situ measurement techniques (e.g., X-ray Diffraction, Extended X-ray Absorption Spectroscopy, Raman and IR Spectroscopy) under high pressure. These techniques can provide a lot of useful information about the kinetics and thermodynamics of nucleation, particle formation and the growth process in the early stage after their generation. Such a measurement toolbox has to be completed by modeling the nucleation and particle growth by either condensation or coagulation on a molecular scale. In addition, the outcome from the above-named research will extend our limited knowledge on reactions at solid-SCF interfaces.

10.2 Synthesis of metal oxide nanoparticles

In contrast to the formation and functionalization of submicron organic particles with $scCO_2$ by means of RESS (rapid expansion of supercritical solution), SAS (supercritical antisolvent) or PGSS (particles from gas-saturated solutions) processes, which can already be carried out continuously and in industrial dimensions, the processes for continuous production and functionalization of metal oxide particles in scH_2O still require considerable research and development. Although it can be stated that the properties of nanoparticles synthesized by CHTS can be controlled through the selection of the suitable process parameters (i.e., pressure, temperature, type of salt and its concentration in H_2O) and in particular through the type of the mixer and reactor geometry.

Currently, however, there is a lack of reliable methods for process design and process control to guarantee constant product quality, particle size and morphology.

For this reason, further systematic experimental and theoretical investigations must be carried out to establish the relationships between the above-named process parameters and the achieved product properties (size, shape and structure as well as surface properties).

One of the major limitations of all scH_2O-based particle formation processes is often the lack of fundamental and reliable multicomponent phase equilibria data of the complex binary and/or ternary systems involved in the different processes. Thus, for the design of the CHTS process and the determination of their best operating conditions reliable experimental thermodynamic data (e.g., pressure-density-temperature behavior, heat capacity) and transport properties (e.g., binary diffusion coefficient) of the systems involved are essential.

For this purpose, the increased use and further development of microreactor systems is absolutely necessary. Such systems should enable to map exactly the multiphase flows that occur and to describe the individual physical-chemical processes during nanoparticle synthesis precisely. Furthermore, under the given demanding reaction conditions of the CHTS, reliable in situ and/or online measurement techniques, such as Electrochemical Impedance Spectroscopy, Small Angle X-ray Scattering, Extended X-ray Absorption Fine Structure and/or X-ray Absorption Near-Edge Structure are necessary for process analysis and/or control. Such an integration of the sensor technology into the high pressure and temperature microreactor system represents a great challenge. However, it should then be possible, at least for selected model reactions, to map the individual substeps during the multistage nanoparticle synthesis with spatial and temporal resolution and thus to identify and understand the essential reaction mechanisms.

Furthermore, the results of such work would also make an important contribution to a deeper understanding of the basic chemical mechanisms, phase transition phenomena and nucleation and growth processes in reactive flow systems under supercritical reaction conditions, and thus also provide answers to the numerous open questions.

In **conclusion**, the points named above hold for the broad range of formation and processing of nanostructured materials that exhibit promising properties for applications in fields such as catalysis, sensing, energy, drug delivery, electronics, and optics, which are of increasing economic and ecologic interest.

From a more general point of view, it must be noted that the theoretical work on both the SFRD and CHTS process published so far in the literature often only considers partial aspects and not the complete particle formation process.

The progress in such particle formation processes can be enhanced by the combination of the findings received from simulation and theoretical analysis coupled with the comparison of the experimental results. Such an approach can be very helpful for a deeper understanding and improved interpretation of achieved product properties as well as the process and product optimization.

Up to date and based on the various experimental and modeling results published in the literature one can conclude that there is a need for

- Comparison between catalysts prepared by SFRD and those produced by conventional methods in terms of activity, light-off and light-out temperature, long-term stability, reaction rate and conversion.
- Design and development of new tailor-made precursors and different type of salts.
- Kinetic and thermodynamic adsorption behavior of precursors and their mixtures on porous substrates from SCFs.
- Phase behavior and thermodynamics of the educts (precursor or salt) and their mixtures in $scCO_2$ or scH_2O and development of predictive methods.
- Understanding the mechanisms for nucleation and growth of nanoparticles in SCFs using both computational chemistry and fast online and in situ measurement techniques such as Powder X-ray Diffraction, Extended X-ray Absorption Spectroscopy, Raman and IR Spectroscopy under high pressure.

In addition, and with regard to process and product optimization a further development of the following techniques and tools is needed:

- Molecular dynamics studies of binary and ternary interactions in supercritical mixtures.
- Transformation from batch to continuous processes.
- Knowledge transfer from lab-scale findings to pilot plants.
- Investigations on nanomaterial handling for technological and security considerations.

Index

Note: Page numbers followed by "*f*" indicate figures, and "*t*" indicate tables.

FSC
www.fsc.org
MIX
Papier | Fördert
gute Waldnutzung
FSC® C083411

Kolibri 360 GmbH
Ferdinand-Jühlke-Straße 7
99095 Erfurt, Deutschland
produktsicherheit@kolibri360.de